T0336823

THE BETHE WAVEFUNCTION

Michel Gaudin's book *La fonction d'onde de Bethe* is a uniquely influential masterpiece on exactly solvable models of quantum mechanics and statistical physics. Available in English for the first time, this translation brings his classic work to a new generation of graduate students and researchers in physics. It presents a mixture of mathematics interspersed with powerful physical intuition, retaining the author's unmistakably honest tone.

The book begins with the Heisenberg spin chain, starting from the coordinate Bethe Ansatz and culminating in a discussion of its thermodynamic properties. Delta-interacting bosons (the Lieb–Liniger model) are then explored, and extended to exactly solvable models associated with a reflection group. After discussing the continuum limit of spin chains, the book covers six- and eight-vertex models in extensive detail, from their lattice definition to their thermodynamics. Later chapters examine advanced topics such as multicomponent delta-interacting systems, Gaudin magnets and the Toda chain.

MICHEL GAUDIN is recognized as one of the foremost experts in this field, and has worked at Commissariat à l'Énergie Atomique (CEA) and the Service de Physique Théorique, Saclay. His numerous scientific contributions to the theory of exactly solvable models are well known, including his famous formula for the norm of Bethe wavefunctions.

JEAN-SÉBASTIEN CAUX is a Professor in the theory of low-dimensional quantum condensed matter at the University of Amsterdam. He has made significant contributions to the calculation of experimentally observable dynamical properties of these systems.

Tu ergo, Domine, fecisti ea, qui pulcher es: pulchra sunt enim; qui bonus es: bona sunt enim; qui es: sunt enim. Nec ita pulchra sunt, nec ita bona sunt, nec ita sunt, sicut Tu conditor eorum; quo comparata, nec pulchra sunt, nec bona sunt, nec sunt. Scimus haec, gratias Tibi, et scientia nostra scientiae Tuae comparata ignorantia est.

Saint AUGUSTINE

THE BETHE WAVEFUNCTION

MICHEL GAUDIN

Translated from the French original
'La fonction d'onde de Bethe' (1983)
by Jean-Sébastien Caux

CAMBRIDGE
UNIVERSITY PRESS

CAMBRIDGE
UNIVERSITY PRESS

University Printing House, Cambridge CB2 8BS, United Kingdom

One Liberty Plaza, 20th Floor, New York, NY 10006, USA

477 Williamstown Road, Port Melbourne, VIC 3207, Australia

314-321, 3rd Floor, Plot 3, Splendor Forum, Jasola District Centre, New Delhi - 110025, India

79 Anson Road, #06-04/06, Singapore 079906

Cambridge University Press is part of the University of Cambridge.

It furthers the University's mission by disseminating knowledge in the pursuit of education, learning and research at the highest international levels of excellence.

www.cambridge.org
Information on this title: www.cambridge.org/9781107045859

© Dunod, Paris 1983

This publication is in copyright. Subject to statutory exception and to the provisions of relevant collective licensing agreements, no reproduction of any part may take place without the written permission of Cambridge University Press.

English edition published 2014

A catalogue record for this publication is available from the British Library

ISBN 978-1-107-04585-9 Hardback

Cambridge University Press has no responsibility for the persistence or accuracy of URLs for external or third-party internet websites referred to in this publication, and does not guarantee that any content on such websites is, or will remain, accurate or appropriate.

Contents

Foreword

It is my pleasure to welcome this translation of the original French version of my book *La fonction d'onde de Bethe* into English.

The theory of exactly solvable models, perhaps more than any other subfield of many-body physics, has the distinct advantage of providing solid, reliable and long-lasting knowledge. This latter characteristic perhaps explains why my original text, which is by now over three decades old, is still used by members of the scientific community, despite too many mistakes and neglects on my part. This is why the present work of J.-S. Caux is more than a translation, for by his revision he has drawn 'new from old' thanks to his style and rigour.

Despite all the developments which the field has known in the intervening period, and which are of course not treated or mentioned here, the fact probably remains that much of what is presented has not been too deprecated in the years since the original version appeared. I am convinced that this translation will bring to a much larger readership an accurate image of the status of the knowledge on these fascinating models at the moment of publication of the original. The initiative and merit belong to the translator, to whom I express my gratitude.

Michel Gaudin
Paris, April 2013

Translator's note

Michel Gaudin's 1983 book *La fonction d'onde de Bethe* remains, to the people who have read it, a uniquely influential masterpiece on the subject of exactly solvable models of quantum mechanics. As a beginning Ph.D. student in the mid-1990s, I remember coming across this rather intriguing work, which stood out by being already out of print, and in French. My experience with this book, which sparked my own personal interest in integrable models, is similar to that of many members of the community. Not only the contents but also the *style* of presentation are truly unique: a mixture of mathematical economy interspersed with powerful physical intuition, written in the author's unmistakably humble and honest tone.

Most of the motivation for performing this translation came from a desire to share the high-quality, timeless knowledge contained in the original book with a new generation of researchers. Besides, it is likely that the translation of this work into Russian in the late 1980s significantly contributed to establishing that country's strong and ongoing tradition in the field of exactly solvable models; the sad fact that the book remained inaccessible for other readers seemed too great an omission not to correct. Despite the fact that many important developments have taken place in the field of exactly solvable models since the original publication, the foundations of the theory of integrable models remain essentially unchanged, and the work of Michel Gaudin still stands as a unique 'snapshot' of the field as it was at an important moment in its history.

Finally, one of my hopes with this endeavour was to help make the many unknown (often rediscovered, but too often uncredited) contributions of Michel Gaudin more widely and justly recognized by the wider community.

I am grateful to my team in Amsterdam, namely Rianne van den Berg, Giuseppe Brandino, Michael Brockmann, Jacopo De Nardis, Sebas Eliëns, Davide Fioretto, Thessa Fokkema, Rob Hagemans, Jorn Mossel, Miłosz Panfil, Olya Shevchuk, Rogier Vlijm and Bram Wouters for pointing out many typographical errors as this translation was being prepared. I also thank Stijn de Baerdemacker, Patrick Dorey,

Fabian Essler, Holger Frahm, Dimitri Gangardt, Paul Johnson, Austen Lamacraft, Barry McCoy, Giuseppe Mussardo and Maxim Olshanii for their encouragement and feedback.

I am particularly indebted to Vincent Pasquier, who helped this project from the very beginning and provided much useful feedback throughout.

I also thank Professor James McGuire for his feedback on the text and for sharing his reminiscences of his friendship and collaboration with Michel Gaudin. These helped put the project in its proper historical and personal perspective.

Last, but without doubt most importantly, I am extremely grateful to Michel Gaudin himself for his support of the project, his suggestions for improvements and for the extreme kindness with which he always welcomed my queries. It has been a privilege to be in direct contact with such an inspiring individual, both as a researcher and a human being.

Introduction

From the exact solution of the Ising model by Onsager in 1944 up to that of the hard hexagon model by Baxter in 1980, the statistical mechanics of two-dimensional systems has been enriched by a number of exact results. One speaks (in quick manner) of exact models once a convenient mathematical expression has been obtained for a physical quantity such as the free energy, an order parameter or some correlation, or at the very least once their evaluation is reduced to a problem of classical analysis. Such solutions, often considered as singular curiosities upon their appearance, often have the interest of illustrating the principles and general theorems rigorously established in the framework of definitive theories, and also enabling the control of approximate or perturbative methods applicable to more realistic and complex models. In the theory of phase transitions, the Ising model and the results of Onsager and Yang have eminently played such a reference role. With the various vertex models, the methods of Lieb and Baxter have extended this role and the collection of critical exponents, providing new useful elements of comparison with extrapolation methods, and forcing a refinement of the notion of universality. Intimately linked to two-dimensional classical models (but of less interest for critical phenomena), one-dimensional quantum models such as the linear magnetic chain and Bethe's famous solution have certainly contributed to the understanding of fundamental excitations in many-body systems. One could also mention the physics of one-dimensional conductors. All these theoretical physics questions, while undoubtedly representing the main motivation for research in solvable models, are not the main object of the present study, since their exposition would require a deep and diversified knowledge of physical theories. Here, the main focus is on methods particular to a family of models, in other words the techniques for their solution.

Despite an apparent diversity of definitions and methods, the models which have up to now been solved have intimate relationships. They can often be seen as special or limiting cases of a more general model encompassing them. This is the case for example of the ice model, various ferroelectrics, the discrete lattice gas or Ising model, all joined in the quadratic lattice eight-vertex model defined by Fan and Wu. Aside systems of dimers on planar lattices (periodic or not) treated by Pfaffian methods, exactly solvable models of two-dimensional statistical mechanics are brought back to one-dimensional quantum problems via the transfer matrix method; the static configurations of these nearest-neighbour interacting two-dimensional systems present a topological analogy with the spatio-temporal trajectories of a dynamical system in one space dimension. Whether one seeks to obtain the free energy of the classical system or the excitations of the quantum one, one is led to the calculation of a largest eigenvalue, a second-largest or an entire spectrum, in other words one must diagonalize a matrix, be it a Hamiltonian or a transfer matrix. The prototype of such a Hamiltonian is that of the chain of spin-1/2 atoms with nearest-neighbour exchange interactions, for which in 1931 Bethe discovered the eigenfunctions and spectrum by induction. These eigenfunctions or amplitudes are of a fundamentally original structure and take the form of sums over the various operations of a finite permutation group or, more generally, a reflection group. One could baptize them 'Bethe sums'; one rather speaks of Bethe's Ansatz or hypothesis. These amplitudes describe a non-diffractive 'scattering', in other words a totally elastic collision of the particles or spin waves concerned. This is accompanied by the conservation of as many 'quasi-momenta' as there are particles; the existence of these constants of motion leads to the complete integrability of the Hamiltonian system.

Bethe's method provides the unifying thread between the chapters of this book. Chapter 1 exposes the technique he used in his famous article from 1931, aiming to obtain the wavefunctions and energy spectrum of the Heisenberg–Ising anisotropic magnetic chain. The results of Bethe and Griffiths on the asymptotic localization of the roots of the coupled equations for the spectrum permit the classification of states and, in Chapter 2, the treatment of the chain's thermodynamics at arbitrary temperature. Yang's approach to the thermodynamics of bosons in one dimension was followed, giving a probably correct evaluation of the entropy which would still deserve a rigorous proof. The study of limiting cases contained in Chapter 3, including high- and low-temperature limits as well as the Ising limit, provides convincing evidence for the exactness of the results. Chapters 4 and 5 are dedicated to the δ-interacting Bose gas, which can be seen as a continuous limit of the Heisenberg–Ising spin chain. An extension of Bethe's technique is described therein, which views the N-body problem as that of the diffusion of a multidimensional wave by a kaleidoscope or ensemble of infinitely thin concurrent refracting plates invariant under the reflection group they generate. This is applied to the solution of the bosonic problem defined on a finite segment and to that of the open magnetic chain. Thereafter, a conjecture on the norm of Bethe wavefunctions is presented, constituting a prelude to the open question of correlation functions. Chapter 6 is an incursion into $(1 + 1)$-dimensional quantum field theory aiming to examine the connection between the integrable models of Luttinger and Thirring with a certain continuum limit of the anisotropic magnetic chain.

Vertex models on quadratic lattices are treated in Chapter 7 using Lieb's method for the diagonalization of the transfer matrix of the general neutrality condition-preserving six-vertex model. Only a sketchy presentation is given of the thermodynamics of the various models of ferroelectrics, and one must consult the extended review of Lieb and Wu to go any further. The solution of the (self-conjugate) eight-vertex model is described in Chapters 8 and 9, following for the most part Baxter's method. There again, the integrability of the transfer matrix or of the derived Hamiltonian with three anisotropy constants is linked to the existence of ternary relations between transfer micromatrices. These ternary relations, which can also be understood as star–triangle relations, constitute remarkable representations of permutations and lead to the existence of one-parameter commuting families, and thus to integrability.

The general theory of n-component δ-interacting particles occupies three chapters. The Bethe hypothesis for the real-space wavefunction is proven in Chapter 10; Yang's method intimately relates the realization of symmetry conditions to that of periodicity conditions. The analogy with the inverse scattering method is already manifest and Zamolodchikov's S-matrix algebra is similar to that of Yang's operators. Chapter 11 gives three pathways towards the solution of the fermionic system with two internal states. The first reduces the problem to an inhomogeneous six-vertex model. The second summarizes the algebraic method used in the author's thesis; the third is Faddeev's method giving an operatorial form of Bethe's sum, this being without doubt the most compact representation and that most susceptible to future developments. Chapter 12 exposes Sutherland's general solution for arbitrary wavefunction symmetry, together with various corollaries concerning in particular a remarkable basis for representation of permutations. Chapter 13 contains mixed results concerning certain completely integrable spin systems. Baxter's most recent solution to the ternary relations, which led him to that of the hard hexagon model, is described there. Chapter 14 is devoted to the study of the Toda chain as an integrable classical and quantum system, though it is not associated with Bethe's wavefunction in its primitive form.

Being concerned with a field subject to rapid and varied developments, the ensemble of these 14 chapters does not pretend to be either a synthesis or a review of results on models pertaining in one way or another to Bethe's method. The latter in any case now tends to be viewed in the framework of integrable models as a technique derived from the inverse scattering method. My point of view here is more concrete than general and remains that of a first internal publication in 1972 on Bethe's method for exactly solvable models, of which the present book is only a relative extension and update. It is limited to the treatment of finite or extensive systems of statistical mechanics, ignoring beautiful aspects of quantum field theory with the exception of the few elementary considerations of Chapter 6. I have simply ordered the questions arousing my interest for the benefit of those who might feel attracted by the ingeniosity of the construction and the unifying character of Bethe's method. The mathematical tools are simple, and on the subject of rigour, one can justly be more demanding; one must then refer to the original publications and to the numerous recent works on these questions.

The stimulating atmosphere of the Service de Physique Théorique of the Centre d'Études Nucléaires de Saclay has allowed me to bring this manuscript to fruition, which I publish thanks to the supporting encouragements of my colleagues C. Itzykson, G. Ripka, M. L. Mehta and R. Balian. My interest in the subject was initiated by J. des Cloizeaux, to whom I owe more than I can say, and was renewed by Professor C. N. Yang whose intuition did not refrain from exerting itself in this one-dimensional field. My thanks go to all those who have taught me something on the subject, in particular to E. Brézin, B. McCoy, J. B. McGuire, B. Derrida, M. Takahashi, D. Chudnovsky and G. Chudnovsky; they go to Dr C. K. Lai with whom I have had the pleasure to collaborate at the State University of New York at Stony Brook, and to J. M. Maillard who has agreed to proofread a number of chapters.

This book owes its publication to the Commissariat à l'Énergie Atomique, with the help of its edition and documentation services and, in the first instance, M. R. Dautray who has given it a warm welcome within the scientific collection he directs.

1

The chain of spin-1/2 atoms

1.1 Model for a one-dimensional metal

In a famous contribution to the theory of metals published in 1931, Bethe studies one of the simplest quantum models one can imagine for N mutually interacting atoms: a linear chain of two-level atoms interacting with their nearest neighbours, constituting a one-dimensional crystal. If the interaction between two neighbouring atoms is represented graphically by a link, the chain obviously realizes the simplest connected configuration. In order to avoid edge effects, one considers a closed or cyclic chain. The problem of the open chain will be treated in Chapter 5.

The levels of each atom are those of the two-spin states of a spin-1/2 electron outside a completely filled electronic level. In the absence of interactions between atoms, in other words for a sufficiently large value of the lattice spacing, the degeneracy of the 2^N levels is complete. Bethe considers calculating the states and energies of the atomic chain to first order in the electronic interaction, in the framework of the theory of Slater. In a first approximation, the Coulomb forces and eventually also the spin-dependent magnetic forces have the following consequences:

(a) they separate the average value of the energy of each pair of neighbouring atoms into parallel ↑↑ and antiparallel ↑↓ states;
(b) they exchange the spin states of electrons localized on neighbouring atoms.

The non-vanishing elements of the energy matrix are thus intuitively defined. More rigorously, if one first neglects the magnetic moments coming from spin, degenerate perturbation theory gives the Hamiltonian (Heisenberg, 1928; Dirac, 1967)

$$H = -J \sum_{n=1}^{N} P_{n,n+1} \tag{1.1}$$

in which P_{12}, P_{23}, ... are operators exchanging the orbital states, and J represents Slater's exchange integral. In view of Pauli's principle, H in the basis of completely antisymmetric states is equivalent to the so-called isotropic spin Hamiltonian

$$H = \frac{J}{2} \sum_n \vec{\sigma}_n \cdot \vec{\sigma}_{n+1} \tag{1.2}$$

since one has, in this basis, the representation

$$P_{12} = -\frac{1}{2}(1 + \vec{\sigma}_1 \cdot \vec{\sigma}_2).$$

The spin vector of the index n atom of the chain is $\vec{S}_n = \frac{1}{2}\vec{\sigma}_n$, where $\sigma^{x,y,z}$ are the spin-1/2 Pauli matrices. For the closed chain, sites $N + n$ and n are identified: $\vec{S}_{N+1} \equiv \vec{S}_1$.

If one must take into account average magnetic forces between electrons along a certain anisotropy direction, one considers the so-called Heisenberg–Ising Hamiltonian depending on a parameter Δ; this anisotropy parameter takes the value 1 for the isotropic case in which only exchange interactions occur

$$H_\Delta = 2J \sum_{n=1}^{N} S_n^x S_{n+1}^x + S_n^y S_{n+1}^y + \Delta \left(S_n^z S_{n+1}^z - \frac{1}{4} \right). \tag{1.3}$$

A constant has been added to ensure that the ferromagnetic reference state $|S^z = \frac{1}{2}N\rangle$ has zero energy. The H_Δ Hamiltonian commutes with the total spin projection on the anisotropy axis $S^z = \sum_n S_n^z$. For the case $\Delta = 1$, isotropy implies $[H, \vec{S}] = 0$.

More generally, one considers the three-parameter Hamiltonian defining the XYZ model (Bonner, 1968; Sutherland, 1970; Baxter, 1971b)

$$H_{\Delta,\Gamma} = 2 \sum_n J_x S_n^x S_{n+1}^x + J_y S_n^y S_{n+1}^y + J_z S_n^z S_{n+1}^z \tag{1.4}$$

depending on two anisotropy parameters Δ and Γ such that

$$J_x : J_y : J_z = 1 - \Gamma : 1 + \Gamma : \Delta. \tag{1.5}$$

Its spectrum will be determined in Chapter 8 following Baxter's method.

Concerning the physics of the magnetic chain and a list of useful references, the reader can consult the article of Bonner and Fisher (1964) and Bonner's thesis (1968); for the classical (non-quantum) chain, see the article of Fogedby (1980).

1.2 Bethe's method

1.2.1 Equation for the amplitudes

The method invented by Bethe for the diagonalization of H (1.2) is also simply applicable to H_Δ (1.3) (Orbach, 1958). The important distinguishing feature with respect to the three-parameter Hamiltonian (1.4) is the conservation of the magnetic component S^z. An eigenstate $|M\rangle$ of H_Δ belonging to spin $S^z = \dfrac{N}{2} - M$, where M is the spin deviation with respect to the ferromagnetic state $|S^z = \dfrac{N}{2}\rangle$, is a superposition of states of basis

$$|n_1 n_2 \ldots n_M\rangle = S_{n_1}^- S_{n_2}^- \ldots S_{n_M}^- |0\rangle \qquad (1.6)$$

characterized by a sequence of integer coordinates of atoms with magnetic component $-\dfrac{1}{2}$ (denoted $\downarrow$) along the chain. One has

$$|M\rangle = \sum_{\{n\}} a(n_1 n_2 \ldots n_M)|n_1 n_2 \ldots n_M\rangle. \qquad (1.7)$$

The sum goes over the C_M^N sets $\{n\}$ of M increasing integers varying from 1 to N

$$1 \le n_1 < n_2 < \ldots < n_M \le N. \qquad (1.8)$$

A typical configuration of the chain is represented in Figure 1.1. It can also be characterized by the coordinates of the antiparallel links, alternating the two types $-\bullet- \equiv \uparrow\downarrow$ and $-\bullet- \equiv \downarrow\uparrow$.

The off-diagonal elements of H_Δ

$$\langle\{n'\}| \sum_n S_n^+ S_{n+1}^- + S_n^- S_{n+1}^+ |\{n\}\rangle \qquad (1.9)$$

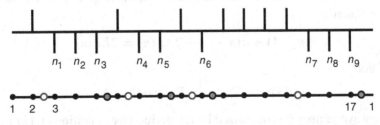

Figure 1.1 The figure represents the base state $|n_1 n_2 \ldots n_9\rangle$ for $N = 17$, $M = 9$. The transitions operated by the non-diagonal part of H_Δ number 8, which exactly equals the number of antiparallel links; $3 \to 2, 5 \to 6, 7 \to 6, 8 \to 9, 10 \to 9$ and $11, 15 \to 14, 17 \to 18 \equiv 1$.

lead to the transitions

$$|n_1 n_2 \ldots n_M\rangle \rightarrow |n'_1 n'_2 \ldots n'_M\rangle \qquad (1.10)$$

in which configurations $\{n'\}$ are obtained from $\{n\}$ by the exchange of a single pair of antiparallel neighbours

$$n'_1 = n_1, \quad n'_2 = n_2, \quad \ldots, \quad n'_\alpha = n_\alpha \pm 1, \quad \ldots, \quad n'_M = n_M \qquad (1.11)$$

provided that the set $\{n'\}$ thus defined is an admissible configuration with respect to (1.8). In order to make inequalities (1.11) valid on the whole defining interval of the n_α coordinates, this interval shall however be extended beyond N by identifying sites n and $N + n$. The periodicity condition on the amplitudes is thus

$$a(n_1 n_2 \ldots n_M) \equiv a(n_2 n_3 \ldots n_M, n_1 + N). \qquad (1.12)$$

These have been defined up to now only for sets of increasing integers with $n_M - n_1 < N$.

The diagonal elements of H_Δ are

$$2\Delta \langle \{n\}| \sum_m \left(S_m^z S_{m+1}^z - \frac{1}{4} \right) |\{n\}\rangle$$

$$= -\Delta \times \text{number of antiparallel links } \uparrow\downarrow \text{ or } \downarrow\uparrow$$

$$= -\Delta \times \text{number of allowed transitions } |\{n\}\rangle \rightarrow |\{n'\}\rangle. \qquad (1.13)$$

The equation for the amplitudes $a(\{n\})$ and the energy levels $2JE$ of Hamiltonian H_Δ is thus

$$\sum_{\{n'\}} (a(\{n'\}) - \Delta a(\{n\})) = 2E a(\{n\}) \qquad (1.14)$$

in which the sum is over all allowed transitions defined by (1.11).

1.2.2 Decoupling of the eigenvalue equation

Let us start with the elementary case of one downturned spin $M = 1$ leading to a difference equation

$$a(n + 1) + a(n - 1) - 2\Delta a(n) = 2E a(n) \qquad (1.15)$$

with solution

$$a(n) = e^{ikn}, \qquad E = \cos k - \Delta, \qquad (1.16)$$

where the wavenumber k is quantized by the periodicity condition (1.12)

$$e^{ikN} = 1 \quad \Leftrightarrow \quad kN = 2\pi\lambda, \qquad \lambda = 0, 1, 2, \ldots, N - 1. \qquad (1.17)$$

Going to the $M = 2$ case, Bethe obtains a solution defined by two momenta k_1 and k_2:

$$a(n_1 n_2) = e^{\frac{i}{2}\psi_{12} + ik_1 n_1 + ik_2 n_2} + e^{-\frac{i}{2}\psi_{12} + ik_2 n_1 + ik_1 n_2}. \tag{1.18}$$

The remarkable fact in this spin wave scattering is that the wavefunction $a(n_1 n_2)$ coincides with its asymptotic expression which, in one dimension, only contains two elastic terms, the direct and exchange terms. This result generalizes, using appropriate phase shifts, when superimposing M spin waves with distinct momenta forming a set $\{k\} \equiv \{k_1 k_2 \ldots k_M\}$.

The success of this scheme formally comes from decoupling the system (1.14), which is achieved as follows.

(a) The definition of the $a(\{n\})$ is extended to values of n such that equality is allowed:

$$n_1 \le n_2 \le n_3 \le \ldots \le n_M. \tag{1.19}$$

(b) One assumes that equations (1.14) are still verified if the sum is no longer constrained to the allowed configurations (distinct n_α) only, but is freely extended to all transitions (1.11). One thus obtains second-order difference equations with constant coefficients

$$\sum_{\alpha=1}^{M}(a(n_1 \ldots n_\alpha + 1 \ldots n_M) + a(n_1 \ldots n_\alpha - 1 \ldots n_M)$$
$$-2\Delta a(n_1 \ldots n_\alpha \ldots n_M)) = 2Ea(n_1 n_2 \ldots n_M). \tag{1.20}$$

(c) Since equalities (1.14) and (1.20) are by hypothesis simultaneously verified, the terms of the left-hand side of (1.20) which do not appear in (1.14) have a vanishing total contribution. These terms come from two possible sources

• $n'_\alpha = n_{\alpha+1} = n_\alpha + 1$,
• $n'_\beta = n_{\beta-1} = n_\beta - 1$,

for certain values of the α and β indices.

Let us examine the two terms associated with two neighbouring down spins $\downarrow$ on sites n_α and $n_{\alpha+1} = n_\alpha + 1$. A first forbidden term in (1.14) comes from $n'_\alpha = n_\alpha + 1$, since n'_α coincides with $n_{\alpha+1} = n'_{\alpha+1}$, i.e.

$$a(\ldots n_\alpha + 1, n_\alpha + 1, \ldots) - \Delta a(\ldots n_\alpha, n_\alpha + 1, \ldots).$$

A second term in (1.14) comes from $n'_{\alpha+1} = n_{\alpha+1} - 1 = n_\alpha$, i.e.

$$a(\ldots n_\alpha, n_\alpha, \ldots) - \Delta a(\ldots n_\alpha, n_\alpha + 1, \ldots).$$

Collecting the two forbidden terms associated with the neighbouring pair $(n_\alpha, n_\alpha + 1)$, equations (1.14) and (1.20) will be identical provided one can impose the constraints

$$a(\ldots n_\alpha + 1, n_\alpha + 1, \ldots) + a(\ldots n_\alpha, n_\alpha, \ldots)$$
$$-2\Delta a(\ldots n_\alpha, n_\alpha + 1, \ldots) = 0, \tag{1.21}$$

which must be verified for all values of coordinates subject to conditions (1.19).

1.2.3 The Bethe wavefunction

The difference equation (1.20) is separable in each coordinate and admits factorized solutions $\exp(ik_1 n_1 + ik_2 n_2 + \ldots + ik_M n_M)$ associated with a set $\{k\}$. The corresponding energy is thus

$$E = \sum_{\alpha=1}^{M} (\cos k_\alpha - \Delta). \tag{1.22}$$

Bethe attempts to fulfil the complementary constraints (1.21) by expressing the amplitude as a sum over the $M!$ identical energy waves obtained by permuting the momenta in set $\{k\}$. Bethe's sum (or Bethe's Ansatz) is the following:

$$a(n_1 \ldots n_M) = \sum_{P \in \pi_M} A(P) \exp\left(i \sum_{\alpha=1}^{M} k_{P\alpha} n_\alpha \right). \tag{1.23}$$

The $A(P)$ coefficients are to be determined from (1.21), which is written as

$$\sum_{P} A(P) \left\{ e^{i(k_{P\alpha} + k_{P(\alpha+1)})} - 2\Delta e^{ik_{P(\alpha+1)}} + 1 \right\} \times e^{ik_{P1}n_1 + \ldots + i(k_{P\alpha} + k_{P(\alpha+1)})n_\alpha + \ldots} = 0 \tag{1.24}$$

and must be verified for all $\{n\}$ and each α such that $n_{\alpha+1} = n_\alpha + 1$. In the sum, one can then associate terms $A(P)\ldots$ and $A(PP_{\alpha,\alpha+1})\ldots$ which have the same $\{n\}$ dependence, and sum over the subset $P/P_{\alpha,\alpha+1}$ of permutations of π_M, where $P_{\alpha,\alpha+1} \equiv (\alpha, \alpha+1)$ represents the permutation of two neighbours α and $\alpha+1$. To simplify the notation, take $(\alpha, \alpha+1) = (34)$. It is sufficient to have the following relations for $A(P)$:

$$A(P)(e^{i(k_{P3}+k_{P4})} - 2\Delta e^{ik_{P4}} + 1) + A(PP_{34})(e^{i(k_{P4}+k_{P3})} - 2\Delta e^{ik_{P3}} + 1) = 0. \tag{1.25}$$

Defining the antisymmetric phases $\psi_{\alpha\beta} = -\psi_{\beta\alpha}$ by the relation

$$e^{i\psi_{\alpha\beta}} = -\frac{e^{i(k_\alpha+k_\beta)} - 2\Delta e^{ik_\alpha} + 1}{e^{i(k_\alpha+k_\beta)} - 2\Delta e^{ik_\beta} + 1}, \tag{1.26}$$

condition (1.25) is written as

$$A(PP_{\alpha,\alpha+1}) = A(P)e^{-i\psi_{P\alpha,P(\alpha+1)}}, \forall P \text{ and } \alpha. \tag{1.27}$$

The unique solution up to a trivial factor is

$$A(P) = \exp\left(\frac{i}{2}\sum_{\alpha<\beta}\psi_{P\alpha,P\beta}\right). \tag{1.28}$$

Note that the relation between the phases and momenta can be written

$$\cot\frac{\psi_{\alpha\beta}}{2} = \Delta\frac{\sin(k_\alpha - k_\beta)/2}{\cos(k_\alpha + k_\beta)/2 - \Delta\cos(k_\alpha - k_\beta)/2}. \tag{1.29}$$

1.2.4 Periodicity conditions

If we were dealing with an infinite one-dimensional crystal, we would choose the momenta $\{k\}$ arbitrarily (real, modulo 2π), the phases would be determined by (1.20), and we would have thus constructed a continuous set of eigenstates of the infinite chain. However, in view of the thermodynamic limit and in order to address the questions of classification and completeness, it is necessary to consider a finite system; thus, the idea of treating the periodic chain. It so happens that periodicity conditions (1.12) are compatible with the structure (1.23) of the Bethe sum. These conditions are written as

$$A(P) = A(PC)e^{ik_{P1}N}, \tag{1.30}$$

in which C designates the cyclic permutation $C\alpha = \alpha + 1$, or $C = (1, 2, \ldots, M)$. With the help of (1.28), we have

$$A(PC) = \exp\left(\frac{i}{2}\sum_{1\le\alpha<\beta\le M}\psi_{P(\alpha+1),P(\beta+1)} + \frac{i}{2}\sum_{1\le\alpha<M}\psi_{P(\alpha+1),P1}\right)$$

$$= \exp\left(\frac{i}{2}\sum_{1\le\alpha<\beta<M}\psi_{P\alpha,P\beta} + i\sum_{1<\alpha\le M}\psi_{P\alpha,P1}\right).$$

Relation (1.30) is thus written as

$$e^{ik_{P1}N+i\sum_{1<\alpha\le M}\psi_{P\alpha,P1}} = 1, \quad \forall P. \tag{1.31}$$

There must thus be a set of integers λ_α such that

$$Nk_\alpha = 2\pi\lambda_\alpha + \sum_\beta\psi_{\alpha,\beta}, \quad \alpha = 1, 2, \ldots, M, \tag{1.32}$$

with $\psi_{\alpha,\alpha} = 0$ by convention.[1]

---

[1] Note that this entails that $\psi_{\alpha,\beta}$ is discontinuous at $k_\alpha - k_\beta = 0$. See the paragraph preceding equation (1.59).

These are the coupled equations used to determine the admissible sets of momenta $\{k\}$, and consequently the energies and amplitudes. We note that this is a high-degree algebraic system in the e^{ik} variables. The complexity of system (1.32) renders the problems of existence and classification particularly acute, and these will be treated in Section 1.5 (see Yang and Yang, 1966a–d).

1.3 Parameters and quantum numbers

1.3.1 Orbach's parametrization

Before defining the sets $\{\lambda\}$ and classifying the solutions, it is necessary to fix the branches of k and ψ. We shall take

$$0 \leq \Re k_\alpha \leq 2\pi, \qquad -\pi \leq \Re \psi_{\alpha,\beta} \leq \pi. \tag{1.33}$$

In contrast, relation (1.29) between the momenta and the phases can be parametrized in a way which is useful for our further purposes; we take

$$\cot \frac{\psi_{\alpha,\beta}}{2} = \Delta \frac{\cot \frac{k_\alpha}{2} - \cot \frac{k_\beta}{2}}{\cot \frac{k_\alpha}{2} \cot \frac{k_\beta}{2} (\Delta - 1) + (\Delta + 1)} \tag{1.34}$$

which leads us to distinguishing the regions $|\Delta| \gtrless 1$.

(a) $\Delta > 1$.

We take

$$\Delta = \cosh \Phi, \qquad 0 < \Phi < \infty \tag{1.35}$$

and we have, as a function of parameters φ_α,

$$\begin{cases} \cot \dfrac{k_\alpha}{2} = \coth \dfrac{\Phi}{2} \tan \dfrac{\varphi_\alpha}{2} \\ \cot \dfrac{\psi_{\alpha,\beta}}{2} = \coth \Phi \tan \dfrac{\varphi_\alpha - \varphi_\beta}{2}, \end{cases} \quad \text{or} \quad \begin{aligned} e^{ik_\alpha} &= \dfrac{e^{i\varphi_\alpha} - e^{\Phi}}{e^{i\varphi_\alpha + \Phi} - 1}, \\ &-\pi < \Re\varphi_\alpha < \pi. \end{aligned} \tag{1.36}$$

(b) $-1 < \Delta < 1$.

We take

$$\Delta = \cos \Theta, \qquad 0 < \Theta < \pi \tag{1.37}$$

and we obtain, in terms of the parameters θ_α,

$$\begin{cases} \cot \dfrac{k_\alpha}{2} = \cot \dfrac{\Theta}{2} \tanh \dfrac{\theta_\alpha}{2}, \\ \cot \dfrac{\psi_{\alpha,\beta}}{2} = \cot \Theta \tanh \dfrac{\theta_\alpha - \theta_\beta}{2}. \end{cases} \tag{1.38}$$

One must take note that the real domain of k is covered by the θ domain

$$-\infty < \Re\theta_\alpha < +\infty, \qquad \Im\theta_\alpha = 0 \text{ or } \pi.$$

(c) The boundary case $\Delta = 1$ is obtained from either side by taking

$$x_\alpha = \lim_{\Phi \to 0} \frac{\varphi_\alpha}{\Phi} = \lim_{\Theta \to 0} \frac{\theta_\alpha}{\Theta},$$

$$\cot \frac{k_\alpha}{2} = x_\alpha, \qquad \cot \frac{\psi_{\alpha,\beta}}{2} = \frac{1}{2}(x_\alpha - x_\beta). \tag{1.39}$$

(d) Finally, the region $\Delta < -1$ is brought back to the study of the region $\Delta > 1$ by virtue of a proposition demonstrated in the following subsection, see (1.42).

Proposition: The H_Δ and $H_{-\Delta}$ Hamiltonians are equivalent.

To close this subsection, let us write down the form that the coupled equations (1.31) take in terms of the φ or θ parameters:

$$\left(\frac{\sin \frac{1}{2}(\varphi_\alpha + i\Phi)}{\sin \frac{1}{2}(\varphi_\alpha - i\Phi)} \right)^N = \prod_{\beta \neq \alpha} \frac{\sin \frac{1}{2}(\varphi_\alpha - \varphi_\beta + 2i\Phi)}{\sin \frac{1}{2}(\varphi_\alpha - \varphi_\beta - 2i\Phi)}. \tag{1.40}$$

Let us also mention the expression for the energy levels:

$$E = -\sum_\alpha \frac{\sin^2 \Theta}{\cosh \theta_\alpha - \cos \Theta} \quad \text{or} \quad -\sum_\alpha \frac{\sinh^2 \Phi}{\cosh \Theta - \cos \varphi_\alpha}, \tag{1.41}$$

depending on the case. We go from one to the other by the continuations $\varphi = i\theta$, $\Phi = i\Theta$.

1.3.2 Quantum numbers

We first have the following symmetry in the Δ parameter, allowing us to restrict our study to the domain $\Delta > 0$:

$$U H_\Delta U^{-1} = -H_{-\Delta}, \tag{1.42}$$

with the S^z-conserving unitary operator

$$U = \exp\left(i\pi \sum_{n=1}^N n S_n^z \right). \tag{1.43}$$

The spectra of H_Δ and $H_{-\Delta}$ are thus inverted with respect to each other.

Besides the S^z magnetic component, the total momentum K is another conserved quantum number of the periodic chain. The one-site translation operator $T = e^{iK}$ commutes with H by virtue of the property

$$T^{-1} \vec{S}_n T = \vec{S}_{n+1}. \tag{1.44}$$

T is thus the circular site permutation operator along the chain; it admits the classical representation

$$T = \operatorname*{Tr}_{\tau} P_{01} P_{02} \dots P_{0N} \equiv C^{-1} \tag{1.45}$$

$$C : \alpha \to \alpha + 1$$

with

$$P_{0n} = \frac{1}{2}(1 + \vec{\tau} \cdot \vec{\sigma}_n). \tag{1.46}$$

We deduce from (1.23)

$$T|M\rangle = e^{i(k_1 + \dots + k_M)}|M\rangle. \tag{1.47}$$

In other words

$$K = \sum_{\alpha=1}^{M} k_\alpha = \frac{2\pi}{N} \sum_\alpha \lambda_\alpha. \tag{1.48}$$

The U transformation commutes with K if M is even, but shifts K by π if M is odd, N being assumed even.

1.3.3 Total spin

For the value $\Delta = 1$, there exists an additional absolute quantum number which is the total spin S such that $\vec{S}^2 = S(S+1)$.

> **Proposition:** If the numbers k_α solving (1.32) for $\Delta = 1$ are all different from zero (modulo 2π), then the eigenstate $|M\rangle$ belongs to the sector with total spin $S = S^z = \frac{N}{2} - M$.

Proof:

$$|M\rangle = \sum_{\{n\}} a(n_1 \dots n_M) S_{n_1}^- \dots S_{n_M}^- |0\rangle. \tag{1.49}$$

The condition $S^+|M\rangle = 0$ which characterizes the $S = S^z$ state is

$$\sum_{n=1}^{n_2-1} a(n n_2 \dots n_M) + \sum_{n=n_2+1}^{n_3-1} a(n_2 n n_3 \dots n_M) + \dots$$

$$+ \sum_{n=n_M+1}^{N} a(n_2 n_3 \dots n_M n) = 0. \tag{1.50}$$

Let us substitute in the first term of (1.50) the expressions for the amplitudes (1.23), perform the sums over n under the hypothesis $z_j = e^{ik_j} \neq 1$, and regroup terms of the same nature to obtain

$$\sum_{j=1}^{M-1}\sum_{P} A(P) \left(\frac{1}{1-z_{Pj}} - \frac{z_{P(j+1)}}{1-z_{P(j+1)}} \right) e^{i(k_{P1}n_2+...+(k_{Pj}+k_{P(j+1)})n_{j+1}+...+k_{PM}n_M)}$$

$$+\sum_{P} A(P) \left(\frac{-z_{P1}}{1-z_{P1}} e^{i(k_{P2}n_2+...+k_{PM}n_M)} \right. \qquad (1.51)$$

$$\left. + \frac{1}{1-z_{PM}} e^{i(k_{P1}n_2+...+k_{P(M-1)}n_M+k_{PM}(N+1))} \right) = 0.$$

After shifting $P \to PC$ in the last sum, the two terms of the last parentheses in (1.51) compensate each other by virtue of the periodicity condition (1.30). As far as the first sum is concerned, for each value of j we can join permutations P and $P_{(j,j+1)}$, which gives the typical sum

$$A(P) \left(\frac{1}{1-z_{P3}} - \frac{z_{P4}}{1-z_{P4}} \right) + A(P(34)) \left(\frac{1}{1-z_{P4}} - \frac{z_{P3}}{1-z_{P3}} \right) = 0.$$

This sum vanishes by virtue of (1.25) for $\Delta = 1$. $\qquad\square$

States with $S - S^z \neq 0$ are obviously obtained from the $S = S^z$ states by repeated application of the total spin-lowering operator $S^- = \sum_n S_n^-$. They can also be given directly by the coupled system of equations of order $M = N/2 - S^z$ by admitting that certain k can vanish. If there exist m momenta k equal to zero, which is made possible by choosing the corresponding integers λ equal to zero, the state $|M\rangle$ has total spin $S^z + m$; we have, in fact, by (1.39) that $\psi_{\alpha,\beta} = 0$ if $k_\alpha = 0$.

While this quantum number has been defined only for the value $\Delta = 1$, we will by convention call states of total spin S the states of H_Δ which are continuously corresponding to the spin S states of the isotropic Hamiltonian. This continuity principle will be used in what follows; it finds its justification in the fact that the energy and the amplitudes are algebraic functions of Δ for finite N. It is therefore sufficient to avoid the eventual branch points in order to establish continuity.

1.3.4 Quantum numbers of real states

In the case of the isotropic interaction, Bethe showed that the solutions involving real $\{k\}$ do not exhaust the list of eigenstates of given M, of which there are C_M^N. Before presenting his results on complex roots, we shall briefly describe the real solutions by use of the continuity principle in Δ, starting from values of this parameter for which the system is simpler, namely $\Delta = 0$ and ∞.

For $\Delta = 0$, we fall back on a particular case of the XY model

$$H_{\Delta=0,\Gamma} = 2 \sum_{n=1}^{N} (1 - \Gamma) S_n^x S_{n+1}^x + (1 + \Gamma) S_n^y S_{n+1}^y$$

$$= \sum_n S_n^+ S_{n+1}^- + S_{n+1}^+ S_n^- + \Gamma (S_n^+ S_{n+1}^+ + S_n^- S_{n+1}^-), \quad (1.52)$$

which is equivalent to a system of free fermions (Lieb, Schultz and Mattis, 1961). The Jordan–Wigner transformation

$$S_n^+ = a_n e^{-i\pi \sum_{m>n} a_m^\dagger a_m}, \quad S_n^z = \frac{1}{2} - a_n^\dagger a_n, \quad (1.53)$$

in which the a_n, $a_n^\dagger$ are fermionic operators ($\{a_n, a_m^\dagger\} = \delta_{nm}, \ldots$), gives us exactly, for M odd,

$$\begin{cases} H_{\Delta=0} = \displaystyle\sum_{n=1}^{N} a_{n+1}^\dagger a_n + a_n^\dagger a_{n+1}, \\ M = \displaystyle\sum_n a_n^\dagger a_n. \end{cases} \quad (1.54)$$

(For M even, there is a boundary effect changing the sign of the $n = N$ term in sum (1.54), creating a surmountable complication.) From (1.54), we deduce the complete spectrum

$$E_{\Delta=0} = \sum_{\alpha=1}^{M} \cos k_\alpha, \quad (1.55)$$

with

$$k_\alpha = \frac{2\pi}{N} \left(J_\alpha + \frac{N}{2} \right), \quad (1.56)$$

in which $\{J\}$ is a set of distinct integers (modulo N). For M even we would have J_α a half-odd integer.

Examining the amplitudes (1.23) for $\Delta = 0^+$, we deduce from (1.26) and (1.27)

$$A(P) = \text{antisymmetric character of } P,$$

and, therefore, the determinant expression of the amplitude

$$a(n_1 \ldots n_M) = |e^{ikn_1}, e^{ikn_2}, \ldots, e^{ikn_M}|. \quad (1.57)$$

The periodicity condition is thus written as

$$e^{ik_\alpha N} = -(-)^M, \quad (1.58)$$

giving rule (1.56) for the quantum numbers $\{J\}$.

However, the limiting form of the coupled equations (1.32) links the J and λ in a subtle manner, since, from (1.29), we have

$$\psi_{\alpha\beta} = \pi \varepsilon (\sin k_\alpha - \sin k_\beta), \qquad \varepsilon(x) = x/|x|$$

and the order of k is that of $\sin k$ only in the interval $\pi/2 < k < 3\pi/2$, $\forall \alpha$; alternatively $-\pi/2 < k_\alpha < \pi/2$, in other words if all the k are in the same semi-band of energy, positive ($\cos k_\alpha > 0$) or negative. In particular, the antiferromagnetic ($J > 0$) ground state is obtained by completely filling the semi-band $\cos k_\alpha < 0$. For example, if we have $N/2$ odd, we obtain the only two possible sequences

$$\{J\} \equiv \left\{ -\frac{M-1}{2}, -\frac{M-3}{2}, \ldots, \frac{M-1}{2} \right\} \tag{1.59}$$

and

$$\left\{ -\frac{M+1}{2}, -\frac{M-1}{2}, \ldots, \frac{M-3}{2} \right\}. \tag{1.60}$$

The limit form of (1.32) for $\Delta = 0$ gives us the relation between J_α and λ_α:

$$\frac{N}{2} + J_\alpha = \lambda_\alpha - \alpha + \frac{M+1}{2}, \qquad \alpha = 1, \ldots, M, \tag{1.61}$$

in other words the first sequence

$$\{\lambda\} = \{1, 3, 5, \ldots, 2M - 1\} \tag{1.62}$$

corresponding to total spin $S = 0$, and the second sequence

$$\{\lambda\} = \{0, 2, 4, \ldots, 2M - 2\} \tag{1.63}$$

corresponding to spin $S = 1$. These are the only two states of the $J > 0$ case in which the relation between the λ and the limiting momenta is simple. It was possible to show (Yang and Yang, 1966b) that the $S = 0$ state defined above corresponds by continuity to the ground state in the whole region $\Delta > -1$.

In the $J < 0$ case, the lowest-lying $S^z = 0$ state corresponds to the filling of all $\cos k_\alpha > 0$, in other words to the sequence

$$\left\{ \frac{N}{2} + J \right\} = \left\{ -\frac{M-1}{2}, -\frac{M-3}{2}, \ldots, \frac{M-1}{2} \right\},$$

or equivalently

$$\lambda_\alpha = 0, \qquad \alpha = 1, 2, \ldots, M. \tag{1.64}$$

This is thus the state of total spin $S = N/2$.

1.3.5 Ising limit

Let us examine the other simple limit $\Delta = \infty$, in which the Hamiltonian is equivalent to the Ising one

$$H_{\text{Ising}} = \lim_{\Delta \to \infty} \frac{H_\Delta}{\Delta} = 2 \sum_n \left(S_n^z S_{n+1}^z - \frac{1}{4} \right). \tag{1.65}$$

The limiting eigenvalues of H_Δ/Δ are the negative integers $-Q, 0 \leq Q \leq M$. The degeneracy of the (M, Q) subspace is easily computed thanks to the representation of antiparallel pairs in Figure 1.1. We find

$$g(M, Q) = \frac{N}{M} C_Q^M C_{Q-1}^{N-M-1}. \tag{1.66}$$

We could consider the problem of diagonalizing the perturbation $\Delta^{-1} \sum_n S_n^+ S_{n+1}^-$ +h.c. in the (M, Q) subspace. In Section 1.5 we shall study the large-Δ asymptotic behaviour of the energies and amplitudes.

If we confine ourselves to the real states, we put $K_\alpha = \lim k_\alpha$ and, according to (1.29), we obtain

$$\lim_{\Delta \to \infty} \psi_{\alpha,\beta} = K_\alpha - K_\beta - \pi \varepsilon (K_\alpha - K_\beta), \tag{1.67}$$

from which we obtain the limiting amplitudes

$$\lim_{\Delta \to \infty} a(n_1 \ldots n_M) = \left| e^{iKP_\alpha(n_\alpha - \alpha + \frac{M+1}{2})} \right|. \tag{1.68}$$

The limit of system (1.32) gives us

$$K_\alpha = \frac{2\pi}{N - M} \left(J_\alpha - \sum_\beta J_\beta \right), \tag{1.69}$$

with

$$N/2 + J_\alpha = \lambda_\alpha - \alpha + (M + 1)/2.$$

Therefore, amplitude (1.68) is non-vanishing only if the K_α are distinct (modulo 2π), and thus also the J_α. In the $S^z = 0$ case, the 2π interval for K is saturated by M successive integers, and we see that the possible sequences are the two defined previously in (1.59) and (1.60). For $S^z = 0$, the only two real states in the $\Delta = \infty$ limit are defined from the same sequence of quantum numbers as the simple states, total spin $S = 0$ and $S = 1$ ground states, obtained in the $\Delta = 0^+$ limit.

We shall provide in Section 1.5 a quick proof of the existence of the corresponding solutions to the coupled equations in the $\Delta \geq 1$ domain (Griffiths, 1964).

For arbitrary values of S^z, this subset of real states is defined by increasing J sequences or increasing λ sequences such that $\lambda_{\alpha+1} - \lambda_\alpha \geq 2$. The number of such real solutions is thus exactly equal to the degeneracy of the $Q = M$ subspace, namely

$$g(M, M) = \frac{N}{M} C_{M-1}^{N-M-1},\tag{1.70}$$

which equals 2 for $M = N/2$.

1.4 Asymptotic positioning of complex momenta

1.4.1 Limit equations

The existence of complex solutions to system (1.31) is already manifest for $M = 2$, in which two spin waves can form a bound state, in the sense that the amplitude $a(n_1 n_2)$ decreases exponentially with $|n_2 - n_1|$ provided this difference remains finite with respect to N. In this section, the intuitive method employed by Bethe for the asymptotic positioning ($N \gg 1$) of the complex roots is extended to all values of $\Delta \geq 1$. As a first step we evaluate exactly the imaginary parts of the unknowns φ_α (1.36) of the following auxiliary system:

$$\Re \left(N k_\alpha - 2\pi \lambda_\alpha - \sum_\beta \psi_{\alpha,\beta} \right) = 0,\tag{1.71}$$

$$\Im \left(\sum_\beta \psi_{\alpha,\beta} \right) = \infty,\tag{1.72}$$

differing from the original system (1.32) only by the replacement $N = \infty$ in the imaginary part of the first member. We show as a second step that the exact roots are very close to those of (1.71) for the exponential in N approximation.

1.4.2 Limit strings

In this subsection we simply show that the proposed complex structure is compatible with the limit form (1.72). We suppose that, in the complex plane of the φ variable, the roots φ_α are grouped into various finite series which are symmetric with respect to the real axis, and which we call 'strings'. A string is characterized by an abscissa φ and an integer n called the order of the string $C_n(\varphi)$:

$$\Delta > 1, \quad C_n(\varphi) = \{\varphi_\nu = \varphi + i\nu\Phi\},$$

with

$$\nu = -(n-1), -(n-3), \ldots, n-3, n-1\tag{1.73}$$

without restriction on n;

$$-1 < \Delta < 1, \qquad C_n(\theta) = \{\theta_\nu = \theta + i\nu\Theta\}$$

with the restriction on n (justified in Subsection 1.5.3),

$$(n-1)\Theta < \pi. \tag{1.74}$$

The general index α used so far to label the momenta k_α is now expanded in such a way that $\varphi_\alpha = \{\varphi, \nu\}$. The different strings will be denoted $C_n(\varphi), C_{n'}(\varphi'), \ldots$ We define the total momentum of a string, its quantum numbers and its energy, by

$$k = \sum_{\nu \in C} k_\nu, \qquad \lambda = \sum_\nu \lambda_\nu; \qquad E_n(\varphi) = \sum_\nu (\cos k_\nu - \Delta) \tag{1.75}$$

and the mutual scattering phase of two strings C and C' by

$$\psi(\varphi, \varphi') = \sum_{\substack{\mu \in C \\ \nu \in C'}} \psi_{\mu,\nu}. \tag{1.76}$$

If we define the real quantities $k_p(\varphi)$ and $\psi_p(\varphi - \varphi')$ by relations analogous to (1.36)

$$\begin{cases} \cot \dfrac{k_p}{2} = \coth \dfrac{p\Phi}{2} \tan \dfrac{\varphi}{2}, \\ \cot \dfrac{\psi_p}{2} = \coth \dfrac{p\Phi}{2} \tan \dfrac{\varphi - \varphi'}{2}, \quad -\pi < \psi_p < \pi, \end{cases} \tag{1.77}$$

we obtain after an easy calculation the total momentum of $C_n(\varphi)$

$$k = k_n(\varphi),$$

and the total energy

$$E_n(\varphi) = \frac{-\sinh\Phi \sinh n\Phi}{\cosh n\Phi - \cos\varphi} = (\cos k_n - \cosh n\Phi)\frac{\sinh\Phi}{\sinh n\Phi} = \sinh\Phi \frac{dk_n}{d\varphi}. \tag{1.78}$$

Finally, the mutual phase of $C_n(\varphi)$ and $C_m(\varphi')$

$$\psi(\varphi, \varphi') = \sum_p [nmp]\psi_p(\varphi - \varphi'), \tag{1.79}$$

in which the symmetric symbol $[nmp]$ is defined as

$$[nmp] = \begin{cases} 1 & \text{if } p = |m-n| \text{ or } m+n, \\ 2 & \text{if } p = |m-n|+2, |m-n|+4, \ldots, m+n-2, \\ 0 & \text{otherwise.} \end{cases} \tag{1.80}$$

We have the exception $[nn0] = 0$. This symbol verifies the sum rules

$$\sum_p [mnp] = M(n, m) = \begin{cases} 2n & n < m \\ 2m & m < n \\ 2n - 1 & m = n \end{cases}$$

$$\sum_p p[mnp] = 2mn. \tag{1.81}$$

If we add term by term all equations (1.32) belonging to a given string, we obtain as many equations as there are strings for the determination of the abscissas φ of each string of arbitrary given order:

$$Nk_n(\varphi) = 2\pi\lambda + \sum_{C'_m(\neq C_n)} \sum_p [nmp]\psi_p(\varphi - \varphi'). \tag{1.82}$$

Let us suppose that there exist integers $\lambda, \lambda', \lambda'', \ldots$ and $n, n', \ldots$ such that system (1.82) admits a real solution. From this, we obtain a solution to the auxiliary system (1.71), the momenta k_α being computed using (1.73) and (1.36). The $\psi_{\alpha,\beta}$ phases are given by the formula

$$e^{i\psi_{\mu,\nu}(\varphi - \varphi')} = e^{2\Phi} \frac{e^{i(\varphi - \varphi') - (\mu - \nu + 2)\Phi} - 1}{e^{i(\varphi - \varphi') - (\mu - \nu - 2)\Phi} - 1}, \tag{1.83}$$

with the exception of the two singular phases $\psi_{\mu,\mu\pm2}(\varphi, \varphi)$ pertaining to a given string $C(\varphi)$, which are not determined by (1.83).

Let us show that their real part is determined by relations (1.71); adding term by term all equations pertaining to $C_m(\varphi)$ from index $\nu = -(m - 1)$ up to $\nu = \mu$, we obtain

$$N \sum_{-(m-1)}^{\mu} \Re k_\nu = 2\pi \sum_{-(m-1)}^{\mu} \lambda_\nu + \Re\psi_{\mu,\mu+2}(\varphi, \varphi) + \text{determined phases.} \tag{1.84}$$

The real part of each exceptional phase $\psi_{\mu,\mu+2}(\varphi, \varphi)$ is thus determined from the original equations, and the compatibility of equations (1.84) for $\mu = m - 1$ is ensured precisely by relations (1.82). As far as the imaginary part of $\psi_{\mu,\mu\pm2}$ is concerned, one can see that (1.72) expresses the fact that it is infinite by virtue of (1.83).

1.4.3 Asymptotic solution $(N \gg 1)$

Let us briefly indicate here how the preceding string structure can by justified by anticipating some results valid in the thermodynamic limit $N \to \infty$ established in Chapter 3.

For N finite but large, we seek solutions φ_α in the neighbourhood of expressions (1.73)

$$\varphi_\nu = \varphi + i\nu\Phi + i\delta_\nu, \qquad \delta_{-\nu} = \delta_\nu^*, \tag{1.85}$$

with $\delta_\nu = O(\delta)$ where δ should be a quantity much smaller than the mean level spacing N^{-1} of the abscissas, that is $\delta \ll N^{-1}$. By continuity, all phases will be displaced by a quantity $O(\delta)$, except the exceptional phases which, according to (1.83), take the now defined values

$$\Im\psi_{\mu,\mu+2}(\varphi, \varphi) = -\ln\frac{|\delta_{\mu+2} - \delta_\mu|}{2\sinh 2\Phi} + O(\delta). \tag{1.86}$$

The real part is modified by $O(\delta)$. Thus the coupled equations give us, for $0 < \mu < m - 1$,

$$\frac{2}{N}\Im\{\psi_{\mu,\mu+2}(\varphi, \varphi) + \psi_{\mu,\mu-2}(\varphi, \varphi)\} = -2\chi_\mu(\varphi) + O(\delta), \tag{1.87}$$

with the definition of χ_μ

$$-2\chi_\mu(\varphi) = \ln\frac{\cosh(\mu - 1)\Phi - \cos\varphi}{\cosh(\mu + 1)\Phi - \cos\varphi}$$
$$+\frac{1}{N}\sum_{C'_n}\ln\left\{\frac{\cosh(\mu + n + 1)\Phi - \cos(\varphi - \varphi')}{\cosh(\mu - n - 1)\Phi - \cos(\varphi - \varphi')}\frac{\cosh(\mu + n - 1)\Phi - \cos(\varphi - \varphi')}{\cosh(\mu - n + 1)\Phi - \cos(\varphi - \varphi')}\right\}, \tag{1.88}$$

in which the sum runs over all strings $C'_n(\varphi')$, including $C_m(\varphi)$ but excluding the factor corresponding to $n - 1 = \mu$. The sum appearing in the second term of (1.88) can only be evaluated in the limit $N \to \infty$, except for the case $\Phi = \infty$ ($\Delta = \infty$) for which we have, in the notation of Section 1.4,

$$2\chi_\mu \approx \frac{2\Phi}{N}(N - \sum_n M(\mu, n)\nu_n) = \frac{2\Phi}{N}Q_\mu. \tag{1.89}$$

Here, ν_n is the number of strings of order n. We here anticipate some results of Section 2.2, which will be based on the real system (1.82) uniquely and express its solutions in terms of positive densities of strings $\tilde{\rho}_n(\varphi)$ in the thermodynamic limit. The essential result of Appendix A is the positivity of $\chi_\mu(\varphi)$,

$$\chi_\mu(\varphi) = \int_{-\pi}^{+\pi}\Xi(\varphi - \varphi')(\tilde{\rho}_{\mu-1}(\varphi') + \tilde{\rho}_{\mu+1}(\varphi'))d\varphi' \tag{1.90}$$

with

$$\Xi(\varphi) = \sum_k \frac{\tanh k\Phi}{2k}e^{ik\varphi} = \sum_{-\infty}^{\infty}\ln\coth\frac{\pi}{4\Phi}|\varphi + 2k\pi|, \tag{1.91}$$

which is manifestly positive.

We deduce, using (1.87), that the exceptional phases are of order N:

$$\Im \psi_{m-3,m-1} \approx N \chi_{m-1} > 0,$$
$$\Im \psi_{m-5,m-3} \approx N(\chi_{m-1} + \chi_{m-3}) > 0, \ldots,$$
$$\Im \psi_{-1,+1} \approx N(\chi_{m-1} + \chi_{m-3} + \ldots + \chi_1) > 0.$$

Using (1.86) we can thus bound the quantities δ_μ:

$$|\delta_{m-1} - \delta_{m-3}| \approx 2 \sinh 2\Phi e^{-N\chi_{m-1}},$$
$$|\delta_{m-3} - \delta_{m-5}| \approx 2 \sinh 2\Phi e^{-N(\chi_{m-1}+\chi_{m-3})}, \ldots$$

The δ_μ are thus exponentially small in N. In the limit $\Phi = 0$ or $\Delta = 1$, we simply find

$$\chi_\mu(\varphi) = 2\pi \times \text{ density of available states for strings } C_\mu(\varphi).$$

The consistency of the proposed structure is thus probable in the $\Delta \geq 1$ region, provided the solutions to system (1.82) exist, which is proven for a large class of states in the following section.

The preceding analysis, on which the finite-temperature thermodynamics of the chain will be based (Chapter 2), is not in contradiction to other results (Woynarovich, 1982; Babelon, de Vega and Viallet, 1983) showing that the description of states in the vicinity of the antiferromagnetic ground state does not in general involve strings of order greater than two. Our argument is in fact based on the positivity of the $\tilde{\rho}_n$ (hole density) and thus requires a macroscopic number of vacancies in the antiferromagnetic 'vacuum'. It is thus not valid for this system at absolute zero temperature.

1.5 State classification and counting

1.5.1 Existence of a class of solutions

We shall prove the existence of a class of real solutions of system (1.32) (Griffiths, 1964), for which the quantum numbers are obtained by the continuity principle, Section 1.4. Griffiths' method is actually extended directly to equations (1.82), if we restrict to only one type of string of given order n, the real case corresponding to $n = 1$.

This class is defined by ν strings of equal length n, for which the quantum numbers verify

$$\begin{cases} 0 \leq \lambda_1 < \lambda_2 < \ldots < \lambda_\nu \leq N \\ \lambda_{\alpha+1} - \lambda_\alpha > M(n,n) = 2n-1, \quad \alpha = [1, \nu-1] \\ \lambda_\nu - \lambda_1 < N - 2n + 1, \quad\quad\quad M = \nu n. \end{cases} \tag{1.92}$$

Proof: The set $\{k\}$ being viewed as a point of $\mathbb{R}_\nu$, equations (1.32) or (1.82) for the total momenta of each string can be written in the form $k_j = \mathscr{F}_j\{k\}$, where the mapping $k \to k' = \mathscr{F}\{k\}$ is defined by

$$\mathscr{F}_j\{k\} \equiv \frac{2\pi}{N}\lambda_j + \frac{1}{N}\sum_{l=1}^{\nu}\psi_{j,l}. \tag{1.93}$$

According to (1.36) or (1.77), the mutual phase of two strings C_j and C_l, namely $\psi_{j,l}$, is an increasing function of k_j, continuous on each interval $]0, k_l[$ and $]k_l, 2\pi[$, but discontinuous in k_l. On the compact $\mathscr{C}$ of $\mathbb{R}_\nu$ defined by the inequalities

$$\mathscr{C} : k_{\alpha+1} - k_\alpha \geq \frac{2\pi}{N}, \qquad k_\nu - k_1 \leq 2\pi - \frac{2\pi}{N}, \tag{1.94}$$

we thus have $\psi_{j+1,l} > \psi_{j,l}, \forall l \neq j, j+1$.

We deduce herewith, by subtracting two consecutive relations,

$$k'_{j+1} - k'_j \geq \frac{2\pi}{N}(\lambda_{j+1} - \lambda_j + \frac{1}{\pi}\psi_{j+1,j}). \tag{1.95}$$

However, for $n = 1$, we have $|\psi_{j,j+1}| \leq \pi$ and for each value of n, using (1.79) and (1.81),

$$|\psi_{j+1,j}| \leq M(n,n)\pi.$$

From hypothesis (1.92) and (1.95), we deduce

$$k'_{j+1} - k'_j \geq \frac{2\pi}{N}, \qquad j = [1, \nu - 1]. \tag{1.96}$$

In contrast, we have

$$\begin{cases} k_l > k_1 \Rightarrow \psi_{1,l} \geq 0 \Rightarrow k'_1 \geq 0, \\ k_l < k_\nu \Rightarrow \psi_{\nu,l} \leq 0 \Rightarrow k'_\nu \leq 2\pi, \end{cases} \tag{1.97}$$

from which we obtain the inequality

$$k'_\nu - k'_1 = \frac{2\pi}{N}\left(\lambda_\nu - \lambda_1 + \frac{\psi_{\nu,1}}{\pi}\right) + \frac{1}{N}\sum_{l=2}^{\nu-1}\psi_{\nu,1} - \psi_{1,l}$$
$$\leq \frac{2\pi}{N}(N - M(n,n) - 1 + M(n,n)) = 2\pi - \frac{2\pi}{N}. \tag{1.98}$$

From (1.94) and (1.96), etc. comes the fact that the image of the compact $\mathscr{C}$ under $\mathscr{F}$ is in $\mathscr{C}$. However, the mapping $\mathscr{F}$ is continuous on $\mathscr{C}$, which is connected and convex; it thus admits at least one fixed point in $\mathscr{C}$, which verifies $k = \mathscr{F}\{k\}$ (Brouwer's theorem). $\qquad\square$

1.5.2 Asymptotic form of the coupled equations

An immediate corollary of the preceding argument is the non-crossing of the momenta k_j or of the abscissas φ_j of same-length strings in the whole domain $\Delta > 1$. We shall make the stronger hypothesis that non-crossing is generally applicable between all strings. Despite the fact that we are dealing with an asymptotic system ($N \gg 1$ but finite), we shall still use the principle of continuity in Δ to obtain a classification of the set of levels, by studying the Ising limit $\Delta = \infty$, hoping for a kind of uniformity in the double limit. Before going any further, let us provide a representation of equations (1.82) which involves continuous functions only and whose writing is systematic.

Let v_m be the number of strings of order m ($m = 1, 2, \ldots$); the constraint on spin is written

$$\sum_{n>0} n v_n = M. \tag{1.99}$$

The total number of strings is $Q = \sum_n v_n$, since by (1.78) each string contributes a partial energy $E = -1$ in the limit $H_{\text{Ising}} = \lim(H_\Delta / \Delta)$ defined in (1.65).

The abscissa of the ith string of order n is denoted $\varphi_{n,i}$, $i = 1, 2, \ldots, v_n$. We use the odd continuous functions $f_{p\Phi}(\varphi)$, such that

$$\tan \frac{1}{2} f_{p\Phi}(\varphi) = - \coth \frac{p\Phi}{2} \tan \frac{\varphi}{2}, \tag{1.100}$$

which allow us to write, following (1.77),

$$\begin{cases} k_{n,i} = f_{n\Phi}(\varphi_{n,i}) + \pi, \\ \psi_p(\varphi - \varphi') = f_{p\Phi}(\varphi - \varphi') + \pi\varepsilon(\varphi - \varphi'). \end{cases} \tag{1.101}$$

Our system of equations now takes a form which is convenient for the thermodynamic limit:

$$N f_{n\Phi}(\varphi_{n,i}) = 2\pi J_{n,i} + \sum_{m,p} \sum_{j=1}^{v_m} [nmp] f_{p\Phi}(\varphi_{n,i} - \varphi_{m,j}), \quad i = 1, 2, \ldots, v_n. \tag{1.102}$$

The still undetermined quantum numbers $J_{n,i}$ are related to the λ of system (1.82) in the following way. Set

$$\lim_{\Delta \to \infty} k_{n,i} = \lim_{\Delta \to \infty} (\pi - \varphi_{n,i}) = K_{n,i}. \tag{1.103}$$

In the non-crossing hypothesis

$$\varepsilon(\varphi - \varphi') = -\varepsilon(K - K').$$

We thus have

$$J_{n,i} = \lambda_{n,i} - \frac{N}{2} - \frac{1}{2} \sum_{m,j} M(n,m)\varepsilon(K_{n,i} - K_{m,j}). \tag{1.104}$$

The $J_{n,i}$ are integer or half-integer according to whether $N + \nu_n$ is odd or even. The allowed sets remain to be determined, which we will do by examining the Ising limit. Result (1.92) already allows us to conclude that $J_{n,i+1} - J_{n,i} > 0$.

1.5.3 Bound states and strings

With the objective of understanding the Ising limit ($\Delta = \infty$), let us show that each string of order n gives rise to a bound state of n spin waves, characterized by a factorization of the amplitude and a typical exponential decay.

Let us consider two successive momenta k_μ and $k_{\mu+2}$, belonging to two successive indices in a given string C_n. In sum (1.23) for the amplitude, let us consider the permutations to be acting on the coordinates $n_1, n_2, \ldots, n_M$. For convenience of notation, we shall apply the indexation in strings of k_α to that of the coordinates n_α. The transposition $P_{\mu,\mu+2}$ determines two classes of permutations, and we can rewrite the amplitude in a succinct notation

$$a(n_1 n_2 \ldots n_M) = \sum_{P(\bar{P}\mu < \bar{P}(\mu+2))} \exp\{\ldots\}\left[\exp\left(\frac{i}{2}\psi_{\mu,\mu+2} + ik_\mu n_{\bar{P}\mu} + ik_{\mu+2}n_{\bar{P}(\mu+2)}\right)\right.$$
$$\left. + \exp\left(-\frac{i}{2}\psi_{\mu,\mu+2} + ik_\mu n_{\bar{P}(\mu+2)} + ik_{\mu+2}n_{\bar{P}\mu}\right)\right]. \tag{1.105}$$

In fact, for each P, the ratio of the modulus of the two terms is

$$\exp\left\{-\left\{\Im\psi_{\mu,\mu+2} - \Im(k_\mu - k_{\mu+2})(n_{\bar{P}(\mu+2)} - n_{\bar{P}\mu})\right\}\right\}$$

which, according to (1.90), is of order $\exp(-(cst)N)$ provided that the distance between two $\downarrow$ coordinates remains finite with respect to N.

Under this hypothesis, only one class of permutations contributes to the definition of the asymptotic amplitude. If we apply this result to all neighbouring pairs of indices within a given string, the dominant amplitude becomes a sum over the set of permutations which completely reverses the order of the indices of each string:

$$n_{\bar{P}(\overline{m-1})} > \ldots > n_{\bar{P}(m-3)} > n_{\bar{P}(m-1)} \quad (\bar{\mu} = -\mu). \tag{1.106}$$

We can see that relative to each string, the amplitude contains a factor

$$\exp\left\{\sum_{\mu>0} \Im k_\mu (n_{\bar{P}\bar{\mu}} - n_{\bar{P}\mu})\right\} \tag{1.107}$$

in which, according to (1.86), we have $\Im k_\mu < 0$ if $\mu > 0$.

The bound state behaviour is thus manifest in (1.107), and the interpretation of this factor as the wavefunction of a bound state, or string of ↓ spins, is particularly clear when the group of coordinates $n_{\bar{P}_\mu}$ forms an isolated cluster.

1.5.4 The Ising limit

We consider the limit of the Bethe amplitudes for $\Delta \to \infty$, and first of all the limit of a spin-bound state or string. For $\Phi \gg 1$, we have the equivalence $|\Im k_\mu| = \Phi$ and the amplitude of the bound state is dominated by the term in which the argument $\sum_{\mu>0}(n_{\bar{P}_\mu} - n_{\bar{P}_\mu})$ is minimal, in other words in which the sequence (1.106) is made up of successive integers. It is sufficient to renormalize the amplitude by a factor $\exp(m^2\Phi/2)$ in order to keep the only dominant term in the Ising limit. In this limit a string or bound state of spins is a train of n consecutive ↓ spins. Let us call $N_{m,i}$ and $K_{m,i}$ the centre of mass and momentum of the string $C'_{m,i}$. We show (Gaudin, 1972) that the limit amplitude of the system of strings has the form

$$a(\ldots N_{m,1} \ldots N_{m,v_m} \ldots) = \prod_{m>0} \left| e^{iK_{m,j}\hat{N}_{m,i}} \right|_{v_m} \tag{1.108}$$

with the definition

$$\hat{N}_{m,i} = N_{m,i} + \frac{1}{2}\sum_n \sum_{j=1}^{v_n} M(n,m)\varepsilon(N_{n,j} - N_{m,i}), \tag{1.109}$$

which correctly gives back result (1.68) in the real case $n = 1$.

It remains to determine the limit values of the string momenta. The limit of (1.102) gives us

$$NK_{n,i} = 2\pi\left(J_{n,i} + \frac{N}{2}\right) + \sum_{m,p}\sum_{j=1}^{v_m}[nmp](K_{n,i} - K_{m,j}). \tag{1.110}$$

Introducing the quantities

$$K_m = \sum_i K_{m,i}, \qquad J_m = \sum_i J_{m,i}, \tag{1.111}$$

and the positive integers

$$Q_n = N - \sum_{m>0} M(n,m)v_m, \tag{1.112}$$

relations (1.110) can be written as

$$2\pi\left(J_{n,i} + \frac{N}{2}\right) = Q_n K_{n,i} + \sum_m M(n,m)K_m, \tag{1.113}$$

in which the J_n are related to the K_n by

$$2\pi \left(J_n + \frac{1}{2} N v_n \right) = Q_n K_n + v_n \sum_m M(n, m) v_m, \qquad (1.114)$$

allowing us to re-express the K_n, and subsequently the $K_{n,i}$, in terms of the quantum numbers.

The matrix

$$\| Q_n \delta_{nm} + v_n M(n, m) \|$$

associated with this linear system has the curious property of having $Q_1 - v_1$, $Q_2 - v_2, \ldots$ as eigenvalues, except the last one which should be replaced by N. Despite this, inversion is not straightforward; relation (1.113) however determines the set $J_{n,i}$ up to an overall shift, and this will suffice for the thermodynamics of the chain (Chapter 2).

1.5.5 *Quantum numbers of the $S = S^z$ class*

According to (1.108), the momenta $K_{n,i}$ for given n must be distinct (modulo 2π). Therefore, according to (1.113), the $J_{n,i}$ form a sequence of distinct integers (or half-odd integers) over an interval of length Q_n. Up to an as yet undetermined overall shift, we can choose $\{J\}$ in the following manner:

$$-\frac{1}{2}(Q_n - 1) \le J_{n,1} < J_{n,2} < \ldots < J_{n v_n} \le \frac{1}{2}(Q_n - 1). \qquad (1.115)$$

We thus define, for a given number of strings of each order, a set of $Z(N; v_1 v_2 \ldots)$ distinct states with

$$Z(N; v_1 v_2 \ldots) = C_{v_1}^{Q_1} C_{v_2}^{Q_2} \ldots C_{v_n}^{Q_n} \ldots \qquad (1.116)$$

Proposition: The total number of states defined by inequalities (1.115) is given by the remarkable formula (Bethe, 1931)

$$\sum_{v_1 + 2v_2 + 3v_3 + \ldots = M} Z(N; v_1 v_2 \ldots) = C_M^N - C_{M-1}^N, \qquad (1.117)$$

which is exactly the number of eigenstates belonging to total spin $S = S^z$, according to the convention of Subsection 1.3.3.

Proof: We define

$$Z(N, M, Q) = \sum_{\sum v_n = Q, \sum n v_n = M} Z(N; v_1 v_2 \ldots). \qquad (1.118)$$

According to definition (1.112) of Q_n,

$$Q_n(N; v_1 v_2 \ldots) = Q_{n-1}(N - 2Q; v_2 v_3 \ldots)$$

and consequently

$$Z(N; \nu_1\nu_2\ldots) = C_{\nu_1}^{N-2Q+\nu_1} Z(N - 2Q; \nu_2\nu_3\ldots).$$

From this we deduce the recursion relations

$$Z(N, M, Q) = \sum_{\nu_1=0}^{Q-1} C_{\nu_1}^{N-2Q+\nu_1} Z(N - 2Q, M - Q, Q - \nu_1), \qquad (1.119)$$

for which we can verify that the following solution:

$$Z(N, M, Q) = \frac{N - 2M + 1}{N - M + 1} C_Q^{N-M+1} C_{Q-1}^{M-1} \qquad (1.120)$$

satisfies the initial conditions $Z(N, 1, 1) = N - 1$. Identity (1.117) is finally obtained by summing both sides of (1.120) over Q. □

We will have noticed the link between the expression for $Z(N, M, Q)$ counting only the $S = S^z$ states as conjectured in the following subsection, and the total degeneracy $g(M, Q)$ of the (M, Q) states (equation (1.66)).

1.5.6 Quantum numbers of the $S - S^z = \bar{m} > 0$ states

When Δ takes the value 1, an eigenstate of magnetic component S^z differing from the total spin S is obviously of the form $(S^-)^{\bar{m}}|S = S^z\rangle$ with $\bar{m} = S - S^z$. The amplitudes $a(\{n\})$ can thus be considered as built from $M = (N/2) - S^z$ momenta k_α of which $\bar{m}$ are equal to zero, for $\Delta = 1$. This implies choosing the corresponding integers λ to be zero. When Δ is close to 1, our continuity principle leads to the splitting of the zero momenta into a cluster of distinct momenta, approaching zero with Φ, and *a priori* belonging to a set of strings $C_{m_j}(\varphi_j)$ such that $\bar{m} = \sum_j m_j$.

According to (1.77), the abscissas of the strings with limit momentum different from zero also tend to zero with Φ. In contrast, the abscissas φ_j can have a nonzero limit $\bar{\varphi}_j$. The limit form of the coupled equations (1.82) divided by Φ is then

$$(N - 2M + 2\bar{m}) \cot\frac{\bar{\varphi}_j}{2} = \sum_l 2m_l \cot\frac{\bar{\varphi}_j - \bar{\varphi}_l}{2}, \qquad \bar{m} = \sum_j m_j, \qquad \bar{\varphi}_j \in \mathbb{R} \neq 0.$$
$$(1.121)$$

Let us show that equations (1.121) have no solution except in the case of a single string $C_{\bar{m}}$ of abscissa $\bar{\varphi} = \pm\pi$. In fact, the above equations give the two-dimensional electrostatic equilibrium positions on the circle of positive charges $2m_1$ at $\bar{\varphi}_1$, $2m_2$ at $\bar{\varphi}_2$, etc. in the presence of a fixed negative charge $N - 2M + 2\bar{m} = 2S \geq 2\bar{m}$ located at the origin of angles on the circle. There exists thus at most one exceptional string, of order $\bar{m}$, denoted $C_{\bar{m}}(\pi)$, continuously corresponding to the $k_\nu = 0$ group ($\nu = -(\bar{m} - 1), \ldots, \bar{m} - 1$) of the isotropic limit.

According to the continuity principle, if the limit phase $\varphi_{\bar{m},l}$ is equal to $+\pi$, we shall have $\varphi_{n,i} < \varphi_{\bar{m},l}$ ($\forall n, i$) and consequently $K_{n,i} > K_{\bar{m},l}$, which justifies the choice of sign $\varphi = +\pi$. We thus deduce from (1.104) and from $\lambda_{\bar{m},l} = 0$ that the quantum number $J_{\bar{m},l}$ of the exceptional string is

$$J_{\bar{m},l} = -\frac{1}{2}(Q_{\bar{m}} + 2\bar{m} - 1) = -\frac{1}{2}\bar{Q}_{\bar{m}}$$

with

$$\bar{Q}_n = Q_n(\nu_1, \ldots, \nu_{\bar{m}} - 1, \ldots).$$

By virtue of (1.117), we would obtain the correct number of states for $M \to M - \bar{m}$ if we proposed the following choices for the quantum numbers of the $S - S^z = \bar{m}$ states:

$$-\frac{1}{2}(Q_{\bar{m}} - 1) - \bar{m} = J_{\bar{m},1} < J_{\bar{m},2} < \ldots < J_{\bar{m},\nu_{\bar{m}}} \leq \frac{1}{2}(Q_{\bar{m}} - 1) - \bar{m}.$$

1.5.7 Remark on strings in the $|\Delta| \leq 1$ region

The principle of continuity and level (momenta!) non-crossing which allowed us to reach the preceding asymptotic results seems not to be applicable to the crossing of the $\Delta = 1$ (and -1) point which is a level degeneracy point, nor to be extendable to the $|\Delta| < 1$ region, except for the real states, for example the two ground states followed in the whole $\Delta \geq -1$ region. The asymptotic structure of the complex roots in the $|\Delta| < 1$ region was studied by Takahashi and Suzuki (1972) and displays a curiously discontinuous behaviour as a function of the angle $\Theta = \text{acos } \Delta$. The maximal order of the different strings is determined by the continuous arithmetic fraction development of Θ/π. We can already anticipate this non-uniformity with respect to Δ in the limit $N \to \infty$ when examining by perturbation theory the complicated relationship between the momenta and quantum numbers for the free fermion gas in the vicinity of $\Delta = 0$ ($\Theta = \pi/2$).

2

Thermodynamic limit of the Heisenberg–Ising chain

2.1 Results for the ground state and elementary excitations

For each eigenstate of the chain that is considered, we give the quantum numbers, the energy per site and the excitation energy in the limit of the infinitely long chain. The calculation methods are hereafter very briefly described using a single example (Bethe, 1931; Hulthén, 1938; Walker, 1959; des Cloizeaux and Gaudin, 1966; Yang and Yang, 1966a–d; Johnson, Krinsky and McCoy, 1973).

2.1.1 Antiferromagnetic ground state

This is the only state for which the quantum numbers have been determined rigorously and for which the associated solution has been shown to be unique using a variational method (Yang and Yang, 1966b). Restricting ourselves to the case of an even number of sites $N = 2M_0$, the ground state quantum numbers for region $\Delta > -1$ are

$$S^z = 0, \quad S = 0, \quad \text{total momentum } q = \begin{cases} \pi & (M_0 \text{ odd}) \\ 0 & (M_0 \text{ even}) \end{cases}$$

$$\{\lambda\} \equiv \{1, 3, 5, \ldots, N - 1\}$$

or

$$\{J\} \equiv \left\{ -\frac{M_0 - 1}{2}, -\frac{M_0 - 3}{2}, \ldots, \frac{M_0 - 1}{2} \right\}, \quad \nu_1 = M_0. \tag{2.1}$$

The energy per site $\varepsilon = \lim E/N$ admits the following expressions:

- $-1 < \Delta < 1$,

$$\varepsilon_1(\Delta) = -\sin \Theta \int_0^\infty \left(1 - \frac{\tanh \omega \Theta}{\tanh \omega \pi} \right) d\omega \tag{2.2}$$

27

- $\Delta > 1$,

$$\varepsilon_2(\Delta) = -\sinh\Phi \sum_{k=-\infty}^{+\infty} (1 + e^{2|k|\Phi})^{-1} \qquad (2.3)$$

- $\Delta < -1$,

$$\varepsilon = 0,$$

The ground state is ferromagnetic and the energy is vanishing.

The two analytical functions $\varepsilon_1(\Delta)$ and $\varepsilon_2(\Delta)$ are distinct, but are perfectly joined from either side of the $\Delta = 1$ point on the real axis, their value and that of all their derivatives being equal when $\Delta = 1$. The ε_2 function is cut by an essential singularity line in $[-1, +1]$, the ε_1 function in $[1, \infty[$. Between the continuations, we have the relation

$$\varepsilon_2(\Delta) - \varepsilon_1(\Delta) = 2\pi i \sum_{k=1}^{\infty} (1 + e^{(2k-1)\pi^2/\Phi})^{-1},$$

$$\Phi = i\Theta, \qquad \Delta = \cosh\Phi, \qquad 0 < \Re\Theta < \pi. \qquad (2.4)$$

2.1.2 Antiferromagnetic state quasi-degenerate with the ground state

Quantum numbers

$$S^z = 0, \quad \text{total spin } S = 1, \quad \text{total momentum } q = \begin{cases} 0 & (M_0 \text{ odd}) \\ \pi & (M_0 \text{ even}) \end{cases}$$

$$\{\lambda\} = \{0, 2, 4, \dots, N - 2\} \qquad (2.5)$$

$$\{J\} = \left\{ -\frac{M_0 + 1}{2}, -\frac{M_0 - 1}{2}, \dots, \frac{M_0 - 3}{2} \right\}, \qquad \bar{Q}_1 = M_0 + 1.$$

The difference in energy per site between this state and the ground state vanishes in the thermodynamic limit.

2.1.3 Antiferromagnetic spin wave $S^z = \pm 1$

(des Cloizeaux and Pearson, 1962)

$$S = 1, \qquad M = M_0 - 1, \qquad \bar{m} = 0, \qquad Q_1 = M_0 + 1.$$

$$\{J\} = \left\{ -\frac{M_0 - 2}{2}, -\frac{M_0 - 4}{2}, \dots, \boxed{J_0}, \dots, \frac{M_0 - 2}{2}, \frac{M_0}{2} \right\}. \qquad (2.6)$$

$$\underset{\text{hole}}{\downarrow}$$

Total momentum q: $|q| = (\pi/2) - 2\pi J_0/N$.

Dispersion curve:

$$-1 < \Delta \leq 1,$$
$$\eta_1(q) = \frac{\pi}{2} \frac{\sin \Theta}{\Theta} |\sin q|. \tag{2.7}$$

In parametric form for $\Delta > 1$:

$$\eta_1(q) = \sinh \Phi (\mathrm{Dn}\varphi_0 + \mathrm{Dn}\pi),$$
$$\frac{\pi}{2} - |q| = \int_0^{\varphi_0} \mathrm{Dn}\varphi d\varphi, \tag{2.8}$$

with the following definition of the elliptic function $\mathrm{Dn}\varphi$ of periods 2π and $4i\,\Phi$ (or $K'/K = \Phi/\pi$):

$$\mathrm{Dn}\varphi = \sum_{k=-\infty}^{\infty} \frac{e^{ik\varphi}}{2\cosh k\Phi} = \frac{K}{\pi}\mathrm{dn}\left(\frac{K\varphi}{\pi}; k\right), \tag{2.9}$$

in Jacobi's notation for the modulus k and complete elliptic integral K.

The elimination of the φ_0 parameter between the energy η_1 and momentum q can be explicitly performed by noticing the equality

$$\int_0^{\varphi_0} \mathrm{Dn}\varphi d\varphi = am\left(\frac{K\varphi_0}{\pi}; k\right),$$

which gives

$$\eta_1(q) = K' \frac{\sinh \Phi}{\Phi}(\sqrt{1 - k^2\cos^2 q} + k'). \tag{2.10}$$

In the $\Delta = 1$ limit, $\Phi = 0$, $K' = \pi/2$, $k = 1$ and we fall back on formula (2.7) for $\Theta = 0$.

2.1.4 Antiferromagnetic spin wave $S^1 = 0$, $S = 1$

$$M = M_0, \qquad \bar{m} = 1.$$

In the set (2.5) of quantum numbers, we admit two holes such that two real roots of the coupled equations are replaced by a complex pair:

$$v_1 = M_0 - 2, \qquad v_2 = 1.$$

$$\begin{cases} Q_1 = N - M(1,1)v_1 - M(1,2)v_2 \equiv M_0, & \bar{Q}_1 = M_0 + 1, \\ Q_2 = N - M(2,1)v_1 - M(2,2)v_2 \equiv 1, \end{cases} \tag{2.11}$$

from which we obtain the sets

$$\{J_1\} = \left\{ -\frac{M_0 + 1}{2}, -\frac{M_0 - 1}{2}, \ldots, \boxed{J_0}, \ldots, \frac{M_0 - 5}{2} \right\},$$

$$\underset{\text{hole}}{\downarrow}$$

$$\{J_2\} = \{0\}. \tag{2.12}$$

The coupled equations are written

$$N f_2(\varphi) = \sum_{j=1}^{M_0 - 2} f_3(\varphi - \varphi_j) + f_1(\varphi - \varphi_j),$$

$$N f_1(\varphi_i) = 2\pi J_1 + \sum_{j=1}^{M_0 - 2} f_2(\varphi_i - \varphi_j) + f_3(\varphi_i - \varphi) + f_1(\varphi_i - \varphi). \tag{2.13}$$

According to the method described in the next section, we deduce a dispersion curve identical to (2.10); we thus obtain a degenerate triplet (Johnson *et al.*, 1973).

2.2 Calculation method for the elementary excitations

2.2.1 Ground state at a given magnetization

Quantum numbers

$$\{J\} = \left\{ -\frac{M - 1}{2}, \ldots, \frac{M - 1}{2} \right\}, \qquad \frac{S^z}{N} = \sigma. \tag{2.14}$$

In the $N = \infty$ limit, we suppose that the abscissas φ_j solving equation

$$N f_1(\varphi_J) = 2\pi J + \sum_{J'} f_2(\varphi_J - \varphi_{J'}) \tag{2.15}$$

admit a limit density $\rho(\varphi)$ over a certain interval $[-\tilde{\varphi}, +\tilde{\varphi}]$:

$$2\pi \, d(J/N) = -\rho(\varphi) d\varphi, \tag{2.16}$$

$$2\sigma = 1 - \frac{1}{\pi} \int_{-\tilde{\varphi}}^{+\tilde{\varphi}} \rho(\varphi) d\varphi. \tag{2.17}$$

After dividing both terms of (2.15) by N, taking the limit and differentiating with respect to φ, we obtain an inhomogeneous integral equation for the root density $\rho(\varphi)$:

$$\frac{\sinh \Phi}{\cosh \Phi - \cos \varphi} = \rho(\varphi) + \frac{1}{2\pi} \int_{-\tilde{\varphi}}^{+\tilde{\varphi}} \frac{\sinh 2\Phi}{\cosh 2\Phi - \cos(\varphi - \varphi')} \rho(\varphi') d\varphi' \tag{2.18}$$

and the formula for the energy per site

$$\varepsilon = \lim \frac{E}{N} = \frac{1}{2\pi} \int_{-\tilde{\varphi}}^{+\tilde{\varphi}} \frac{\sinh^2 \Phi}{\cosh \Phi - \cos \varphi'} \rho(\varphi') d\varphi'. \qquad (2.19)$$

From this we could derive the zero-temperature magnetic susceptibility $\dfrac{\partial \varepsilon}{\partial \sigma}$. Solving equation (2.18) and eliminating $\tilde{\varphi}$ represent a complicated problem in general, the case of the total ground state ($\sigma = 0$) being an exception. In this case, integrating both terms in (2.18) from $-\pi$ to π and comparing with (2.17) gives $\tilde{\varphi} = \pi$. We can then easily find the solution by introducing the Fourier series for ρ and we obtain

$$\rho(\varphi) = \sum_k \frac{e^{ik\varphi}}{2 \cosh k\Phi} = \mathrm{D}n\varphi > 0, \qquad (2.20)$$

from which we reproduce expression (2.3) for the energy in region $\Delta > 1$.

2.2.2 $S = S^z = 1$ spin wave

We write the $(M_0 - 1)$ equations pertaining to the set (2.6) of quantum numbers (denoted $\hat{J}$), for the $\hat{\varphi}_J$ variables. We subtract term by term ($\hat{J} \leftrightarrow J$) the $(M_0 - 1)$ equations for $\hat{\varphi}_J$ from the corresponding equations for φ_J in the ground state. We obtain for $J' \neq J$ and $J' \neq J_0$:

$$N(f_1(\hat{\varphi}_J) - f_1(\varphi_J)) = -\pi - f_2(\varphi_J - \varphi_{J_0})$$
$$+ \sum_{J'(\neq J, J_0)} (f_2(\hat{\varphi}_J - \hat{\varphi}_{J'}) - f_2(\varphi_J - \varphi_{J'})), \quad (2.21)$$

and, for the excitation energy,

$$\eta_1 = \sum_{J \neq J_0} (\cos \hat{k}_J - \cos k_J) - (\cos k_{J_0} - \Delta). \qquad (2.22)$$

Assuming the existence of the limit

$$\lim_{N \to \infty} N(\hat{\varphi}_J - \varphi_J) = g(\varphi) \quad (\varphi = \varphi_J) \qquad (2.23)$$

we deduce the limit form of equations (2.21):

$$-\frac{\sinh \Phi}{\cosh \Phi - \cos \varphi} g(\varphi) = -\pi - f_2(\varphi - \varphi_0)$$
$$-\frac{1}{2\pi} \int_{-\pi}^{+\pi} \frac{\sinh 2\Phi}{\cosh 2\Phi - \cos(\varphi - \varphi')} [g(\varphi) - g(\varphi')] \rho(\varphi') d\varphi' \qquad (2.24)$$

and that of the excitation energy

$$\eta_1 = -\frac{1}{2\pi}\int_{-\pi}^{+\pi}\frac{d}{d\varphi}\frac{\sinh^2\Phi}{\cosh\Phi - \cos\varphi}g(\varphi)\rho(\varphi)d\varphi - (\cos k_0 - \cosh\Phi). \quad (2.25)$$

We have denoted the parameters of the J_0 hole by φ_0 and k_0. Taking (2.18) into account, we extract from (2.24) the integral equation for the unknown function

$$h(\varphi) = g(\varphi)\rho(\varphi) \quad (2.26)$$

in the form

$$h(\varphi) + \frac{1}{2\pi}\int_{-\pi}^{+\pi}\frac{\sinh 2\Phi}{\cosh 2\Phi - \cos(\varphi - \varphi')}h(\varphi')d\varphi' = \pi + f_2(\varphi - \varphi_0). \quad (2.27)$$

To invert (2.27), we make use of the Fourier transform on $[-\pi, +\pi]$ but one should first notice that the second term is not 2π-periodic since f_2 is a continuous function. The function

$$\pi + \pi\lfloor\frac{\varphi + \pi}{2\pi}\rfloor + f_2(\varphi - \varphi_0),$$

in which $\lfloor\ldots\rfloor$ is the 'integer part of' function, is periodic and coincides with $(\pi + f_2)$ in $[-\pi, +\pi]$. Only the expansion of the derivative $h'(\varphi)$ will prove to be useful:

$$h'(\varphi) = \sum_k h_1(k)e^{ik\varphi}. \quad (2.28)$$

The transformed equation is written

$$h_1(k)(1 + e^{-2|k|\Phi}) = e^{ik\pi} - e^{-2|k|\Phi - ik\varphi_0} \quad (2.29)$$

and expression (2.25) for the energy gives us

$$\eta_1 = \sinh\Phi\sum_k h_1(k)e^{-|k|\Phi} + \frac{\sinh^2\Phi}{\cosh\Phi - \cos\varphi_0}. \quad (2.30)$$

Substituting for $h_1(k)$ from (2.29), we obtain

$$\eta_1 = \sum_k\frac{\sinh\Phi}{2\cosh k\Phi}(e^{ik\pi} + e^{-ik\varphi_0}) = \sinh\Phi(\mathrm{Dn}\varphi_0 + \mathrm{Dn}\pi), \quad (2.31)$$

using definition (2.20) of the $\mathrm{Dn}\varphi$ function.

The second formula (2.8) comes from the definition (2.6) of J_0 and from (2.16):

$$|q| = (\pi/2) + \int_0^{\varphi_0}\rho(\varphi)d\varphi.$$

For the $S = 1$, $S^z = 0$ spin wave, involving a complex pair (2.13), the method is analogous; we would have on the right-hand side of (2.27) the supplementary terms $f_1(\ ) + f_3(\ )$. The effect of these two terms is to exactly compensate the energy of the string, leading again to (2.30) irrespective of the string parameter φ.

2.3 Thermodynamics at nonzero temperature: Energy and entropy functionals ($\Delta \geq 1$)

2.3.1 Method (Yang and Yang, 1969)

We shall apply to the solution of the problem of the thermodynamics of the spin chain, the method with which Yang and Yang obtained the thermodynamics of the system of bosons in one dimension (see Subsection 4.4.3). These authors think that this method gives the exact solution for the thermodynamics problem, avoiding the impossible calculation of $\operatorname{tr} e^{-\beta N}$ for the finite-length chain. The success of the method is no doubt due to the linearity of the energy as a functional of the occupation densities, a property which is shared by solvable one-dimensional systems in the $N \to \infty$ limit. Furthermore, according to the description of quantum numbers given in Section 2.1, these can be considered as the quantum numbers of a system of independent fermions. All thus happens as if we had to calculate the partition function of a system of free fermions. In order to obtain the equilibrium densities, it suffices to determine the energy and entropy functionals and to minimize the free energy.

As shown by the expression[1] (see (1.78))

$$E = - \sinh \Phi \sum_{n,i} \frac{\sinh n\Phi}{\cosh n\Phi - \cos \varphi_{n,i}}, \qquad (2.40)$$

the energy of the system is of the order of the number of strings in the system. Strings of macroscopic order cannot be present in macroscopic numbers and therefore do not contribute to the energy per site. It thus suffices to introduce densities of strings of order 1, 2, 3, etc. More precisely, let us consider the class of states of the system for which there exist $N\rho_n(\varphi)d\varphi$ strings of abscissas $\varphi_{n,i}$ in the interval between φ and $\varphi + d\varphi$, where $\rho_n(\varphi)$ ($n = 1, 2, \ldots$) represents a 2π-periodic positive measure. The energy per site ε is a linear functional of the set of densities and can be written, according to (2.40),

$$\varepsilon = \lim \frac{E}{N} = - \sinh \Phi \sum_{n=1}^{\infty} \int_{-\pi}^{+\pi} \frac{\sinh n\Phi}{\cosh n\Phi - \cos \varphi} \rho_n(\varphi)d\varphi. \qquad (2.41)$$

The magnetization per site obviously depends on the existence of strings of macroscopic order, in particular of the exceptional string $\bar{m} = S - S^z$, and satisfies the inequalities

$$\sigma_z = \frac{S^z}{N} \leq \frac{S}{N} \leq \frac{1}{2} - \sum_{n=1}^{\infty} n \int_{-\pi}^{+\pi} \rho_n(\varphi)d\varphi. \qquad (2.42)$$

Equality takes place in the absence of macroscopic strings.

---

[1] Equations (2.32)–(2.39) are not present in the original, whose equation numbering we strictly follow.

2.3.2 String density of states. Hole densities

What is the limit distribution of the quantum numbers J determining such a class of states of densities $\rho_1, \rho_2, \ldots$? The answer is given by the limit form of the coupled equations (1.102):

$$\lim_{\varphi_{n,i} \to \varphi} 2\pi \frac{J_{n,i}}{N} \equiv 2\pi F_n(\varphi) = f_{n\Phi}(\varphi) - \sum_{m,p} [nmp] \int_{-\pi}^{+\pi} f_{p\Phi}(\varphi - \varphi')\rho_m(\varphi')d\varphi',$$

$$(2.43)$$

with the inequalities coming from (1.115) and (1.112):

$$|F_n(\varphi)| \leq \frac{1}{2} - \frac{1}{2}\sum_{m=1}^{\infty} M(n, m) \int_{-\pi}^{+\pi} \rho_m(\varphi)d\varphi. \qquad (2.44)$$

According to our general hypothesis of continuity in Φ and of non-crossing (for finite systems), the correspondence between the F_n and the ρ_n must be bijective and the $F_n(\varphi)$ functions must be decreasing. According to the definition of $J_{n,i}[i = 1, \nu_n]$, the function

$$r_n(\varphi) = -\frac{dF_n}{d\varphi} \qquad (2.45)$$

is a positive measure which represents the maximum available density of states for strings of order n. We thus have

$$0 \leq \rho_n(\varphi) \leq r_n(\varphi). \qquad (2.46)$$

Since ρ_n represents the actual density of strings, the difference

$$\tilde{\rho}_n = r_n - \rho_n \geq 0 \qquad (2.47)$$

can be called the density of 'holes' of strings. The densities ρ_n and r_n are related by the equations derived from (2.43):

$$2\pi r_n(\varphi) = \frac{\sinh n\Phi}{\cosh n\Phi - \cos\varphi} - \sum_{m,p} [nmp] \int_{-\pi}^{\pi} \frac{\sinh p\Phi}{\cosh p\Phi - \cos(\varphi - \varphi')}\rho_m(\varphi')d\varphi',$$

$$n = 1, 2, 3, \ldots \qquad (2.48)$$

whose solutions must verify the inequalities (2.46) and also, according to (2.44),

$$\int_{-\pi}^{+\pi} r_n(\varphi)d\varphi = 1 - \sum_{m=1}^{\infty} M(n, m) \int_{-\pi}^{+\pi} \rho_m(\varphi)d\varphi. \qquad (2.49)$$

2.3.3 Fourier series analysis

The convolutions (2.48) give rise to very simple relations after Fourier transformation. We define the coefficients

$$\rho_n(k) = \int_{-\pi}^{+\pi} \rho_n(\varphi)e^{-ik\varphi}d\varphi = \rho_n^*(-k) \tag{2.50}$$

and similarly for $r_n(k)$ and $\tilde{\rho}_n(k)$. The transformed equations (2.48) are immediately written as

$$r_n(k) = e^{-|k|n\Phi} - \sum_{\substack{m>0 \\ p>0}}[nmp]\rho_m(k)e^{-|k|p\Phi} \tag{2.51}$$

and we sum over p using the definition (1.80) of the $[nmp]$ symbols. For convenience, we define coefficients $\rho_m(k)$ for negative indices

$$\rho_{-m}(k) = -\rho_m(k) \tag{2.52}$$

and for (2.51) we thus obtain

$$\tilde{\rho}_n(k) = r_n(k) - \rho_n(k) = e^{-|k|n\Phi} - \coth|k|\Phi \sum_{m=-\infty}^{+\infty} e^{-|k||n-m|\Phi}\rho_m(k). \tag{2.53}$$

This convolution form allows for an easy inversion of this infinite matrix and we simply find the recursion

$$r_n(k) = \frac{1}{2\cosh k\Phi}(\tilde{\rho}_{n+1}(k) + \tilde{\rho}_{n-1}(k)) \tag{2.54}$$

valid for $n > 0$, provided we set

$$\tilde{\rho}_0(k) = 1. \tag{2.55}$$

In the φ variable, relations (2.54) are written

$$r_n(\varphi) = \frac{1}{2\pi} \int_{-\pi}^{+\pi} \mathrm{Dn}(\varphi - \varphi')\{\tilde{\rho}_{n-1}(\varphi') + \tilde{\rho}_{n+1}(\varphi')\}d\varphi' \tag{2.56}$$

in which the function $\mathrm{Dn}(\varphi)$ has been defined in (2.9).

The constraints on the spin can also be expressed easily. The second equality (2.42) gives us

$$\sigma = \lim \frac{S}{N} = \frac{1}{2} - \sum_{n>0} n\rho_n(0) \quad (k = 0) \tag{2.57}$$

and relations (2.49)

$$r_n(0) = 1 - \sum_{m>0} M(n,m)\rho_m(0). \tag{2.58}$$

We curiously find that these relations (2.58) are the same as (2.54) for $k = 0$, completed by the asymptotic condition originating from (2.58) and (2.57):

$$\lim_{n \to \infty} \tilde{\rho}_n(0) = 2\sigma. \tag{2.59}$$

Relations (2.59), (2.54) and (2.46) thus contain all the information in the coupled equations in the limit $N = \infty$.

2.3.4 Expression for the energy

We here show that the energy depends in the end only on the density $\tilde{\rho}_1$, in other words on the distribution of real roots. Expression (2.41) indeed also equals the Fourier coefficient series

$$\varepsilon = \frac{E}{N} = -\sinh \Phi \sum_{n=1}^{+\infty} \sum_{k=-\infty}^{+\infty} e^{-n|k|\Phi} \rho_n(k). \tag{2.60}$$

After substituting in (2.60) the expressions for $\rho_n = r_n - \tilde{\rho}_n$ given by (2.54), we obtain the remarkable result

$$\varepsilon = \varepsilon_0 + \sinh \Phi \sum_{k=-\infty}^{+\infty} \frac{\tilde{\rho}_1(k)}{2 \cosh k\Phi} \tag{2.61}$$

in which the quantity ε_0 is the ground state energy per site (Walker, 1959)

$$\varepsilon_0 = \lim \frac{E_0}{N} = -\sinh \Phi \sum_{k=-\infty}^{+\infty} \frac{e^{-|k|\Phi}}{2 \cosh k\Phi}. \tag{2.62}$$

We can furthermore write

$$\varepsilon = \varepsilon_0 + \sinh \Phi \int_{-\pi}^{+\pi} \mathrm{Dn}(\varphi) \tilde{\rho}_1(\varphi) d\varphi. \tag{2.63}$$

The positivity of Dn and $\tilde{\rho}_1$ means that ε_0 is indeed the ground state energy.

2.3.5 Expression for the entropy

The expression for the entropy comes directly from the description (1.104) of the quantum numbers. Let us consider a subdivision of the intervals of variation of $J_{n,i}$, defined by an interval of standard length $d\varphi$. By definition of the densities $r_n(\varphi)$ (expression (2.45)), there exist

$$dJ_n = Nr_n(\varphi)d\varphi \quad (Nd\varphi \gg 1) \tag{2.64}$$

possible string states in the interval dJ_n defined by the abscissas φ, $\varphi + d\varphi$. If the density of strings of order n is $\rho_n(\varphi)$, the number of $J_{n,i}$ in dJ_n is

$$dv_n = N\rho_n(\varphi)d\varphi. \tag{2.65}$$

The total number of states of the chain which are compatible with the given densities is thus

$$\mathcal{N} = \prod_n \prod_{d\varphi} C_{dv_n}^{dJ_n}. \tag{2.66}$$

The entropy of the corresponding class of states is thus

$$\mathcal{S} = \ln \mathcal{N}, \tag{2.67}$$

and we obtain exactly for the entropy per site

$$\frac{\mathcal{S}}{N} = \sum_{n=1}^{\infty} \int_{-\pi}^{+\pi} d\varphi \{r_n \ln r_n - \rho_n \ln \rho_n - \tilde{\rho}_n \ln \tilde{\rho}_n\}_\varphi. \tag{2.68}$$

2.4 Thermodynamics at nonzero temperature: Thermodynamic functions

2.4.1 Variation of the free energy

Having obtained the energy and entropy as functionals of the densities ρ_n and $\tilde{\rho}_n$ in Section 2.2, it remains to minimize the free energy

$$F = E - T\mathcal{S} \tag{2.69}$$

taking constraint (2.42) on the magnetic component of total spin into account. Starting again from the remark following equation (2.40), the strings of macroscopic order do not contribute to the free energy and we shall choose the spin constraint in form (2.59).

It is natural to represent the densities as Fermi functions

$$\rho_n = \frac{r_n}{e^{\frac{\varepsilon_n}{T}} + 1}, \qquad \tilde{\rho}_n = \frac{r_m}{e^{-\frac{\varepsilon_m}{T}} + 1} \tag{2.70}$$

in terms of the maximal density $r_n(\varphi)$ and of a still arbitrary pseudo-energy function $\varepsilon_n(\varphi, T)$, which is a means of parametrizing (2.46) and (2.47). Taking expressions (2.63) for the energy and (2.68) for the entropy into account, we write the free energy in the form

$$F_1 = \frac{F}{N} = \frac{E_0}{N} + \sinh \Phi \int_{-\pi}^{+\pi} \mathrm{Dn}(\varphi)\tilde{\rho}_1(\varphi)d\varphi$$

$$+ \sum_{n=1}^{\infty} \int_{-\pi}^{+\pi} \{\tilde{\rho}_n(\varphi)\varepsilon_n(\varphi) - r_n(\varphi)\ln(1 + e^{\frac{\varepsilon_n}{T}})\}d\varphi \tag{2.71}$$

or still, in terms of Fourier coefficients,

$$F_1 = \frac{E_0}{N} + \sinh \Phi \sum_k \frac{\tilde{\rho}_1(k)}{2 \cosh k\Phi} + \sum_{n>0} \sum_k \{\tilde{\rho}_n(k)\varepsilon_n(k) - L_n(k)r_n(k)\} \quad (2.72)$$

with the definition

$$L_n(k) = \frac{T}{2\pi} \int_{-\pi}^{+\pi} \ln(1 + e^{\frac{\varepsilon_n(\varphi)}{T}}) e^{ik\varphi} d\varphi. \quad (2.73)$$

Since relations (2.54) explicitly express the ρ in terms of the hole densities $\tilde{\rho}$, we shall consider the latter as independent variables when calculating the variation of the free energy.

$$\delta F_1 = \sinh \Phi \sum_k \frac{\delta\tilde{\rho}_1(k)}{2 \cosh k\Phi} + \sum_{n>0} \sum_k \{\varepsilon_n(k)\delta\tilde{\rho}_n(k) - L_n(k)\delta r_n(k)\}. \quad (2.74)$$

The variations $\delta\tilde{\rho}_n(k)$ $(n = 1, 2, 3, \ldots)$ being independent, we derive from the stationarity of F_1 the set of equations for the ε_n:

$$\varepsilon_n(k) = \frac{1}{2 \cosh k\Phi}[L_{n+1}(k) + L_{n-1}(k)], \quad n > 1, \quad (2.75)$$

$$\varepsilon_1(k) = \frac{1}{2 \cosh k\Phi}(L_2(k) - \sinh \Phi). \quad (2.76)$$

In φ space we obtain the recurrent sequence of nonlinear equations

$$\frac{\varepsilon_n}{T} = \text{Dn} * \{\ln(1 + e^{\frac{\varepsilon_{n+1}}{T}}) + \ln(1 + e^{\frac{\varepsilon_{n-1}}{T}})\}, \quad n > 1, \quad (2.77)$$

$$\frac{\varepsilon_1}{T} = \text{Dn} * \ln(1 + e^{\frac{\varepsilon_2}{T}}) - \frac{1}{T} \sinh \Phi \text{Dn} \quad (2.78)$$

in which we have used the convolution notation

$$(\text{Dn} * f)(\varphi) = \frac{1}{2\pi} \int_{-\pi}^{+\pi} \text{Dn}(\varphi - \varphi') f(\varphi') d\varphi'. \quad (2.79)$$

2.4.2 Behaviour of the pseudo-energies for $n \to \infty$

Clearly, we did not need to take condition (2.59) into account in order to obtain the set of equations (2.77). In fact, this condition only concerns the asymptotic in n behaviour of $\tilde{\rho}_n(0)$; we thus herewith avoid having to introduce a multiplier.

In contrast, this behaviour is linked to that of r_n and ε_n: it is thus necessary to study which asymptotic behaviour of ε_n is compatible with equations (2.77).

From the positivity of $\text{Dn}(\varphi)$, we deduce from (2.77)

$$\frac{\varepsilon_n}{T} > 0, \quad n > 0. \quad (2.80)$$

The existence of an integrable limit $\bar{\varepsilon}(\varphi)$ for the sequence $\varepsilon_n(\varphi)$ is impossible. In the contrary case we would have the contradiction

$$\frac{\bar{\varepsilon}(0)}{T} = \frac{1}{2\pi} \int_{-\pi}^{+\pi} \ln\left(1 + e^{\frac{\bar{\varepsilon}(\varphi)}{T}}\right) d\varphi > \frac{1}{T}\bar{\varepsilon}(0). \tag{2.81}$$

It is thus natural to suppose that

$$\lim_{n\to\infty} \frac{\varepsilon_n(\varphi)}{T} = +\infty, \qquad -\pi \leq \varphi \leq \pi. \tag{2.82}$$

With this hypothesis our equations (2.77) are linearized in the $n = \infty$ limit and we deduce the asymptotic behaviour

$$\frac{1}{T}\varepsilon_n(k) \sim e^{-|k|n\Phi}, \qquad k \neq 0$$

$$\frac{1}{T}\varepsilon_n(0) \sim \begin{cases} \lambda n + \mu, & k = 0 \\ \ln n, & \text{if} \quad \lambda = 0 \end{cases} \tag{2.83}$$

in which λ and μ are two constants.

We thus conclude that

$$\lim_{n\to\infty} \frac{1}{T}\frac{\varepsilon_n(\varphi)}{n} = \lambda \geq 0. \tag{2.84}$$

In the absence of any proof concerning the solutions of (2.77), we conjecture that providing a positive number λ and a T is sufficient to completely determine a unique sequence $\varepsilon_n(\varphi, T, \lambda)$ verifying (2.77), (2.83) and (2.84); these functions are periodic and analytic in the strip $|\Im\varphi| < \Phi$.

2.4.3 The free energy

We are now in a position to calculate the free energy, taking into account the constraint on spin: $\lim_{n\to\infty} \tilde{\rho}_n(0) = 2\sigma > 2\sigma_z$. Substituting in the expansion (2.72) of the free energy the expressions (2.54) for the $r_n(k)$, we obtain, taking (2.75) into account

$$F_1 = \frac{E_0}{N} + \sum_{k=-\infty}^{+\infty} -\frac{L_1(k)}{2\cosh k\Phi} + \sigma \lim_{n\to\infty} \frac{\varepsilon_{n+1}(k) - \varepsilon_n(k)}{\cosh k\Phi} \tag{2.85}$$

or in other words, taking (2.83) and (2.84) into account, the expression for the free energy

$$F_1 = \frac{E_0}{N} - \frac{T}{2\pi} \int_{-\pi}^{+\pi} d\varphi \mathrm{Dn}(\varphi) \ln\left(1 + e^{\frac{\varepsilon_1(\varphi)}{T}}\right) + \sigma\lambda T \tag{2.86}$$

as a function of the three parameters T, σ and λ.

Up to this point the minimization of the free energy in the space of densities $\tilde{\rho}_n$ leaves us still in the presence of two parameters σ and λ. Minimization in σ is immediate: since the product λT is positive, the minimum is attained for $\sigma = \sigma_z$. The minimization with respect to λ remains, giving in general

$$\sigma = \sigma_z = \frac{1}{2\pi T} \int_{-\pi}^{+\pi} d\varphi \mathrm{Dn}(\varphi) \frac{\frac{\partial \varepsilon_1}{\partial \lambda}}{1 + e^{-\frac{\varepsilon_1}{T}}}. \tag{2.87}$$

Eliminating the parameter λ between expressions (2.86) and (2.87) finally gives the free energy as a function of temperature T and magnetization σ (we shall write σ instead of σ_z). It follows from (2.86) and (2.87) that the quantity λT is the intensive variable conjugate to the magnetization

$$B = \frac{\partial F_1}{\partial \sigma}(\sigma, T) = \lambda T \tag{2.88}$$

and must thus be interpreted as the value of the magnetic field under which the magnetization per site takes the value σ.

It must be noted that the free energy of the chain is formally identical to that of a system of independent quasi-particles of spin-1/2 fermionic type, of which the dispersion curve would be

$$\varepsilon_1 = \varepsilon_1(\varphi, T, \lambda),$$
$$\frac{dq}{d\varphi} = -\mathrm{Dn}(\varphi) \tag{2.89}$$

for a given temperature and magnetic field. The function ε_1 depends uniquely on the solution of the system (2.77) and (2.78).

2.4.4 Equilibrium distribution of strings

It now remains to find the maximal densities $r_n(\varphi)$ in order to completely determine the densities of strings and string holes at equilibrium. To this end we have the sequence of equations (2.56) supplemented by conditions (2.55) and (2.59):

$$r_n = \mathrm{Dn} * \left(\frac{r_{n+1}}{1 + e^{-\varepsilon_{n+1}/T}} + \frac{r_{n-1}}{1 + e^{-\varepsilon_{n-1}/T}} \right), \qquad n > 1,$$

$$r_1 = \frac{1}{2\pi} \mathrm{Dn} + \mathrm{Dn} * \frac{r_2}{1 + e^{-\varepsilon_2/T}}, \tag{2.90}$$

$$\lim_{n \to \infty} r_n(0) = 2\sigma. \tag{2.91}$$

We notice that taking the derivative of system (2.77) and (2.78) with respect to T^{-1} exactly gives system (2.90) and (2.91). By adjusting the behaviour for $n = \infty$, we thus obtain the formulas

$$r_n(\varphi) = -\frac{1}{2\pi} \frac{1}{\sinh \Phi} \frac{\partial}{\partial T^{-1}} (T^{-1} \varepsilon_n(\varphi))_\lambda. \tag{2.92}$$

In particular, the parameter μ of expression (2.83) for the asymptotic behaviour of $\varepsilon_n(\varphi)$ verifies the equation

$$\frac{\partial \mu}{\partial T} = 2\sigma \frac{\sinh \Phi}{T^2}. \tag{2.93}$$

The preceding relations find an application in the calculation of the differential of the free energy:

$$d(T^{-1}F_1) = -\frac{1}{2\pi} \int_{-\pi}^{+\pi} d\varphi \operatorname{Dn}(\varphi) \frac{\frac{\partial(T^{-1}\varepsilon_1)}{\partial T^{-1}}}{1 + e^{-\varepsilon_1/T}} dT^{-1} + \lambda d\sigma. \tag{2.94}$$

By means of the first relation (2.92), we obtain

$$d(T^{-1}F_1) = \sinh \Phi \int_{-\pi}^{+\pi} d\varphi \operatorname{Dn}(\varphi) \tilde{\rho}_1(\varphi) dT^{-1} + \lambda d\sigma, \tag{2.95}$$

in other words the expected relation for the internal energy

$$E - E_0 = N \frac{\partial(F_1/T)}{\partial T^{-1}}. \tag{2.96}$$

2.4.5 Domain of applicability

The expression for the free energy obtained in Section 2.3 is that of the antiferromagnetic chain $(2J = +1)$ in the region $\Delta > 1$. The calculation is also applicable in the same region for the ferromagnetic Hamiltonian $(2J = -1)$; it is sufficient to change the sign of the temperature and of the free energy to obtain

$$F_{\text{ferro}}(T, \Delta) = -F(-T, \Delta), \qquad T > 0, \qquad \Delta > 0. \tag{2.97}$$

The only point where the sign of T might intervene explicitly is in the minimization of the free energy with respect to σ, σ being fixed; but finally the result is given by (2.97).

The solution can futhermore be extended to the region $\Delta < -1$ because of the existence of the canonical transformation U, formula (1.31):

$$\operatorname{tr} e^{-\frac{H_\Delta}{T}} = \operatorname{tr} e^{-\left(\frac{H_{-\Delta}}{-T}\right)}. \tag{2.98}$$

In summary, the expressions for the free energy in the region $|\Delta| > 1$ are in the four cases

$$\begin{cases} F(|\Delta|, T), & J\Delta > 0 \\ -F(|\Delta|, -T), & J\Delta < 0. \end{cases} \tag{2.99}$$

Let us finish with a remark on the limiting case $\Delta = 1$ (Takahashi, 1971). This case can be obtained as the limit $\Delta = 1 + 0$ ($\Phi = 0$). It is sufficient to perform the change of variables $x = \varphi/\Phi$ and replace in the equations and formulas

$$\frac{1}{2\pi}\int_{-\pi}^{+\pi} Dn(\varphi - \varphi')d\varphi' \ldots \quad \text{with} \quad \frac{1}{4}\int_{-\infty}^{+\infty}\frac{dx'}{\cosh\frac{\pi}{2}(x - x')}\ldots$$

and

$$\sinh\Phi Dn(\varphi) \quad \text{with} \quad \frac{\pi}{2}\frac{1}{\cosh\frac{\pi x}{2}}.$$

Appendix A

Our aim is to evaluate the function $\chi_n(\varphi)$ given by expression (1.88) in the asymptotic limit $N \gg 1$. We hypothesize the condensation of the various string abscissas as formulated in Subsection 2.2.1. We thus obtain, as a function of the densities $\rho_n(\varphi)$,

$$-2\chi_\mu(\varphi) = \ln\frac{\cosh(\mu - 1)\Phi - \cos\varphi}{\cosh(\mu + 1)\Phi - \cos\varphi} + \sum_{n=1}^{\infty}\int_{-\pi}^{+\pi}d\varphi'\rho_n(\varphi')$$

$$\ln\frac{\cosh(\mu + n + 1)\Phi - \cos(\varphi - \varphi')}{\cosh(\mu - n - 1)\Phi - \cosh(\varphi - \varphi')}\frac{\cosh(\mu + n - 1)\Phi - \cos(\varphi - \varphi')}{\cosh(\mu - n + 1)\Phi - \cosh(\varphi - \varphi')}.$$
$$(A.1)$$

Applying the Fourier tranform method described in Subsection 2.2.2,[2] we obtain directly

$$|k|\chi_\mu(k) = e^{-|k|\mu\Phi} - \frac{1}{2}\sum_{n=-\infty}^{+\infty}\rho_n(k)(e^{-|k|\|\mu-n+1|\Phi} + e^{-|k|\|\mu-n-1|\Phi}), \qquad (A.2)$$

in which we have set $\rho_{-n}(k) = -\rho_n(k)$.

Distinguishing three sorts of terms $\mu - n \geq 0$, we write (A.2) in the form

$$|k|\chi_\mu(k) = \sinh|k|\Phi(e^{-|k|\mu\Phi} + \rho_\mu(k)) - \sum_{n=-\infty}^{+\infty}\rho_n(k)\cosh k\Phi e^{-|k|\|\mu-n|\Phi} \quad (A.3)$$

which, with the help of equation (2.53), is simply written

$$\chi_\mu(k) = \frac{\sinh k\Phi}{k}r_\mu(k), \qquad \mu > 0. \qquad (A.4)$$

---

[2] Correcting the original, in which Subsection 2.2.3 (which does not exist) is referred to.

It can be checked that the formula is valid for $k = 0$. We note the simplicity of the result in the limit $\Phi = 0$ ($\Delta = 1$), which yields

$$\chi_\mu(x) = r_\mu(x) > 0 \quad \left(x = \lim \frac{\varphi}{\Phi}\right). \tag{A.5}$$

For $\Delta > 1$, things are somewhat less simple and one must write, using (2.54),

$$\chi_\mu(k) = \frac{\tanh k\Phi}{2k}(\tilde{\rho}_{\mu-1}(k) + \tilde{\rho}_{\mu+1}(k)) \tag{A.6}$$

in order to obtain expression (1.90) of $\chi_\mu(\varphi)$ and show that it is positive.

3

Thermodynamics of the spin-1/2 chain:
Limiting cases

3.1 The Ising limit

The $\Delta = \infty$ limit, or Ising limit, provides an interesting test of the calculation presented in Section 2.3, since the thermodynamic functions of the one-dimensional Ising model can easily be obtained using a direct method. To avoid reintroducing the constant J which appears in Hamiltonian (1.6), putting $2J = \dfrac{1}{\Delta}$, and then taking the limit $\Delta = \infty$, we shall change the temperature scale by defining a new inverse temperature β such that

$$\beta T = \Delta. \tag{3.1}$$

The limit Hamiltonian H_∞ is that of Ising:

$$\lim_{\Delta \to \infty} \frac{H_\Delta}{T} = \beta H_\infty = \beta \sum_{n=1}^{N} \left(S_n^z S_{n+1}^z - \frac{1}{4} \right). \tag{3.2}$$

The eigenvalues of H_Δ are the negative integers $-Q$, where Q is precisely the number of strings of the corresponding eigenstates of H_Δ. This comes from the fact that in the limit $\Delta = \infty$, a string of order n gives rise to a bound state of n consecutive $\downarrow$ spins, as we have seen in Subsection 1.5.4. A convenient way to calculate the thermodynamics of the Ising chain is to introduce the partition function

$$Z = \sum_{M,Q} e^{\beta Q - \gamma M} g(M, Q) \tag{3.3}$$

in which γ is the magnetic field and $g(M, Q)$ is the degeneracy of the $S^z = \dfrac{N}{2} - M$, $H_\infty = -Q$ subspace

$$g(M, Q) = \frac{N}{M} C_Q^M C_{Q-1}^{N-M-1}. \tag{3.4}$$

Applying the saddle-point method results in

$$\frac{Q}{N} = \frac{1}{2}e^{\beta}\left[e^{\beta} + \sinh^2\frac{\gamma}{2}\right]^{-\frac{1}{2}}\left[\left(e^{\beta} + \sinh^2\frac{\gamma}{2}\right)^{+\frac{1}{2}} + \cosh\frac{\gamma}{2}\right]^{-\frac{1}{2}}, \quad (3.5)$$

$$2\sigma = 1 - \frac{2M}{N} = \sinh\frac{\gamma}{2}\left[e^{\beta} + \sinh^2\frac{\gamma}{2}\right]^{-\frac{1}{2}}, \quad (3.6)$$

$$\frac{1}{N}\ln Z = \ln\left\{\left[e^{\beta} + \sinh^2\frac{\gamma}{2}\right]^{\frac{1}{2}} + \cosh\frac{\gamma}{2}\right\} - \frac{\gamma}{2}. \quad (3.7)$$

The free energy F_1 must thus be such that in the limit we obtain

$$\lim_{\Delta\to\infty}\frac{F_1}{T} = -\ln\left\{\left[e^{\beta} + \sinh^2\frac{\gamma}{2}\right]^{1/2} + \cosh\frac{\gamma}{2}\right\} + \gamma\sigma. \quad (3.8)$$

We must now determine the limit form of expression (2.86) and of the recursion relations (2.77) and (2.78). When Φ tends to infinity, the function $Dn(\varphi)$ tends uniformly to $\frac{1}{2}$ and the $\frac{\varepsilon_n}{T}$ become φ independent. Setting $z_n = e^{\varepsilon_n/T}$, we obtain

$$z_n^2 = (1 + z_{n+1})(1 + z_{n-1}), \quad n > 0 \quad (3.9)$$

$$z_0 + 1 = e^{-\frac{\Delta}{T}} = e^{-\beta} \quad (3.10)$$

with the asymptotic behaviour

$$\ln z_n \sim \lambda n. \quad (3.11)$$

The nonlinear recursion (3.9) can be solved by setting

$$\zeta_n = \sqrt{1 + z_n} > 0, \quad (3.12)$$

which gives

$$\zeta_n^2 - \zeta_{n-1}\zeta_{n+1} = 1, \quad \zeta_0 = e^{-\frac{\beta}{2}}. \quad (3.13)$$

This reminds us of reductions in the theory of continuous fractions and leads us to the solution in terms of two parameters λ and μ:

$$\zeta_n = \frac{\sinh\frac{1}{2}(\lambda n + \mu)}{\sinh\frac{\lambda}{2}}, \quad (3.14)$$

with the definition of μ

$$\sinh\frac{\mu}{2} = \sinh\frac{\lambda}{2}e^{-\frac{\beta}{2}}. \quad (3.15)$$

We then obtain for ζ_1

$$\zeta_1 = e^{-\frac{\beta}{2}}\left\{\left[e^{\beta} + \sinh^2\frac{\lambda}{2}\right]^{\frac{1}{2}} + \cosh\frac{\lambda}{2}\right\}. \quad (3.16)$$

Formula (2.86) gives us

$$\lim_{\Delta \to \infty} \frac{F_1}{T} = -\ln \zeta_1 + \sigma \lambda - \frac{\beta}{2} \qquad (3.17)$$

and formula (2.87)

$$\sigma = \frac{\partial}{\partial \lambda} \ln \zeta_1 = \frac{1}{2} \tanh \frac{\mu}{2} = \frac{1}{2} \frac{\sinh \frac{\lambda}{2}}{\left[e^\beta + \sinh^2 \frac{\lambda}{2} \right]^{\frac{1}{2}}}. \qquad (3.18)$$

Comparing expressions (3.18) and (3.6) for the magnetization gives $\lambda = \gamma$. Thereafter expression (3.17) for the free energy becomes identical to the expected limit (3.8). One can easily verify that the limit densities are

$$\tilde{\rho}_n = 2\sigma \coth \frac{1}{2} (\lambda n + \mu) \qquad (3.19)$$

and that relations (2.92) are true. The solution proposed for the thermodynamics of the spin chain thus correctly reproduces the Ising limit.

3.2 The $T = \pm 0$ limits

The study of the $T = 0$ limit of the ferro- and antiferromagnetic systems also provides a test of the consistency of the solution, since the minimal energies of the states of given spin and momentum have already been calculated elsewhere (Griffiths, 1964; Yang and Yang, 1966a). We shall develop the cases $T = +0$ and $T = -0$ in parallel, which allows us to cover both cases of the sign of $J\Delta$ (2.99).

3.2.1 Limit of the recursion equations

Still distinguishing the two possible signs of T, we shall set

$$\lim_{T=0} \varepsilon_n(\varphi) = E_n(\varphi), \qquad (3.20)$$

$$\lim_{T=0} \lambda T = \Lambda. \qquad (3.21)$$

We know that ε shares the sign of T for $n > 1$ and we deduce, using (2.73) and (3.20), that

$$\lim_{T=0} L_n(k) = E_n(k), \qquad n > 1 \qquad (3.22)$$

$$\lim_{T=+0} L_1(k) = \frac{1}{2\pi} \int_{E_1(\varphi) > 0} e^{ik\varphi} E_1(\varphi) d\varphi = L^+(k), \qquad (3.23)$$

$$\lim_{T=-0} L_1(k) = \frac{1}{2\pi} \int_{E_1(\varphi) < 0} e^{ik\varphi} E_1(\varphi) d\varphi = L^-(k). \qquad (3.24)$$

Equations (2.75) and (2.76) take the limit form

$$E_n(k) = \frac{1}{2\cosh k\Phi}\{E_{n+1}(k) + E_{n-1}(k)\}, \quad n > 2$$

$$E_2(k) = \frac{1}{2\cosh k\Phi}\{E_3(k) + L^{\pm}(k)\},$$

$$E_1(k) = \frac{1}{2\cosh k\Phi}\{E_2(k) - \sinh \Phi\} \tag{3.25}$$

with the condition

$$\lim_{n\to\infty} \frac{E_n}{n} = \Lambda. \tag{3.26}$$

From (3.25) and (3.26) one deduces

$$E_n(k) = L^{\pm}(k)e^{-(n-1)|k|\Phi} + \Lambda(n-1)\delta(k), \tag{3.27}$$

$$2E_1(k)\cosh k\Phi = L^{\pm}(k)e^{-|k|\Phi} - \sinh \Phi + \Lambda\delta(k). \tag{3.28}$$

In φ space, we thus obtain integral equations for the function $E_1(\varphi)$ in the regions $E_1(\varphi) > 0$ (antiferromagnetic case) or $E_1(\varphi) < 0$ (ferromagnetic case).

3.2.2 Antiferromagnetic case

The transform of (3.28) (divided by $2\cosh k\Phi$) is written

$$E_1(\varphi) - \int_{E_1>0} d\varphi' R(\varphi - \varphi')E_1(\varphi') = \frac{\Lambda}{2} - \sinh \Phi \mathrm{Dn}(\varphi) \tag{3.29}$$

in which the function $R(\varphi)$ is given by the Fourier series

$$R(\varphi) = \frac{1}{2\pi}\sum_{k=-\infty}^{\infty} \frac{e^{ik\varphi}}{e^{2|k|\varphi} + 1}. \tag{3.30}$$

This is a positive function, as the product of two positive functions $\mathrm{Dn}(\varphi)$ and $\dfrac{\sinh \Phi}{\cosh \Phi - \cos \varphi}$.

The limit forms of (2.86) and (2.87) for the energy and magnetization are

$$\lim_{T=+0} F_1 = \frac{E_0}{N} - \frac{1}{2\pi}\int_{E_1>0} \mathrm{Dn}(\varphi)E_1(\varphi)d\varphi + \Lambda\sigma, \tag{3.31}$$

$$2\sigma = \frac{1}{\pi}\int_{E_1>0} \mathrm{Dn}(\varphi)\frac{\partial E_1}{\partial \Lambda}d\varphi. \tag{3.32}$$

From formula (2.92) we extract the density $r_1(\varphi)$:

$$r_1(\varphi) = \frac{1}{2\pi \sinh \Phi}\left(\Lambda\frac{\partial E_1}{\partial \Lambda} - E_1\right), \tag{3.33}$$

which allows us to rewrite $\lim F_1$ in the form

$$\lim_{T=+0} F_1 = \varepsilon_0 + \sinh \Phi \int_{E_1>0} \mathrm{Dn}(\varphi) r_1(\varphi) d\varphi. \tag{3.34}$$

With the help of expression (3.33) and of the Λ-derivative equation (3.29), we verify that $r_1(\varphi)$ satisfies the integral equation

$$r_1(\varphi) - \int_{E_1>0} d\varphi' R(\varphi - \varphi') r_1(\varphi') = \frac{1}{2\pi} \mathrm{Dn}(\varphi). \tag{3.35}$$

Up to making the nature of the region $E_1 > 0$ precise, equations (3.34) and (3.35) have the form obtained by Griffiths for the case of the isotropic chain ($\Phi = 0$). We can also transform them to obtain the equations given by Yang, which involve the region $E_1 < 0$. Substituting in (3.28) the following decomposition of L^+:

$$L^+(k) = E_1(k) - \frac{1}{2\pi} \int_{E_1<0} e^{ik\varphi} E_1(\varphi) d\varphi, \tag{3.36}$$

we obtain

$$E_1(k) + e^{-2|k|\varphi} \frac{1}{2\pi} \int_{E_1<0} d\varphi e^{ik\varphi} E_1(\varphi) = \Lambda \delta(k) - \sinh \Phi e^{-|k|\Phi}, \tag{3.37}$$

which in the φ variable becomes

$$E_1(\varphi) + \frac{1}{2\pi} \int_{E_1<0} d\varphi' \frac{\sinh 2\Phi}{\cosh 2\Phi - \cos(\varphi - \varphi')} E_1(\varphi') = \Lambda - \frac{\sinh^2 \Phi}{\cosh \Phi - \cos \varphi}. \tag{3.38}$$

By virtue of (3.33) we thus obtain, for the density $r_1 = \rho_1$,

$$r_1(\varphi) + \frac{1}{2\pi} \int_{E_1<0} d\varphi' \frac{\sinh 2\Phi}{\cosh 2\Phi - \cos(\varphi - \varphi')} r_1(\varphi') = \frac{1}{2\pi} \frac{\sinh \Phi}{\cosh \Phi - \cos \varphi}. \tag{3.39}$$

In the same way one obtains, after the Fourier transformation of (3.39), division by $2 \cosh k\Phi$ and summation over k,

$$\varepsilon = \frac{E_0}{N} + \sinh \Phi \int_{E_1>0} d\varphi \mathrm{Dn}(\varphi) r_1(\varphi) = -\int_{E_1<0} \frac{\sinh^2 \Phi}{\cosh \Phi - \cos \varphi} r_1(\varphi) d\varphi. \tag{3.40}$$

In the same way one can transform equation (3.32) into the following:

$$2\sigma = 1 - \frac{1}{\pi} \int_{E_1<0} \frac{\sinh \Phi}{\cosh \Phi - \cos \varphi} \frac{\partial E_1}{\partial \Lambda} d\varphi. \tag{3.41}$$

Multiplying equation (3.38) by $\dfrac{\partial E_1}{\partial \Lambda}$ and integrating over the domain $E_1 < 0$, one obtains the identity allowing us to write the magnetization in its final form

$$\sigma = \frac{1}{2} - \int_{E_1 < 0} r_1(\varphi) d\varphi. \tag{3.42}$$

It now remains to prove the following.

Proposition: The region $E_1(\varphi) < 0$ on $[-\pi, +\pi]$ is a symmetric interval $[-\varphi_1, +\varphi_1]$.

The proof makes use of the integral equation (3.29).

(a) $R(\varphi)$ is decreasing in the interval $[0, \pi]$. Indeed, following the remark made after definition (3.30), one can calculate the derivative $R'(\varphi)$ by using the convolution and parity

$$R'(\varphi) = -\frac{1}{2\pi} \int_0^\pi [\mathrm{Dn}(\varphi - \psi) - \mathrm{Dn}(\varphi + \psi)] \frac{\sinh \Phi \sin \psi}{(\cosh \Phi - \cos \psi)^2} d\psi,$$
$$0 < \varphi < \pi \tag{3.43}$$

but the quantity in square brackets in the integrand is a continuous function of ψ which vanishes only for $\psi = 0$ and $\psi = \pi$, as follows from the parity of $\mathrm{Dn}(\varphi)$ and its decreasing nature on $[0, \pi]$. Since it is positive for small enough ψ $(0 < \varphi < \pi)$, it is thus positive for $0 < \psi < \pi$. One then deduces that $R'(\varphi) < 0$ for $0 < \varphi < \pi$.

(b) If λ_1 is the largest eigenvalue of the positive integral kernel $R(\varphi - \varphi')$ in the domain $E_1 > 0$, then $\lambda_1 \le 1/2$.

It is first of all evident that such a kernel is positive. Consider then $f \in L^2(-\pi, +\pi)$. From (3.30), the largest eigenvalue of R on $[-\pi, +\pi]$ is $\dfrac{1}{2}$. We thus have

$$\int \int_{-\pi}^{+\pi} d\varphi d\varphi' f(\varphi) R(\varphi - \varphi') f(\varphi') \le \frac{1}{2} \left[\int_{-\pi}^{+\pi} d\varphi f(\varphi) \right]^2. \tag{3.44}$$

Let us restrain f to the class $f_1 \in L^2(E_1 > 0)$, normalized. By virtue of a well-known theorem (Mikhlin, 1964), we can write

$$\lambda_1 = \mathrm{Max}_{f_1} \int \int_{E_1 > 0} d\varphi d\varphi' R(\varphi - \varphi') f_1(\varphi) f_1(\varphi') \tag{3.45}$$

and with the help of (3.44), we have $\lambda_1 \le \dfrac{1}{2}$.

(c) The function $E_1(\varphi)$ is increasing on $[0, \pi]$. From result (b) the Liouville series, which defines $E_1(\varphi)$ by iteration of (3.29), is uniformly convergent with respect to φ. $E_1(\varphi)$ is thus differentiable and verifies the equation

$$E_1'(\varphi) - \int_{\substack{E_1 > 0 \\ \pi > \psi > 0}} d\psi [R(\varphi - \psi) - R(\varphi + \psi)] E_1'(\psi) = - \sinh \Phi \mathrm{Dn}'(\varphi) \quad (3.46)$$

in which we have used the fact that E_1' is odd. Reasoning as in paragraph (a) and using the fact that $R(\varphi)$ is decreasing on $[0, \pi]$, the kernel of (3.46) is positive. The Liouville series converges according to (b) and all its terms are positive, since $-\mathrm{Dn}'(\varphi) \geq 0$ on $[0, \pi]$ and the iterated kernels are positive. We deduce that $E_1'(\varphi) > 0, 0 < \varphi < \pi$.

(d) Being continuous, $E_1(\varphi)$ vanishes at most once on $[0, \pi]$. Let φ_1 be such a zero; if it exists, $E_1(\varphi) < 0$ corresponds to the interval $[-\varphi_1, \varphi_1]$. It is easy to determine the values of Λ and of $r_1(\varphi)$ which correspond to the extreme cases $\varphi_1 = 0$ and $\varphi_1 = \pi$.

Let Λ_0 be the energy gap between the ground state and the spin waves (formula (2.31), setting $\varphi_0 = \pi$):

$$\Lambda_0 = 2 \sinh \Phi \mathrm{Dn}(\pi). \quad (3.47)$$

We easily deduce from (3.29) that, for $\Lambda < \Lambda_0$,

$$\begin{cases} E_1(\varphi) &= \dfrac{\Lambda}{2} - \sinh \Phi \mathrm{Dn}(\varphi) < 0, \\ r_1(\varphi) &= \dfrac{1}{2\pi} \mathrm{Dn}(\varphi), \\ \sigma &= 0. \end{cases} \quad (3.48)$$

For $\Lambda = \Lambda_0$, we get precisely $\varphi_1 = 0$. The fact that the magnetization vanishes for $\Lambda < \Lambda_0$ is consistent with the fact that the magnetic field must be greater than Λ_0 in order to excite a state of spin $|S^z| = 1$.

In contrast, we deduce from (3.38) and (3.39), for $\Lambda > \cosh \Phi + 1 = \Delta + 1$,

$$r_1(\varphi) = \frac{1}{2\pi} \frac{\sinh \Phi}{\cosh \Phi - \cos \varphi},$$

$$\sigma = \frac{1}{2} \quad (\varphi_1 = \pi, \text{ value for } \Lambda = \Delta + 1). \quad (3.49)$$

We obtain saturation of the chain for a finite value of the field.

The ferromagnetic case is similarly treatable by setting $T = -0$; according to (2.84) and (3.27), the magnetic field is $|\Lambda| = -\Lambda$. One easily finds

$$E_n(\varphi) = \Lambda n - \frac{\sinh \Phi \sinh n\Phi}{\cosh n\Phi - \cos \varphi} \leq 0. \quad (3.50)$$

Clearly E_n is the energy of a bound state or string of order n in a magnetic field $|\Lambda|$. These strings truly form the elementary excitations of the ferromagnetic system.

3.2.3 Calculations in the vicinity of $T = 0$

A systematic study of the nonlinear equations (2.77) and (2.78) in the vicinity of $T = 0$ and in a magnetic field $B \neq 0$ was performed by Johnson and McCoy (1972). Calculations and results are far from trivial. It is constantly necessary to distinguish the $B = O(T)$ and finite B regions, due to the fact that $F(T, \sigma)$ is defined in terms of the parameter $\lambda = \dfrac{B}{T}$. One needs to consider four regions A, B, C, D and their boundaries:

(A) $J\Delta < 0$, B finite

$$F(T, B) = -\frac{B}{2} - \frac{T^{3/2}}{\sqrt{2\pi}} \exp\left(\frac{1 - B - \Delta}{T}\right) + \ldots \tag{3.51}$$

(B) $J\Delta > 0$, $\varepsilon_1 > 0$ outside the neighbourhood $\varepsilon_1 = 0$, in other words $B > \Delta + 1$,

$$F(T, B) = -\frac{B}{2} - \frac{T^{3/2}}{\sqrt{2\pi}} \exp\left(\frac{1 + \Delta - B}{T}\right) + \ldots \tag{3.52}$$

(C) $J\Delta > 0$, $\varepsilon_1 < 0$

$$F(T, B) = \varepsilon_0 - \left(\frac{k'}{2Kk^2 \sinh \Phi}\right)^{1/2} \exp\left(-\frac{\Lambda_c}{2T}\right) + \ldots, \tag{3.53}$$

$$\Lambda_c = \frac{2k'K}{\pi} \sinh \Phi - H_0. \tag{3.54}$$

(D) $J\Delta > 0$, ε_1 positive or negative, and finite close to the boundaries of D.

The preceding results are meant for $\Delta \neq 1$; the limit $\Delta = 1$ can be taken, except maybe in the neighbourhood of the boundaries of the four regions A, B, C, D. Let us mention in particular the result for $\Delta = 1$, $B = O(T)$ obtained by the previously cited authors after solving a Wiener–Hopf problem:

$$F(T, B) = \varepsilon_0 - \frac{B^2}{2\pi^2} - \frac{T^2}{3}\left(1 + \frac{1}{2(\ln B)^2} + \ldots\right) + \ldots \tag{3.55}$$

In zero field ($B = 0$), the specific heat thus behaves as $\dfrac{2}{3}T$; this result is in perfect agreement with the numerical evaluation performed on finite chains of 10 or 12 atoms (Bonner and Fisher, 1964). It is also in agreement with a theoretical estimate based on the analogy between the anisotropic spin chain and a fermionic spin-0 gas.

Information on the lowest excitations and the direct application of Landau's theory yields the $\frac{2}{3}$ factor (des Cloizeaux, 1966).

The preceding results were deduced for $|\Delta| \geq 1$. The free energy is however an analytic function of Δ on the whole real axis, for $T \neq 0$ (Araki, 1969). Open problems include the analytic continuation of the nonlinear equations ($\Phi \to i\Theta$) and the direct derivation of the thermodynamics of the chain for $|\Delta| < 1$.

3.3 $T = \infty$ limit

We here calculate the first two terms in the series expansion of the free energy in T^{-1}, a series which can be obtained directly by standard perturbation methods; in zero field, one finds

$$F_1 = -T \ln 2 - \frac{\Delta}{4} - \frac{1}{32T}(\Delta^2 + 2) + O(1/T^2). \tag{3.56}$$

Following the method applied for the Ising limit one obtains from the recursion relations (2.77),

$$\lim_{T \to \infty} e^{\varepsilon_n/T} = z_n = n(n+2), \tag{3.57}$$

one expands the following quantities to order $1/T$:

$$\begin{cases} e^{\varepsilon_n(\varphi)/T} = z_n + \dfrac{n+1}{T}\eta_n(\varphi) + O(1/T^2), \\ \ln(1 + e^{\varepsilon_n/T}) = \ln(1 + z_n) + \dfrac{1}{T}\dfrac{\eta_n}{n+1} + \cdots \end{cases} \tag{3.58}$$

After substituting this in equations (2.75), we obtain

$$(n+1)\eta_n(k)2\cosh k\Phi = n\eta_{n+1}(k) + (n+2)\eta_{n-1}(k) \tag{3.59}$$

with

$$\eta_0(k) = -\sinh \Phi.$$

The linear recursion (3.59) admits the only appropriate solution

$$\eta_n(k) = -\sinh \Phi \left[\frac{n}{2}(1 - e^{-2|k|\Phi}) + 1\right] e^{-|k|n\Phi}, \tag{3.60}$$

from which we deduce for $n = 1$

$$L_1(k) = T\delta(k)\ln(1 + z_1) - \frac{\sinh \Phi}{4}[3 - e^{-2|k|\Phi}]e^{-|k|\Phi} \tag{3.61}$$

and, with the help of formula (2.85), for the free energy

$$F_1 = -T \ln 2 + \varepsilon_0 + \frac{\sinh \Phi}{8} \sum_{k=-\infty}^{+\infty} \frac{3 - e^{-2|k|\Phi}}{\cosh k\Phi} e^{-|k|\Phi}. \tag{3.62}$$

Taking expression (2.62) for $\varepsilon_0 = \dfrac{E_0}{N}$ into account, one finally obtains, to order $1/T$:

$$F_1 = -T \ln 2 - \frac{\sinh \Phi}{4} \sum_k e^{-2|k|\Phi} = -T \ln 2 - \frac{\cosh \Phi}{4}, \qquad (3.63)$$

a result compatible with (3.56).

These few tests on limiting cases, and most importantly the cited study of Johnson and McCoy (1972), strengthen the conviction that the strange nonlinear system of recursion equations obtained in Chapter 2 is capable of describing exactly the thermodynamics of the anisotropic chain, including in the limit of zero temperature of the antiferromagnetic case.

4

δ-Interacting bosons

This chapter is dedicated to the model of a gas of bosons in one spatial dimension with a repulsive delta function interaction (Girardeau, 1960; Lieb and Liniger, 1963). In a sense to be defined later, this system can be considered as a limit of the Heisenberg–Ising chain when the lattice spacing tends to zero, the total length of the chain remaining finite. The results mainly concern the spectrum, the wavefunctions and the thermodynamics; they present a simplified character compared with the spin chain due to the absence of complex strings in the repulsive case. Normalization problems are addressed and offer a prelude to the calculation of average values or correlations of which we know nothing, except at the limit of impenetrable bosons (Lenard, 1964; Vaidya and Tracy, 1979; Jimbo *et al.*, 1980). The general problem of delta function interacting identical particles with an arbitrary internal degree of freedom will be treated in Chapters 10, 11 and 12.

4.1 The elementary symmetric wavefunctions

We consider in a first instance the problem of N particles with identical mass with a two-body δ function interaction of intensity $2c$, but situated on an indefinite axis, which defines the geometric conditions of the collisional problem.

In units where $\hbar = 2m = 1$, the Schrödinger equation for the wavefunction ψ, which is completely symmetric in the coordinates $x_1, x_2, \ldots, x_N$, is written

$$-\sum_{j=1}^{N} \frac{\partial^2 \psi}{\partial x_j^2} + 2c \sum_{i<j} \delta(x_i - x_j) = E\psi. \tag{4.1}$$

We shall call an *elementary wavefunction* a solution to (4.1) continuous in $\mathbb{R}_N$, disregarding any conditions at boundaries or infinity. Owing to the symmetry, equation (4.1) decomposes into the wave equation

$$(\Delta_N + E)\psi = 0 \qquad (4.2)$$

in the *fundamental domain D*:

$$D : x_1 < x_2 < \ldots < x_N, \qquad (4.3)$$

and into a set of $N - 1$ conditions at the boundary of D:

$$\left(\frac{\partial \psi}{\partial x_{i+1}} - \frac{\partial \psi}{\partial x_i} \right) |_{x_{i+1}-x_i=0^+} = c\psi|_{x_{i+1}-x_i=0^+}. \qquad (4.4)$$

These last conditions are obtained by integrating the first term of (4.1) with respect to $(x_{i+1} - x_i)$ in a small interval containing $x_{i+1} - x_i = 0$ and taking the symmetry into account.

It turns out that the superposition method used by Bethe for the spin-1/2 chain again permits the construction of the elementary wavefunctions $\psi_{\{k\}}(x)$ depending on N parameters $\{k\} = \{k_1, \ldots, k_N\}$, which are assumed distinct:

$$\psi_{\{k\}}(x) = \sum_{P \in \pi_N} A(P)e^{i\sum_j x_j k_{P_j}}, \qquad x \in \mathbb{R}_N. \qquad (4.5)$$

Equation (4.2) is satisfied in the open set D with energy expressed as

$$E = \sum_{j=1}^{N} k_j^2. \qquad (4.6)$$

The gluing conditions (4.4) give us the relations

$$(i(k_{P_{j+1}} - k_{P_j}) - c)A(P) + (i(k_{P_j} - k_{P_{j+1}}) - c)A(PP_{j,j+1}) = 0, \qquad (4.7)$$

which determine $A(P)$ uniquely.[1] We find the rational expression

$$A(P) = \prod_{i<j} \left(1 + \frac{ic}{k_{P_i} - k_{P_j}} \right). \qquad (4.8)$$

The function $\psi_{\{k\}}(x)$ represents an interpolation in the coupling parameter c between a *permanent*, for $c = 0$, and a determinant, for $c^{-1} = 0$.

Solution of the time-dependent Schrödinger equation. Outside the interaction regions $x_j - x_l = 0$, the wavefunction can be written in the following way:

$$\psi_{\{k\}}(x) = \prod_{j<l} \left(c\varepsilon(x_l - x_j) + \frac{\partial}{\partial x_l} - \frac{\partial}{\partial x_j} \right) \left| e^{ikx_1}, e^{ikx_2}, \ldots, e^{ikx_N} \right|, \qquad (4.9)$$

in terms of the application of an antisymmetric differential operator on a Slater determinant.

---

[1] TN: up to an overall factor.

Let us suppose that we know a solution of the time-dependent wave equation

$$\Delta_N \varphi_t + i \frac{\partial \varphi_t}{\partial t} = 0 \tag{4.10}$$

in which $\varphi_t(x)$ is a generalized function on $\mathbb{R}_N$, antisymmetric in the x_j, continuously differentiable on the boundary of D and with polynomial divergence at infinity. In these conditions, the symmetric function

$$\psi_t(x) = \prod_{j<l} \left(c\varepsilon(x_l - x_j) + \frac{\partial}{\partial x_l} - \frac{\partial}{\partial x_j} \right) \varphi_t(x) \tag{4.11}$$

satisfies the time-dependent Schrödinger equation corresponding to (4.1) in D and its images. In fact, in the open set D the derivatives and sign functions ε commute with each other. The hypotheses made above allow the Fourier decomposition of $\varphi_t(x)$ and thus the decomposition of $\psi_t(x)$ given by (4.11) on the basis of elementary functions $\psi_k(x)e^{-iEt}$.

4.2 Normalization of states in the continuum

Let us begin by pointing out that the elementary wavefunction (4.5) gives the solution to the problem of the scattering of N repulsive bosons, of which the initial and final momenta necessarily constitute the same real set $\{k\}$. Owing to the total symmetry of the wavefunction, the scattering matrix is a simple phase factor which we will determine. We shall come back to the general construction of the S matrix in Chapter 10.

Let us by convention classify k in the following manner:

$$\{k\} \in \bar{D}, \qquad \bar{D} : k_1 > k_2 > \ldots > k_N. \tag{4.12}$$

The wavefunction (4.5) contains a unique *purely incoming* term associated with the identity permutation: $A(I) \exp i (k_1 x_1 + \ldots + k_N x_N)$. Indeed a wave packet formed around k cannot have interacted in the past, since x belongs to D. In the same way, in the expansion (4.5) there exists a single *purely outgoing* term, which is associated with the permutation T that completely reverses the order of k:

$$Tj = (N + 1 - j), \qquad \forall j.$$

Consider

$$A(T) \exp i (k_N x_1 + k_{N-1} x_2 + \ldots + k_1 x_N). \tag{4.13}$$

This term would lead in D to a wave packet without future collisions; see Figure 4.1.

Figure 4.1 Rapidities increasing with abscissa: outgoing state.

The S matrix element is thus

$$S(\{k\}) = \frac{A(T)}{A(I)} = \prod_{j<l} \frac{k_j - k_l - ic}{k_j - k_l + ic}. \tag{4.14}$$

Up to an overall factor $A^{-1}(I)$ (or $A^{-1}(T)$), the functions (4.5) can thus be considered as the incoming or outgoing scattering states.

It seemed interesting at this stage to address the problem of the *normalization* of the continuum states, an easier problem than that for the discrete states of the finite chain, or that in a finite volume (Gaudin, 1971a,b).

Proposition: With the definition of the antisymmetric phases

$$\psi_{jl} = 2 \operatorname{atan} \frac{c}{k_j - k_l}, \tag{4.15}$$

and of the normalization factor

$$G(k) = \prod_{j<l} \left(1 + \frac{c^2}{(k_j - k_l)^2}\right), \tag{4.16}$$

the following transformation function, defined on the domain D, is unitary:

$$\langle k|x \rangle = (2\pi)^{-\frac{N}{2}} G(k)^{-\frac{1}{2}} \psi_{\{k\}}(x) = (2\pi)^{-\frac{N}{2}} \sum_{P \in \pi_N} \exp\left\{\frac{i}{2} \sum_{j<l} \psi_{Pj,Pl} + i \sum_j x_j k_{Pj}\right\} \tag{4.17}$$

which means that we have

$$\int_D \langle k|x \rangle^* \langle k'|x \rangle d^N x = \prod_j \delta(k_j - k_j'), \qquad k, k' \in D$$

and also the closure relation.

4.2.1 *Proof of the closure relation*

Our task is to show the equality

$$I_N \equiv \int_{\mathbb{R}_N} d^N k\, G^{-1}(k) \psi_{\{k\}}(x) \psi_{\{k\}}^*(y) = N!(2\pi)^N \delta(x - y), \qquad x, y \in D. \tag{4.18}$$

(a) Algebraic part.
We shall denote

$$a(j, l) = 1 - \frac{ic}{k_j - k_l}; \quad (x, Pk) = \sum_j x_j k_{P_j}. \tag{4.19}$$

Substituting in the integrand (4.18) the expressions (4.5) for the elementary functions, we obtain

$$I_N = \int_{\mathbb{R}_N} d^N k \sum_{P,Q} G^{-1}(k) \prod_{i<j} a(Pj, Pi) a(Qi, Qj) \exp[i(Pk, x) - i(Qk, y)], \tag{4.20}$$

in which the sum is carried out over all Nth-order permutations P and Q. Substituting expression (4.16) of $G(k)$ in (4.20) and summing over permutations P and $R = Q^{-1}P$, we obtain

$$I_N = \int_{\mathbb{R}_N} d^N k \sum_{P,R} \prod_{i<j} \frac{a(PR^{-1}i, PR^{-1}j)}{a(Pi, Pj)} e^{i(Pk, x - Ry)}. \tag{4.21}$$

Consider now the double product over all pairs of indices in the integrand of (4.21). Through the index change

$$R^{-1}i = i' \tag{4.22}$$

in the numerator, we can write

$$\frac{a(PR^{-1}i, PR^{-1}j)}{a(Pi, Pj)} = \frac{\prod_{Ri'<Rj'} a(Pi', Pj')}{\prod_{i<j} a(Pi, Pj)}. \tag{4.23}$$

There appear $\frac{1}{2}N(N-1)$ pairs in both the numerator and denominator. To each pair $\{i, j\}$ appearing in the denominator on the right-hand side of (4.23) we associate the identical pair $\{i', j'\}$ in the numerator: we thus have two possibilities for each pair.

(1) $i' = i, j' = j$.
In other words

$$i < j, \quad Ri < Rj, \tag{4.24}$$

and the corresponding terms in (4.23) disappear.

(2) $i' = j, j' = i$.
In other words

$$i < j, \quad Ri > Rj. \tag{4.25}$$

We shall then say that pair $\{i, j\}$ is inverted by R and shall write the relation $\{i, j\}_R$.

For example, with the permutation $R = \begin{bmatrix} 1 & 2 & 3 & 4 \\ 3 & 1 & 4 & 2 \end{bmatrix}$, we have

$$\{1, 2\}_R, \qquad \{1, 4\}_R, \qquad \{3, 4\}_R.$$

In the second case, the corresponding factor in (4.23) is

$$\frac{a(Pj, Pi)}{a(Pi, Pj)}, \qquad \{i, j\}_R, \qquad i < j. \tag{4.26}$$

We shall therefore write

$$I_N = \int_{\mathbb{R}_N} d^N k \sum_{P,R} \prod_{\{i,j\}_R} \frac{k_{Pi} - k_{Pj} + ic}{k_{Pi} - k_{Pj} - ic} \exp\left(i \sum_l k_{Pl}(x_l - y_{Rl})\right). \tag{4.27}$$

Changing variables $k_{Pi} \to k_i$ gives us

$$I_N = \sum_{R \in \pi_N} I(R), \tag{4.28}$$

with

$$I(R) = N! \int_{\mathbb{R}_N} d^N k \prod_{\{i,j\}_R} \frac{k_{Pi} - k_{Pj} + ic}{k_{Pi} - k_{Pj} - ic} e^{i \sum_l k_l(x_l - y_{Rl})}. \tag{4.29}$$

(b) Integration.

Our remaining task consists of evaluating the integrals (4.29). Let us first consider the term $I(1)$ associated with the identity. We thus have no inverted pairs and simply get

$$I(1) = N!(2\pi)^N \delta(x_1 - y_1) \ldots \delta(x_N - y_N). \tag{4.30}$$

To obtain result (4.18) it thus suffices to prove that

$$I(R) = 0, \qquad \forall R \neq 1, \qquad x, y \in D. \tag{4.31}$$

Let us consider an arbitrary permutation $R \neq 1$. For each pair $\{i, j\}_R$, we have the implications

$$\begin{cases} i < j & \text{and} \quad x \in D \quad \Rightarrow \quad x_i < x_j, \\ Ri > Rj & \text{and} \quad y \in D \quad \Rightarrow \quad y_{Ri} > y_{Rj}. \end{cases} \tag{4.32}$$

If we set

$$\xi_i = x_i - y_{Ri}, \tag{4.33}$$

we can deduce from (4.32)

$$\xi_i < \xi_j \quad \text{if} \quad i < j \quad \text{and} \quad \{i, j\}_R. \tag{4.34}$$

The quantities ξ_i constitute a partially ordered set with respect to the 'inversion by R' relation. Let us introduce the graph Γ with end points given by $i, j, \ldots$, and with a line joining i and j if $\{i, j\}_R$. Since $R \neq 1$, this graph contains at least one line. The integrand of $I(R)$ is built up by associating a factor with each line and each end point of Γ. The integral $I(R)$ is thus a product of integrals each associated with a connected part of Γ. Let us consider a connected part Γ_c of Γ with L end points and at least one line, and let us call the corresponding integral $I(\Gamma_c)$. Upon eventually changing the variable names k, we have

$$I(\Gamma_c) = \int_{-\infty}^{+\infty} \ldots \int_{-\infty}^{+\infty} dk_1 \ldots dk_L \prod_{\{i,j\}_R \in \Gamma_c} \frac{k_i - k_j + ic}{k_i - k_j - ic} \exp\left(i \sum_{l=1}^{L} k_l \xi_l\right).$$

(4.35)

In the $\xi \in \Gamma_c$ variables, $I(\Gamma_c)$ is a distribution which is the Fourier transform of a translationally invariant rational function; its support is thus the plane

$$\sum_{l=1}^{L} \xi_l = 0.$$

(4.36)

Let us now show that the ξ admit a smallest negative element. There surely exists an end point m of Γ_c such that

$$\xi_m \leq \xi_i, \quad \forall i \in \Gamma_c.$$

(4.37)

Let $\{n\}$ be the set of end points of Γ_c connected to m by a line. Since the line (m, n) exists, we have $\{m, n\}_R$, in other words by definition

$$m < n \quad \text{and} \quad \xi_m < \xi_n$$

(4.38)

or

$$m > n \quad \text{and} \quad \xi_m > \xi_n.$$

(4.39)

The second alternative is eliminated by (4.37) and there only remains (4.38). Suppose that $\xi_m \geq 0$; according to (4.38) and (4.37), this would imply $\sum_{i \in \Gamma_c} \xi_i > 0$, contradicting (4.36). We thus have the result

$$\{n, m\}_R \in \Gamma_c \Rightarrow \xi_m < 0 \quad \text{and} \quad n > m,$$

(4.40)

which allows us to show that $I(\Gamma_c)$ vanishes.

According to a theorem of Schwartz and Gel'fand on generalized functions, the Fourier transform $I(\Gamma_c)$ can be calculated as the limit of the integral taken in the box $-K_i < k_i < K_i, i = [1, L]$ with $\lim K_i = +\infty$. Let us first integrate over the k_m variable. The corresponding factor J_m of $I(\Gamma_c)$ is

$$J_m = \lim_{K_m \to \infty} \int_{-K_m}^{+K_m} dk_m e^{ik_m \xi_m} \prod_{n(>m)} \frac{k_m - k_n + ic}{k_m - k_n - ic}. \tag{4.41}$$

The poles of the integrand are all located in the upper half-plane $k_m = k_n + ic$, $c > 0$ and at infinity we have $\prod_n (\ldots) = 1 + O(k_m^{-1})$. We thus deduce that

$$J_m = 2\pi \delta(\xi_m) + \text{function vanishing for } \xi_m < 0. \tag{4.42}$$

Therefore in the open set $\xi_m < 0$, the distribution J_m is zero. This proves that $I(\Gamma_c) = 0$, and $I(R) = 0$ $(R \neq 1)$ in the open set D.

4.2.2 The normalization integral

The direct calculation of integral (4.17) is interesting and rests on a remarkable algebraic identity verified by the coefficients of the Bethe wavefunctions. We define the generalized function of $\{k\}$ and $\{k'\}$:

$$\mathcal{N} \begin{pmatrix} k_1 & \cdots & k_n \\ k_1' & \cdots & k_N' \end{pmatrix} = \int_D d^N x\, \psi_{\{k\}}^*(x) \psi_{\{k'\}}(x), \tag{4.43}$$

which is a Fourier transform sum characteristic of domain D. With the help of definitions (4.5) and (4.8), we obtain

$$\mathcal{N} = \sum_{P,Q} A^*(P) A'(Q) \int_D e^{i(Qk' - Pk,x)} d^N x. \tag{4.44}$$

Taking

$$Pk - Qk' = q \tag{4.45}$$

we calculate the q distribution

$$J_N(q) = \int_D e^{-i(q,x)} d^N x$$
$$= 2\pi \delta(q_1 + q_2 + \ldots + q_N) i^{N-1} [(q_1 + i0)(q_1 + q_2 + i0) \ldots$$
$$(q_1 + q_2 + \ldots + q_{N-1} + i0)]^{-1}. \tag{4.46}$$

Let us now derive an equality which will prove useful later on. On the one hand, according to the integral definition, we obviously have

$$\sum_P J_N(Pq) = (2\pi)^N \delta(q_1) \delta(q_2) \ldots \delta(q_N). \tag{4.47}$$

On the other hand, a well-known algebraic identity also gives us

$$\sum_P J_N(Pq) = 2\pi i^{N-1} \delta(q_1 + q_2 + \ldots + q_N)$$

$$\times \sum_{j=1}^{N} [(q_1 + i0) \ldots (q_{j-1} + i0)(q_{j+1} + i0) \ldots (q_N + i0)]^{-1}. \quad (4.48)$$

We thus obtain equality of the right-hand sides of (4.47) and (4.48). With the help of (4.45) and (4.46), we write the norm in the form

$$\mathcal{N} = 2\pi i^{N-1} \delta(k_1 + \ldots + k_N - k'_1 - \ldots - k'_N)$$

$$\times \tilde{\Delta} \begin{pmatrix} k_1 + i0, & k_2 + i0, & \ldots, & k_N + i0 \\ k'_1, & k'_2, & \ldots, & k'_N \end{pmatrix}, \quad (4.49)$$

with the following definition of the rational function of $\{k\}$ and $\{k'\}$:

$$\tilde{\Delta} \begin{pmatrix} k \\ k' \end{pmatrix} = \sum_{P,Q} A^*(P) A'(Q) [(k_{P1} - k'_{Q1})(k_{P1} + k_{P2} - k'_{Q1} - k'_{Q2}) \ldots$$

$$(k_{P1} + \ldots k_{P(N-1)} - \ldots - k'_{Q(N-1)})]^{-1}. \quad (4.50)$$

The fundamental identity proved in Appendix B is now used to write $\tilde{\Delta}$ in a form which makes its singularities manifest. Starting from the form (B.6) and (B.7):

$$\tilde{\Delta} \begin{pmatrix} k \\ k' \end{pmatrix} = \sum (k - k') \Delta \begin{pmatrix} k \\ k' \end{pmatrix}$$

$$= \sum_R I(R) d(R) \frac{(k_1 + \ldots + k_N - k'_1 - \ldots - k'_N)}{(k_1 - k'_{R1})(k_2 - k'_{R2}) \ldots (k_N - k'_{RN})}, \quad (4.51)$$

in which $I(R)$ is the parity indicator of R and the $d(R)$ coefficient is given by

$$d(R) = \delta_N^{-1} [(k_1 - k'_{R1} + ic) \ldots (k_N - k'_{RN} + ic)]^{-1} \quad (4.52)$$

with

$$\delta_N = \left| \frac{1}{k_i - k'_j + ic} \right|_N. \quad (4.53)$$

The only singularities of $\tilde{\Delta}$ are thus the poles $k_i = k'_{Ri}$, $\forall R$ and i. Substituting the expansion (4.51) for $\tilde{\Delta}$ in expression (4.49) for $\mathcal{N}$, we obtain

$$\mathcal{N} = 2\pi i^{N-1} \sum_R I(R) d(R)$$

$$\times \sum_{j=1}^{N} \frac{\delta(k_1 + \ldots + k_N - k'_1 - \ldots - k'_N)}{(k_1 - k'_{R1} + i0) \ldots (k_{j-1} - k_{R(j-1)} + i0)(k_{j+1} - k'_{R(j+1)} + i0) \ldots}. \quad (4.54)$$

Using the equality of (4.47) and (4.48), we simply obtain

$$\mathcal{N} = (2\pi)^N \sum_R I(R)d(R)\delta(k_1 - k'_{R1})\ldots\delta(k_N - k'_{RN}). \qquad (4.55)$$

It is clear that the coefficient $d(R)$ must be evaluated at the point $k_i = k'_{Ri}$, $i = [1, N]$. According to (4.52) and (4.53), we obtain at this point

$$d(R) = I(R) \prod_{i<j} \frac{(k_i - k_j)^2 + c^2}{(k_i - k_j)^2} \qquad (4.56)$$

and thus

$$\mathcal{N}\begin{pmatrix} k \\ k' \end{pmatrix} = (2\pi)^N G(k) \sum_R \delta(k_1 - k'_{R1})\ldots\delta(k_N - k'_{RN}). \qquad (4.57)$$

This proves (4.17) when we restrict $\{k\}$ and $\{k'\}$ to D.

4.3 Periodic boundary conditions

4.3.1 Coupled equations

The simplest way to obtain the extensive properties of the gas of repulsive bosons is to put the system in a circle of length L, in other words to impose on the wavefunction the period L in all variables. This is analogous to the ring of spin-1/2 atoms treated in Chapter 1. We thereby obtain the cyclicity conditions

$$\Psi(x_1, x_2, \ldots, x_N) \equiv \Psi(x_2, x_3, \ldots, x_1 + L), \qquad (4.58)$$

$$D : 0 < x_1 < x_2 < \ldots < x_N < L.$$

It turns out that these conditions are compatible with the structure of the elementary wavefunctions defined by the sums (4.5) provided the quasi-momenta are conveniently chosen to satisfy

$$A(PC)e^{ik_{P1}L} = A(P), \qquad \forall P \qquad (4.59)$$

in which C denotes the cyclic permutation $(12\ldots N)$. With the help of (4.8) we then obtain the coupled equations

$$Lk_i = 2\pi I_i - 2\sum_{j=1}^{N} \text{atan}\,\frac{k_i - k_j}{c}, \qquad i = [1, N], \qquad (4.60)$$

in which the numbers I_i are integers or half-odd integers according to whether N is odd or even. We show in the next subsection that there exists a unique real solution to (4.60) for each set of quantum numbers I_i satisfying

$$I_1 < I_2 < \ldots < I_N. \qquad (4.61)$$

4.3.2 *Existence of solutions*

(Yang and Yang, 1969)

The system of equations (4.60) can be written in the form

$$\frac{\partial B}{\partial k_i} = 0, \quad i = [1, N], \tag{4.62}$$

in which the function $B\{k\}$ is defined as

$$B\{k\} = \sum_j \frac{L}{2}k_j^2 - 2\pi I_j k_j + \sum_{j,l} \int_0^{k_j - k_l} dk \text{ atan } k/c. \tag{4.63}$$

The matrix of second derivatives

$$\frac{\partial^2 B}{\partial k_i \partial k_j} = \left(L + \sum_l \frac{2c}{(k_i - k_l)^2 + c^2}\right)\delta_{ij} - \frac{2c}{(k_i - k_j)^2 + c^2} \tag{4.64}$$

is positive definite, since the characteristic polynomial det $\left|\frac{\partial^2 B}{\partial k_i \partial k_j} - x\delta_{ij}\right|$ is a Sylvester determinant, sum of positive terms when $L - x > 0$ with real roots thus satisfying $x \geq L$. The function $B\{k\}$ is therefore convex; it behaves as $\frac{L}{2}\sum_j k_k^2$ near infinity, and thus admits a single extremum, which is in fact a minimum, where equations (4.62) are verified.

It is straightforward to deduce from this that the k_i are distinct when the I_i are also distinct. From equations (4.60) we obtain by subtraction

$$|L(k_i - k_j) - 2\pi(I_i - I_j)| = 2\left|\sum_{l=1}^{N} \text{atan}\frac{k_i - k_l}{c} - \text{atan}\frac{k_j - k_l}{c}\right| \leq \frac{2N}{c}|k_i - k_j|, \tag{4.65}$$

in which we used $|\text{ atan } x - \text{ atan } y| \leq |x - y|$. From (4.65) and $|I_i - I_j| \geq 1$ we deduce

$$|k_i - k_j| \geq \frac{2\pi}{L}\left(1 + \frac{2\rho}{c}\right)^{-1}, \quad \rho = \frac{N}{L}. \tag{4.66}$$

In fact the vector $\{k\}$ is a continuous function of the c parameter, since det $|\partial^2 B/\partial k_i \partial k_j|$ does not vanish. According to inequalities (4.66), the differences $(k_i - k_j)$ cannot vanish when c remains positive. The order of $k_1, k_2, \ldots, k_N$ thus remains unchanged from the $c^{-1} = 0$ case, in which we have

$$\lim_{c\to\infty} k_i = \frac{2\pi}{L}I_i. \tag{4.67}$$

The limiting ($c^{-1} = 0$) wavefunctions are those of a system of fermions with periodic or antiperiodic conditions and thus form a complete set. We thus obtain using (4.61) the complete set of quantum numbers. The k satisfy the inequalities

$$
\begin{cases}
\dfrac{2\pi}{L}I_1 < k_1 < k_2 < \ldots < k_N < \dfrac{2\pi}{L}I_N, \\[2mm]
k_{j+1} - k_j > \dfrac{2\pi}{L}\left(1 + \dfrac{2\rho}{c}\right)^{-1}.
\end{cases}
\tag{4.68}
$$

Note on the $c = 0^+$ limit. Since the k do not cross, we can define the antisymmetric phases ψ_{ij}, $-\pi < \psi < \pi$, such that $\tan(\psi_{ij}/2) = \dfrac{2}{k_i - k_j}$ and write system (4.60) in the form

$$
Lk_i = 2\pi n_i + \sum_j{}' \psi_{ij},
\tag{4.69}
$$

with the integers

$$
n_i = I_i - i + \frac{N+1}{2}
\tag{4.70}
$$

which form a non-decreasing sequence: these are the quantum numbers of the non-interacting ($c = 0$) boson system. One can in fact show that the ψ tend to zero with c. Let us suppose for example the case of the state $n_i = 0$, $i = [1, N]$. In the vicinity of $c = 0$, one finds

$$
\begin{cases}
k_i = \left(\dfrac{2c}{L}\right)^{1/2} q_i + O(c), \\[2mm]
\psi_{ij} = (2cL)^{1/2}\dfrac{1}{q_i - q_j} + O(c),
\end{cases}
\tag{4.71}
$$

in which q_i are the zeroes of the Hermite polynomial $H_N(q)$ (see Appendix D).

4.3.3 Conjecture for the norm

Let $\{k\}$ be the set of quasi-momenta defined by the set of quantum numbers $\{n\}$ as explained in Subsection 4.3.2. We conjecture that the following wavefunction is normalized to unity in D:

$$
\psi_{\{n\}} = \left|\frac{\partial^2 B}{\partial k_i \partial k_j}\right|^{-\frac{1}{2}} \sum_P \exp\left\{\frac{i}{2}\sum_{j<l}\psi_{Pj,Pl} + i\sum_j x_j k_{Pj}\right\},
\tag{4.72}
$$

$$
D : 0 < x_1 < x_2 < \ldots < x_N < L.
$$

The determinant appearing in the normalization factor is that of the matrix (4.64); it is also the Jacobian

$$\frac{D(n_1 n_2 \ldots n_N)}{D(k_1 k_2 \ldots k_N)} (2\pi)^N.$$

This conjecture is verified for $N = 2$, for $c^{-1} = 0$ and for $c = 0$. This last limit is not trivial, in the case where certain quantum numbers are equal. Indeed, let us write the sequence $n_1 \leq n_2 \leq \ldots \leq n_N$ in the form $\ldots (-2)^{\nu_{-2}} (-1)^{\nu_{-1}} (0)^{\nu_0} (1)^{\nu_1} (2)^{\nu_2} \ldots$ in which number p appears ν_p times. We must have

$$\lim_{c \to 0} \psi_{\{n\}} = \left(\prod_p \nu_p! \right)^{-\frac{1}{2}} {}_{|}^{+} e^{ik_i x_j} {}_{|N}^{+}. \tag{4.73}$$

Using result (4.71) we see that the determinant $|\partial^2 B / \partial k_i \partial k_j|$ is not diagonal, but decomposes into blocks of order ν_p. A typical block of order ν gives rise to the factor $\det |N_{ij}|$ with

$$N_{ij} = \left(1 + \sum_l' \frac{1}{(q_j - q_l)^2} \right) - \frac{1}{(q_i - q_j)^2}, \tag{4.74}$$

in which the q_i are the zeroes of $H_\nu(q)$, since they verify the relations

$$q_i = \sum_{l=1}^{\nu} \frac{1}{q_i - q_l}. \tag{4.75}$$

We have the following proposition, which finalizes the proof of (4.73).

Proposition: The matrix $||N_{ij}||$ defined by (4.74) has the numbers $1, 2, \ldots, \nu - 1, \nu$ as eigenvalues. Its determinant is thus $\nu!$. The proof can be found in Appendix D.

This conjecture for the norm of the wavefunction of the bosonic gas extends naturally to the amplitude of the anisotropic chain:

$$\sum_{\{n\}} |a(n_1 n_2 \ldots n_M)|^2 = \frac{D(\lambda_1 \lambda_2 \ldots \lambda_M)}{D(k_1 k_2 \ldots k_M)}.$$

The idea justifying it goes back to Richardson (1965), who uses the Hellmann–Feynman theorem (variation of the energy with respect to a parameter) to calculate certain matrix elements in a problem for which the amplitude has the structure of a simplified Bethe sum (see Subsection 13.1.3). This is worked out in detail in Gaudin, McCoy and Wu (1981).

A proof of this conjecture has recently been provided by Korepin (1982), making use of the inverse scattering method.

4.4 Thermodynamic limit

4.4.1 Ground state

(Lieb and Liniger, 1963)

It is proven (Yang and Yang, 1969):

(a) that the quantum numbers of the ground state are the same throughout the $c > 0$ region,

$$n_i = 0, \quad \forall i \quad \text{or} \quad \{I\} = \left\{ -\frac{N-1}{2}, -\frac{N-3}{2}, \dots, \frac{N-1}{2} \right\}; \qquad (4.76)$$

(b) that in the thermodynamic limit $N \to \infty$, $N/L = \rho$ fixed density, the quasi-momenta k tend to a distribution $\rho(k)$ on a finite interval $[-k_0, +k_0]$;

(c) that the energy per particle $\varepsilon_0 = E/N$ and the density ρ are determined from the Lieb–Liniger equations (limits of (4.60) and (4.6))

$$\rho(k) - \frac{c}{\pi} \int_{-k_0}^{+k_0} \frac{\rho(k')dk'}{(k-k')^2 + c^2} = \frac{1}{2\pi}, \qquad (4.77)$$

$$\rho\varepsilon_0 = \int_{-k_0}^{+k_0} k^2 \rho(k)dk, \qquad (4.78)$$

$$\rho = \int_{-k_0}^{+k_0} \rho(k)dk. \qquad (4.79)$$

The elimination of the auxiliary parameter k_0 between ε_0 and ρ gives the equation for the ground state. One can rather easily show that the energy is an analytic function of c on the real semi-axis $c > 0$. The origin is singular, as already suggested by the asymptotic distribution of the quasi-momenta before the thermodynamic limit, given by (4.71). The distribution of the zeroes q_i of the $H_N(q)(N \gg 1)$ polynomial is $\pi^{-1}(2N - q^2)^{1/2}$ (Szegö, 1939) and gives us the density

$$\rho(k) \propto \frac{1}{2nc}(4c\rho - k^2)^{1/2}, \quad k_0 \propto 2\sqrt{c\rho}. \qquad (4.80)$$

It is rigorously demonstrated, as we indicate in the next subsection, that expression (4.80) constitutes the first-order approximation of the solution to (4.77) in the limit $k_0/c \gg 1$, $|k \pm k_0| \gg c$. The singularity at the origin can be understood thanks to an electrostatic analogy, which we now describe.

4.4.2 The circular capacitor analogy

The integral equation (4.77) is well known in potential theory as *Love's equation* (Sneddon, 1966) for the old problem of the circular capacitor. Let us indeed consider two coaxial conducting discs of unit radius, separated by a distance a and charged to opposite potentials $\pm V_0$. In cylindrical coordinates (r, z), the potential created by an axially symmetric charge distribution $\sigma(r)$ on the lower disc ($z = 0$, for example) admits the following useful representation in terms of an even real function $f(t)$:

$$V(r, z) = \int_{-1}^{+1} \frac{f(t)dt}{[r^2 - (t + iz)^2]^{1/2}}. \tag{4.81}$$

$V(r, z)$ is harmonic by construction outside the lower disc, and continuous and real since f is even.

Calculating the discontinuity of the perpendicular field, one finds that the surface charge density $\sigma(r)$ is linked to $f(t)$ by the Abel transformation

$$\sigma(r) = -\frac{1}{\pi} \frac{d}{dr} \int_r^1 \frac{tf(t)dt}{(t^2 - r^2)^{1/2}}. \tag{4.82}$$

The total charge on the lower disc is

$$Q = \int_{-1}^{+1} f(t)dt. \tag{4.83}$$

In the presence of the upper disc located at height $z = a$, carrying a surface charge density $-\sigma(r)$, the equilibrium condition on the lower disc can be written

$$V(\rho, 0) - V(\rho, a) = V_0, \quad 0 \le \rho \le 1. \tag{4.84}$$

Substituting in (4.84) the expressions for $V(\rho, 0)$ and $V(\rho, a)$ given by (4.81), one takes the *inverse Abel transform* by applying the operator $\int_0^t d\rho (t^2 - \rho^2)^{-1/2} \dots$ and one obtains, at the end of calculations, the Love equation

$$f(t) - \frac{a}{\pi} \int_{-1}^{+1} \frac{f(t')dt'}{(t - t')^2 + a^2} = \frac{V_0}{\pi}. \tag{4.85}$$

This is also the Lieb–Liniger equation after a rescaling, taking $a = c/k_0$ and $V_0 = 1/2$. We thus have the correspondence

$$\rho(k) = f(k/k_0), \tag{4.86}$$

$$\frac{\rho}{k_0} = \text{capacitance} = \int_{-1}^{+1} f(t)dt, \tag{4.87}$$

$$\frac{\varepsilon_0}{k_0^3} = \text{mean square radius of the charge distribution}$$

$$= \int_{-1}^{+1} t^2 f(t)dt / \int_{-1}^{+1} f(t)dt. \tag{4.88}$$

The classic difficulty with the capacitor problem is to find an asymptotic expansion for the capacitance for small disc separation, or, equivalently, for a large disc radius at fixed separation. Irrespective of how large the radius is, edge effects remain and affect the distribution far from the edges. In the limit, close to the edges, one has an electrostatic problem in two dimensions which was solved exactly by Maxwell using the method of conformal transformations. The matching of solutions far from and near the edges leads to Kirchhoff's formula for the capacitance:

$$Q = \frac{1}{4a} + \frac{1}{4\pi} \ln \frac{16\pi}{ea} + o(1). \tag{4.89}$$

This result was confirmed rigorously (Huston, 1963) by studying the integral equation (4.85) directly. This method gives the following expansion of $f(t)$, valid at a distance from the edges ± 1 greater than $O(a)$:

$$f(t) = \frac{1}{2\pi a}(1 - t^2)^{\frac{1}{2}} + \frac{1}{4\pi^2}(1 - t^2)^{-\frac{1}{2}} \left(t \ln \frac{1-t}{1+t} + \ln \frac{16\pi e}{a} \right) + o(1). \tag{4.90}$$

At a distance $O(a)$ from the edges, one must take Maxwell's solution which matches (4.90). It turns out that, to the accuracy considered, formula (4.90) can be used validly for the integration over the whole interval $-1, +1$. One thus obtains from expressions (4.87) and (4.88) for the density and energy of the boson gas the dominant terms

$$\rho = \frac{k_0^2}{4c} + \frac{k_0}{4\pi} \ln \frac{16\pi k_0}{ec} + \cdots, \tag{4.91}$$

$$\varepsilon_0 \rho = \frac{k_0^4}{16c} - \frac{k_0^3}{6\pi} + \frac{k_0^3}{8\pi} \ln \frac{16\pi k_0}{ec} + \cdots \tag{4.92}$$

Eliminating k_0 between equations (4.91) and (4.92) gives the first two terms of the ε_0 function:

$$\varepsilon_0 = c\rho - \frac{4}{3\pi} \rho^{1/2} c^{3/2} + \cdots, \tag{4.93}$$

without giving more information on the nature of the expansion. This result coincides with the perturbative calculation using Bogoliubov's method (Lieb and Liniger, 1963).

4.4.3 Finite temperature thermodynamics

Yang and Yang (1969) obtained the following results.

The pressure P of the gas of repulsive bosons at temperature T is given by the expression

$$P = \frac{T}{2\pi} \int_{-\infty}^{+\infty} dk \ln(1 + e^{-\varepsilon(k)/T})$$ (4.94)

in which the pseudo-energy function $\varepsilon(k, T)$ is the unique solution to the nonlinear integral equation

$$\varepsilon(k) = -A + k^2 - \frac{Tc}{\pi} \int_{-\infty}^{+\infty} \frac{dk'}{(k - k')^2 + c^2} \ln(1 + e^{-\varepsilon(k')/T}).$$ (4.95)

Quantity A is the chemical potential such that

$$\rho = \frac{\partial P}{\partial A} = \text{density.}$$ (4.96)

The excitation spectrum of the system can be considered as resulting from the superposition of elementary excitations of 'particle' and 'hole' types, of energy $\varepsilon(k)$ and momentum $q(k)$:

$$q(k) = \frac{\partial}{\partial A} \int_0^k \varepsilon(k')dk'.$$ (4.97)

Appendix B

Let us consider a set $\{k\}$ of distinct complex numbers together with an analogous set $\{k'\}$.

Let us define

$$A(P) = \prod_{1 \leq i < j \leq N} \left(1 + \frac{\gamma}{k_{Pi} - k_{Pj}}\right), \quad \gamma = ic$$

$A'(P)$ obtained by the substitution $\{k\} \rightarrow \{k'\}$

$\bar{A}(P)$ obtained by the substitution $\gamma \rightarrow -\gamma$

$$\bar{A}(P) = A(PT) \text{ permutation } T = \begin{bmatrix} 12\ldots N \\ N(N-1)\ldots 1 \end{bmatrix}.$$ (B.1)

We define the double summation over permutations P and Q:

$$\Delta_N \begin{pmatrix} k_1 & \cdots & k_N \\ k'_1 & \cdots & k'_N \end{pmatrix} = \sum_P \sum_Q \bar{A}(P)A'(Q)$$

$$\times \left[(k_{P1} - k'_{Q1})(k_{P1} + k_{P2} - k'_{Q1} - k'_{Q2}) \cdots \right.$$
$$\left. (k_{P1} + \ldots + k_{PN} - k'_{Q1} - \ldots - k'_{QN})\right]^{-1}.$$ (B.2)

Clearly Δ_N is a rational function, completely symmetric in the k variables, and separately so in the k' ones. This sum can be expressed as the quotient of two determinants. We have

$$\text{IDENTITY} \qquad \boxed{\Delta_N = \frac{D_N}{\delta_N}} \qquad (\text{B.3})$$

with

$$D_N = \left| \frac{1}{(k_i - k'_j)(k_i - k'_j + \gamma)} \right|_N \qquad (\text{B.4})$$

and the Cauchy determinant

$$\delta_N = \left| \frac{1}{k_i - k'_j + \gamma} \right|_N . \qquad (\text{B.5})$$

Proof: Identity (B.3) was found by generalizing the $N = 2$ case. The proof is made using induction on N.

$N = 2$. Notice that the sum $\sum\limits_{P,Q} \bar{A}(P)A'(Q)\dfrac{1}{k_{P1} - k'_{Q1}}$ is divisible by $k_1 + k_2 - k'_1 - k'_2$. One obtains directly

$$\Delta_2 \begin{pmatrix} k_1 & k_2 \\ k'_1 & k'_2 \end{pmatrix} = \left(1 - \frac{\gamma^2 + \gamma(k_1 + k_2 - k'_1 - k'_2)}{(k_1 - k_2)(k'_1 - k'_2)} \right) \frac{1}{(k_1 - k'_1)(k_2 - k'_2)}$$
$$+ \left(1 + \frac{\gamma^2 + \gamma(k_1 + k_2 - k'_1 - k'_2)}{(k_1 - k_2)(k'_1 - k'_2)} \right) \frac{1}{(k_1 - k'_2)(k_2 - k'_1)} .$$

Subtracting from the first term and adding to the second term the quantity $[(k_1 - k_2)(k'_1 - k'_2)]^{-1}$, we obtain

$$\Delta_2 = \sum_R I(R)d(R)\frac{1}{(k_1 - k'_{R1})(k_2 - k'_{R2})}$$

in which $I(R)$ is the sign of the permutation R and the coefficient

$$d(R) = \frac{(k_1 - k'_{R2} + \gamma)(k_2 - k'_{R1} + \gamma)}{(k_1 - k_2)(k'_1 - k'_2)} .$$

In this form the generalization is immediate and leads to

$$\Delta_N = \sum_R I(R)d(R) \left[(k_1 - k'_{R1})(k_2 - k'_{R2}) \dots (k_N - k'_{RN}) \right]^{-1} \qquad (\text{B.6})$$

with

$$d(R) = \frac{\prod_{i \neq j}(k_i - k'_{Rj} + \gamma)}{\prod_{i<j}(k_i - k_j)(k'_i - k'_j)}. \tag{B.7}$$

With the help of Cauchy's identity, we obtain the aforementioned identity (B.3). The form (B.6), (B.7) is interesting however, since it manifests the true singularities of Δ_N, which are the poles

$$k_i = k'_{Ri}, \quad i = [1, N], \quad \forall R.$$

Induction on N

Definition (B.2) of Δ_N is equivalent to the following recursion relation, obtained by multiplying the two sides of (B.2) by $\sum_{i=1}^{N}(k_i - k'_i)$, setting $PN = r$, $QN = s$ and using the definition of A and A' to order $N - 1$. One thus obtains

$$\sum_i (k_i - k'_i)\Delta_N \begin{pmatrix} k_1 & \cdots & k_N \\ k'_1 & \cdots & k'_N \end{pmatrix} = \sum_{r=1}^{N}\sum_{s=1}^{N} \Delta_{N-1} \begin{pmatrix} \cdots & k_r & \cdots \\ \cdots & k'_s & \cdots \end{pmatrix}$$

$$\times \prod_{j(\neq r)} \left(1 - \frac{\gamma}{k_j - k_r}\right) \prod_{i(\neq s)} \left(1 + \frac{\gamma}{k'_i - k'_s}\right). \tag{B.8}$$

According to the induction hypothesis, let us suppose that identity (B.3) is true up to and including order $N - 1$. We then have

$$\Delta_{N-1} \begin{pmatrix} \cdots & k_r & \cdots \\ \cdots & k'_s & \cdots \end{pmatrix} = \frac{|\text{minor } D_N|_{rs}}{|\text{minor } \delta_N|_{rs}} = \frac{|\text{minor } D_N|_{rs}}{\delta_N}$$

$$\times (-1)^{r+s} \frac{\prod_{j(\neq r)}(k_r - k_j) \prod_{i(\neq s)}(k'_s - k'_i)}{(k_r - k'_s + \gamma) \prod_{j(\neq s)}(k_r - k'_j + \gamma) \prod_{i(\neq r)}(k_i - k'_s + \gamma)} \tag{B.9}$$

in which we have chosen the definition of the minor such that

$$\det |A| = \sum_s (-1)^{r+s} a_{rs} |\text{minor } A|_{rs}. \tag{B.10}$$

We now substitute expression (B.9) for Δ_{N-1} into (B.8). The equality $\Delta_N = \dfrac{D_N}{\delta_N}$ to be proven is written

$$\sum_i (k_i - k'_i)D_N = \frac{1}{\gamma^2} \sum_{r,s} (-1)^{r+s} |\text{minor } D_N|_{rs}(k_r - k'_s + \gamma)e_r e'_s, \tag{B.11}$$

with the notations

$$e_r = \prod_{j=1}^{N} \frac{k_j - k_r - \gamma}{k'_j - k_r - \gamma}, \tag{B.12}$$

$$e'_s = \prod_{i=1}^{N} \frac{k'_i - k'_s + \gamma}{k_i - k'_s + \gamma}. \tag{B.13}$$

It is clear that the right-hand side of (B.11) is a sum of three determinants of order $N + 1$ obtained by adding to $|D_N|$ a line and a column of index 0. One must thus prove that

$$-\gamma^2 \sum (k - k')D_N = |D^{(1)}| - |D^{(2)}| + \gamma|D^{(3)}| \tag{B.14}$$

with the definitions

$$\begin{aligned}
D_{00}^{(\alpha)} &= 0, \quad D_{rs}^{(\alpha)} = |D_N|_{rs}, \quad r, s = [1, N] \\
D_{0s}^{(1)} &= e'_s, \quad D_{r0}^{(1)} = k_r e_r, \\
D_{0s}^{(2)} &= k'_s e'_s, \quad D_{r0}^{(2)} = e_r, \\
D_{0s}^{(3)} &= e'_s, \quad D_{r0}^{(3)} = e_r.
\end{aligned} \tag{B.15}$$

We now notice that the non-vanishing elements of the new line (or column) are linear combinations of the elements of D_N. Using a decomposition into simple elements of the rational fractions of γ, e_r and e'_s, we obtain using (B.12) and (B.13)

$$e_r \equiv 1 - \sum_j \frac{f_j}{k_r - k'_j + \gamma} = \gamma \sum_j \frac{f_j}{(k_r - k'_j)(k_r - k'_j + \gamma)} \tag{B.16}$$

and

$$e'_s \equiv 1 - \sum_i \frac{f'_i}{k_i - k'_s + \gamma} = \gamma \sum_i \frac{f'_i}{(k_i - k'_s)(k_i - k'_s + \gamma)}; \tag{B.17}$$

in other words

$$D_{r0}^{(3)} - \gamma \sum_j f_j D_{rj}^{(3)} = 0. \tag{B.18}$$

In this way, combining lines for $D^{(1)}$ and $D^{(3)}$, and columns for $D^{(2)}$ and $D^{(3)}$, we obtain

$$|D^{(1)}| = -\gamma D_N \left(\sum_i f'_i k_i e_i \right),$$

$$|D^{(2)}| = -\gamma D_N \left(\sum_j f_j k'_j e'_j \right),$$

$$|D^{(3)}| = -\gamma D_N \left(\sum_j f_j e'_j \right) = -\gamma D_N \left(\sum_i f'_i e_i \right). \tag{B.19}$$

After substitution of expressions (B.16) and (B.17) for e_i and e'_j into the right-hand side of (B.19), we obtain

$$\frac{1}{D_N}\left(|D^{(1)}| - |D^{(2)}| + \gamma|D^{(3)}|\right) \tag{B.20}$$

$$= \sum_j (-\gamma(k'_j f_j - k_j f'_j) - \gamma^2 f'_j) + \gamma \sum_{j,i} \frac{-k'_j f_j f'_i + k_i f'_i f_j + \gamma f'_i f_j}{k_i - k'_j + \gamma}$$

$$= -\gamma^2 \sum_i f'_i + \gamma \sum_i (k_i f'_i - k'_i f_i) + \gamma \sum_{i,j} f'_i f_j. \tag{B.21}$$

It remains to evaluate the residues f_i and f'_i using (B.12) and (B.13). One obtains

$$\sum_i f'_i = \text{coefficient of } \frac{1}{\gamma} \text{ in } e'_s = \sum_i (k_i - k'_i).$$

In the same way,

$$\sum_i f_i = \sum_i (k_i - k'_i) = \sum_i f'_i,$$

$$\sum_i k_i f'_i = \frac{1}{2}\sum_i (k'^2_i - k^2_i) - \frac{1}{2}\left(\sum_i (k'_i - k_i)\right)^2,$$

$$\sum_i k'_i f_i = \frac{1}{2}\sum_i (k'^2_i - k^2_i) + \frac{1}{2}\left(\sum_i (k'_i - k_i)\right)^2. \tag{B.22}$$

Substituting (B.21) into (B.22), we finally obtain

$$\frac{1}{D_N}\left(|D^{(1)}| - |D^{(2)}| + \gamma|D^{(3)}|\right) = -\gamma^2 \sum_i (k_i - k'_i),$$

which establishes equality (B.14) and finalizes the proof of identity (B.3).

Limiting cases. The limiting case $c \to 0$ is not trivial and yields an identity due to Borchardt (Muir, 1966) between a permanent and the quotient of two determinants. A known identity allows us to write directly

$$\Delta_N|_{c=0} = {}^+_{\dagger}\frac{1}{k_i - k'_j}{}^+_{\dagger}{}_N,$$

this being Borchardt's equality. The limiting form of (B.3) gives

$$^+_{\dagger}\frac{1}{k_i - k'_j}{}^+_{\dagger}{}_N \left|\frac{1}{k_i - k'_j}\right|_N = \left|\frac{1}{(k_i - k'_j)^2}\right|_N,$$

which can be demonstrated directly. The method is explained in Appendix K, which gives another remarkable expression for Δ_N or for the quotient D_N/δ_N.

Appendix C

With the aim of highlighting the type of difficulty encountered in the direct calculation of the norm or of an average value, we present here the argument for the orthogonality of the eigenstates of the system of bosons on a circle. We want to calculate the scalar product of two states identified by two distinct sets $\{k\} \neq \{k'\}$, which represent two different solutions to the coupled equations:

$$\mathcal{N}\begin{pmatrix} k \\ k' \end{pmatrix} = \int_{0<x_1<...<x_N<L} dx_1 \ldots dx_N \psi^*_{\{k\}}(x)\psi_{\{k'\}}(x). \tag{C.1}$$

(a) $\sum_i \pm \sum k'$.

Orthogonality is manifest for states of differing total momentum. Under a global translation ξ

$$x_j \rightarrow y_j = x_j + \xi \tag{C.2}$$

we have

$$\psi_{\{k'\}}(y) = e^{i\xi \sum k'} \psi_{\{k'\}}(x), \tag{C.3}$$

and as a consequence

$$\mathcal{N}\begin{pmatrix} k \\ k' \end{pmatrix} = \frac{1}{L} \int_0^L d\xi e^{i\xi(\sum k - \sum k')} \int_{\xi<y_1<...<y_N<\xi+L} dy_1 \ldots dy_N \psi^*_{\{k\}}(y)\psi_{\{k'\}}(y). \tag{C.4}$$

By virtue of the periodicity condition, the integral over y does not depend on the choice of origin on the circle. The remaining integral over ξ gives zero if $\sum(k - k') \neq 0$.

(b) Let us thus consider the case of equal total momentum. By cyclic invariance, one can write

$$\mathcal{N}\begin{pmatrix} k \\ k' \end{pmatrix} = L \int_{0<x_2<...<x_N<L} dx_2 \ldots dx_N \psi^*_{\{k\}} \psi_{\{k'\}}(0, x_2, \ldots, x_N)$$

$$= L \sum_{P,Q} A^*(P)A'(Q)I(P, Q) \tag{C.5}$$

with

$$I(P, Q) = \int_{0<x_2<...<x_N<L} e^{-i(q_2 x_2 + ... + q_N x_N)} dx_2 \ldots dx_N. \tag{C.6}$$

We have used here the notation of Subsection 4.2.2: $q = Pk - Qk'$. One easily obtains

$$I(P, Q) = \frac{1}{i^N} \sum_{M=1}^{N} e^{-iL(q_{M+1}+\ldots+q_N)}$$

$$\times \left[q_M(q_M + q_{M-1})\ldots(q_M + \ldots + q_1)(q_M + \ldots + q_1 + q_N)\ldots \right.$$

$$\left. (q_M + \ldots + q_{M+2}) \right]^{-1} \tag{C.7}$$

in which we have taken into account the identity

$$q_1 + q_2 + \ldots + q_N \equiv \sum (k - k') = 0. \tag{C.8}$$

Viewed as a function of $N - 1$ independent variables, the right-hand side of (C.7) is holomorphic; the poles are only apparent. However, if the $\{k\}$ represent discrete solutions to the coupled equations, our calculations will be valid only away from exceptional values (of the L parameter, for example) for which the denominators accidentally vanish. One could then invoke continuity in L of the scalar product. In order to avoid this too vague argument, we shall modify the coupled equations in the following manner:

$$n_j \rightarrow n_j + i\varepsilon \frac{L}{2\pi}, \quad \varepsilon > 0 \tag{C.9}$$

which is equivalent to the translation

$$k_j \rightarrow k_j + i\varepsilon. \tag{C.10}$$

For the $\{k'\}$, we shall take $k'_j \rightarrow k'_j - i\varepsilon$ such that all the denominators of (C.7) have a positive imaginary part.

Let us get rid of the exponential factor in (C.7) using the cyclicity conditions

$$A(P)e^{ik_{PN}L} = zA(PC^{-1}), \quad z = e^{\varepsilon L}.$$

We obtain

$$A^*(P)A'(Q)e^{-i(q_N+\ldots+q_{M+1})L} = z^{2(N-M)}A^*(PC^{M-N})A'(QC^{M-N}). \tag{C.11}$$

Substituting expression (C.7) for $I(P, Q)$ into the right-hand side of (C.5), and taking (C.11) into account, we perform the translations $P \rightarrow PC^{N-M}$, $Q \rightarrow QC^{N-M}$ inside the sums over permutations to obtain

$$\mathcal{N} \begin{pmatrix} k \\ k' \end{pmatrix} = \lim_{\varepsilon \to 0} L \sum_{P,Q} A^*(P)A'(Q) \sum_{M=1}^{N} z^{2(N-M)} \left[q_N(q_N + q_{N-1})\ldots \right.$$

$$\left. (q_N + \ldots + q_2) \right]^{-1} i^{-N}. \tag{C.12}$$

The double summation over P and Q can be performed using the algebraic identity (B.6), (B.7); the crucial point here being that the result is proportional to

$$\sum k - \sum k' = 2Ni\varepsilon.$$

We have

$$\mathcal{N}\begin{pmatrix} k \\ k' \end{pmatrix} = \lim_{\varepsilon \to 0} L2Ni\varepsilon N\Delta\begin{pmatrix} k \\ k' \end{pmatrix}, \tag{C.13}$$

with

$$\Delta\begin{pmatrix} k \\ k' \end{pmatrix} = \left| \frac{1}{k_j - k'_i} \frac{1}{k_j - k'_i + ic} \right| \left| \frac{1}{k_j - k'_i + ic} \right|. \tag{C.14}$$

Away from the values of L (or c) where we have a coincidence $k_j = k'_i$, we deduce

$$\mathcal{N}\begin{pmatrix} k \\ k' \end{pmatrix} = 0. \tag{C.15}$$

It seems difficult to prove that such a coincidence cannot occur for two solutions $\{k\}$ and $\{k'\}$ of equations (4.60). Here again we must invoke continuity to obtain the general result.

(c) Let us finally note that the case of the norm is excluded from the preceding reasoning. This is because some identically vanishing denominators then appear in formulas such as (C.7) for some P and Q. The replacement $k \to k + i\varepsilon$, $k' \to k' - i\varepsilon$ must then be made before formula (C.7), since use was made of $\sum (k - k') = 0$ in the latter's derivation.

Appendix D

Proposition: The order ν matrix

$$N_{ij} = \left(1 + \sum_l' \frac{1}{(q_j - q_l)^2} \right) \delta_{ij} - \frac{1}{(q_i - q_j)^2}, \tag{D.1}$$

in which the quantities q_i are the zeroes of the Hermite polynomial $H_\nu(q)$, has the integers $1, 2, \ldots, \nu - 1, \nu$ as eigenvalues.

Proof: Let us show that matrix N_{ij} is similar to a triangular matrix A_{nm} such that $A_{nn} = n$.

We compute the following sum ($n > 1$):

$$\sum_{j=1}^{\nu} N_{ij} q_j^n = -\sum_j{}' \frac{q_j^n - q_i^n}{(q_j - q_i)^2} + q_i^n = -\sum_j{}' \sum_{\alpha+\beta=n-1} \frac{q_j^\alpha q_i^\beta}{q_j - q_i} + q_i^n$$

$$= -\sum_j{}' \sum_{\alpha+\beta=n-1} \frac{q_j^\alpha - q_i^\alpha}{q_j - q_i} q_i^\beta + (n+1)q_i^n, \qquad (D.2)$$

in which we have used the relation

$$\sum_j{}' \frac{1}{q_j - q_i} = -q_i.$$

We thus obtain

$$\sum_{j=1}^{\nu} N_{ij} q_j^n = -\sum_j{}' \sum_{\lambda+\mu=n-2} (\mu+1) q_j^\lambda q_i^\mu + (n+1)q_i^n. \qquad (D.3)$$

Replacing on the right-hand side of (D.3) the sums $\sum_{j\neq i} q_j^\lambda$ by $-q_i^\lambda +$ numerical function of λ and ν, the right-hand side is equal to $(n+1)q_i^n +$ polynomial in q_i of degree $n - 2$. There thus exist coefficients A_{nm} such that

$$\sum_{j=1}^{\nu} N_{ij} q_j^n = \sum_{m=0}^{n} q_i^m A_{mn}; \qquad A_{nn} = n+1. \qquad (D.4)$$

The above equations are valid for $n = 0, 1, 2, \ldots, \nu - 1$. The $n = 0, 1$ cases are easy to verify.

The equalities (D.4) express the fact that $||N||$ is similar to $||A||$. We thus deduce the corollary

$$|N| = \nu!. \qquad (D.5)$$

5

Bethe wavefunctions associated with a reflection group

The symmetric wavefunctions of the boson system or of the spin chain, which were considered in the preceding chapters, display themselves as 'sums over permutations', or in other words as sums over the operations of a reflection group. This remark is linked to the optical analogy of δ-interacting systems (McGuire, 1964): the Schrödinger equation for N identical particles with a δ-function interaction in one dimension is also the equation for an optical wave in a Euclidean space in N dimensions, reflected and transmitted by $N(N-1)/2$ infinitely thin refracting plates, situated on the bisecting hyperplanes $x_i - x_j = 0$. The Bethe wavefunction is a superposition of all waves transmitted and reflected by the kaleidoscope made from the ensemble of plates. The absence of diffracted waves shall be proven in Chapter 10.

The method of superposition which is suggested by the optical analogy leads to the solution of the problem of bosons confined to an interval $[0, L]$, which allows the calculation of the *boundary energy*. The same method gives the solution to the problem of the open spin chain, with free spins on each end. Lastly, we build the wavefunction of the generalized kaleidoscope (Coxeter, 1948) associated with certain finite reflection groups.

5.1 Bosonic gas on a finite interval

5.1.1 Elementary wavefunctions on a semi-infinite axis

The boundary conditions on the wavefunction of the system defined in the 'box' $[0, L]$ are the following:

$$\psi(x_1 = 0, x_2, \ldots, x_N) \equiv 0, \tag{5.1}$$

$$\psi(x_1, x_2, \ldots, x_N = L) \equiv 0, \tag{5.2}$$

$$\{x\} \in \bar{D}: \quad 0 \le x_1 \le x_2 \le \ldots \le x_N \le L.$$

The idea is to build elementary solutions to the Schrödinger equation (4.1) in $\mathbb{R}_N^+$ ($x_i \geq 0$) which also verify equation (5.1) at the boundary $x_1 = 0$ of the D^+ fundamental region

$$D^+ : \quad 0 < x_1 < x_2 < \ldots < x_N < \infty. \tag{5.3}$$

According to the optical analogy, it is natural to try to build ψ using a superposition of all the elementary $\psi_{\{k\}}$ (formula (4.5)) obtained by reflection on the $x = 0$ surface. Let us consider a sequence $\{|k|\}$ of N distinct positive numbers

$$\{|k|\} : \quad |k_1| < |k_2| < \ldots < |k_N| \tag{5.4}$$

and let us define the 2^N sequences $\{\varepsilon\}$, $\varepsilon_j = \pm 1$:

$$\{k\} = \{k_1, k_2, \ldots, k_N\}, \quad k_j = \varepsilon_j |k_j|. \tag{5.5}$$

All the states $\psi_{\{k\}}$ have the same energy

$$E = \sum_{j=1}^{N} k_j^2 \tag{5.6}$$

and the following wavefunction, formed by superposition:

$$\Psi_{\{|k|\}}(x) = \sum_{\{\varepsilon\}} C(\varepsilon_1, \varepsilon_2, \ldots, \varepsilon_N) \psi_{\{k\}}(x) \tag{5.7}$$

is obviously a solution to equation (4.1). The coefficients $C\{\varepsilon\}$ are determined by condition (5.1), which is written as

$$\sum_{\{\varepsilon\}} \sum_{P} C\{\varepsilon\} \prod_{i<j} \left(1 + \frac{ic}{k_{Pi} - k_{Pj}} \right) e^{ik_{P2}x_2 + \ldots + ik_{PN}x_N} \equiv 0. \tag{5.8}$$

We are still free to sum over ε_{P1}, when P and the other ε are fixed. We thus obtain the 2^{N-1} equations

$$C(\varepsilon_1 \ldots \varepsilon_{P1} \ldots \varepsilon_N) \prod_{\beta(\neq P1)} \left(1 + \frac{ic}{k_{P1} - k_\beta} \right)$$
$$+ C(\varepsilon_1 \ldots - \varepsilon_{P1} \ldots \varepsilon_N) \prod_{\beta(\neq P1)} \left(1 + \frac{ic}{-k_{P1} - k_\beta} \right) = 0, \tag{5.9}$$

of which the unique solution is

$$C(\varepsilon_1 \ldots \varepsilon_N) = \varepsilon_1 \varepsilon_2 \ldots \varepsilon_N \prod_{i<j} \left(1 - \frac{ic}{k_i + k_j} \right), \tag{5.10}$$

in which it is agreed that $k_i + k_j = \varepsilon_i |k_i| + \varepsilon_j |k_j|$. The symmetric elementary solution which vanishes at the origin is thus, for $\{x\} \in D^+$,

$$\Psi_{\{|k|\}}(x) = \sum_{\{\varepsilon\}} \sum_P \prod_j \varepsilon_j \prod_{j<l} \left(1 - \frac{ic}{k_j + k_l}\right)\left(1 + \frac{ic}{k_{Pj} - k_{Pl}}\right) e^{ik_{P1}x_1 + \ldots + ik_{PN}x_N}$$

$$(5.11)$$

with $k_{Pi} = \varepsilon_{Pi} |k_{Pi}|$. This is a sum over the $2^N N!$ elements of the reflection group of the N-dimensional cube.

5.1.2 Energy spectrum

We attempt to satisfy the second boundary condition (5.2) by conveniently choosing $|k|$ of the elementary solution (5.11). One obtains directly the relations

$$\sum_{\varepsilon_{PN}} \varepsilon_{PN} \prod_{j<l} \left(\left(1 - \frac{ic}{k_{Pj} + k_{Pl}}\right)\left(1 + \frac{ic}{k_{Pj} - k_{Pl}}\right)\right) e^{ik_{PN}L} = 0, \qquad (5.12)$$

which must be satisfied for all P and all $\{\varepsilon\}$. Setting $PN = \alpha$, we obtain

$$e^{2ik_\alpha L} = \prod_\beta{}' \frac{k_\beta + k_\alpha + ic}{k_\beta + k_\alpha - ic} \frac{k_\beta - k_\alpha - ic}{k_\beta - k_\alpha + ic} = \prod_{\beta(\neq\alpha)} \frac{(k_\alpha + ic)^2 - k_\beta^2}{(k_\alpha - ic)^2 - k_\beta^2}, \qquad \alpha = [1, N].$$

$$(5.13)$$

In this last form, the system of equations is evidently invariant under all reflections and constitutes the set of coupled equations for the spectrum. Writing k_α instead of $|k_\alpha|$ in order to lighten the notation, the principle of continuity in the coupling c allows us to write

$$k_\alpha L = \pi n_\alpha + \sum_{\beta(\neq\alpha)} \left(\operatorname{atan} \frac{c}{k_\alpha - k_\beta} + \operatorname{atan} \frac{c}{k_\alpha + k_\beta}\right) \qquad (5.14)$$

with $k_\alpha > 0$, and in which n_α are a set of quantum numbers for free bosons in a box

$$1 \leq n_1 \leq n_2 \leq \ldots \leq n_N. \qquad (5.15)$$

System (5.14) is very similar to system (4.69) for the periodic gas and the existence of real solutions could be proved in the same way. In fact, there exists a precise analogy between the system in a box of length L and the system with $2L$ periodicity with the same density. Let us consider the class of $2L$-periodic states with $2N$ particles defined by the quantum numbers

$$\{n\}_{2N} = \{\{-n\}_N, \{+n\}_N\}, \qquad n_j > 0, \qquad (5.16)$$

in which the notation used is self-evident. The set $\{k\}_{2N}$ will have the form

$$\{k\}_{2N} = \{\{-k\}_N, \{k\}_N\}, \qquad k_j > 0. \qquad (5.17)$$

In fact, under this hypothesis the equations of the $2L$-periodic system take the form (5.14). The uniqueness of solutions thus allows us to state the proposition

$$E^{\text{box } L}\{n\}_N = \frac{1}{2}E^{\text{circle } 2L}\{\{-n\}_N, \{n\}_N\}. \tag{5.18}$$

5.1.3 Boundary energy in the ground state

The ground state in the box of length L corresponds to the excited state $\{\{-1\}_N, \{1\}_N\}$ of the periodic gas. This fact allows the calculation of the boundary energy E_b of the gas in a box, in the thermodynamic limit:

$$E_b = \lim_{N \to \infty} (E_N^{\text{box } L} - E_N^{\text{circle } L}). \tag{5.19}$$

This definition coincides with the natural definition

$$E_N = \varepsilon_0 N + E_b + O(L^{-1}), \tag{5.20}$$

since one can relatively easily prove (Gaudin, 1971c) that

$$E_N = \varepsilon_0 N + O(N^{-1}) \tag{5.21}$$

by a simple extension of Yang's method for providing the rigorous proof of equations (4.77), (4.78). The quantity $2E_b$ can thus be calculated as an excitation energy between $\{\{-1\}_N, \{1\}_N\}$ and $\{0\}_{2N}$, and is thus positive.

One obtains by an already described method

$$E_b = 2 \int_{-k_0}^{+k_0} kg(k)dk, \tag{5.22}$$

in which the odd function $g(k)$ solves the integral equation

$$g(k) - \frac{c}{\pi} \int_{-k_0}^{+k_0} \frac{g(k')dk'}{(k-k')^2 + c^2} = \frac{1}{2}\varepsilon(k). \tag{5.23}$$

In order to calculate the leading term of E_b in the limit $c/\rho \ll 1$, one can obtain an approximate solution to (5.23) using the method of Kac and Pollard (1956). In the rescaled variable $x = k/k_0$, we have the approximate equation

$$P \int_{-1}^{+1} \frac{g'(y)dy}{y-x} = -\frac{\pi k_0}{2c}\varepsilon(x), \tag{5.24}$$

which can be inverted to give

$$g(y) = \frac{k_0}{4\pi c} \int_{-1}^{+1} dx \, \varepsilon(x) \ln \frac{1 - xy + \sqrt{(1-x^2)(1-y^2)}}{1 - xy - \sqrt{(1-x^2)(1-y^2)}}. \tag{5.25}$$

Formula (5.22) gives us the dominant term of the boundary energy in the high-density limit

$$E_b = 2k_0^2 \int_{-1}^{+1} yg(y)dy \propto \frac{8}{3}c^{1/2}\rho^{3/2}. \tag{5.26}$$

This is a positive quantity since it can be viewed as an excitation energy of the periodic gas, as seen previously. It seems to possess the same type of singularity and energy per particle.

5.2 The generalized kaleidoscope

The objective of this section is to show that the elementary wavefunctions of the boson gas on a line (expressions (4.5) and (4.8)), or on a half-line (expression (5.11)) are particular cases of more general wavefunctions, which we could call wavefunctions of the generalized kaleidoscope. Such a kaleidoscope is simply a system of thin plates, in a space of arbitrary dimension, which generate a finite set of distinct reflections. This generalization thus corresponds to McGuire's optical analogy. Unfortunately, the new solutions to the wave equation in $\mathbb{R}_N$ are no longer those of a system of N particles with two-body interactions. Some δ-function potentials involving many particles (3,4,6,8) intervene and represent less than realistic forces. The generalization however sheds light on the nature of Bethe wavefunctions, displays their rigid nature and reveals in the end the limitation of the models considered.

5.2.1 Generalization by analogy

The two examples dealt with in Section 4.1 and Subsection 5.1.1 lead by simple analogy to the following generalization. Let G be a reflection group, of finite order, for which the elements g act in the Euclidean vector space $\mathbb{R}_N$ leaving the scalar product (written as (x, y)) invariant. By hypothesis the group is generated by a finite set of generators g_ν, $\nu = [1, r]$, which are the reflections with respect to hyperplanes H_ν, having normal vector n_ν whose direction will be specified, such that we have

$$g_\nu H_\nu = H_\nu, \qquad g_\nu n_\nu = -n_\nu, \qquad g_\nu^2 = 1. \tag{5.27}$$

Let D be a *fundamental region* for G, in other words a convex domain of $\mathbb{R}_N$ with the following properties:

$$gD \cap D = 0, \qquad \forall g \in G, \quad g \neq 1$$
$$\bigcup_g \overline{gD} = \mathbb{R}_N. \tag{5.28}$$

There exists such a domain with the following properties (Weyl, 1935; Coxeter, 1948).

(a) It is limited by r hyperplanes H_v, concurrent at point 0 if the group is finite. The H_v define a set of generators of G.

(b) The normal vectors n_v, $(n_v, n_v) = 1$, verify

$$(n_v, n_\mu) = -\cos \frac{\pi}{p_{\mu v}} \le 0; \qquad p_{\mu v} = 2, 3, 4, \ldots, \infty. \qquad (5.29)$$

(c) The convex domain D is defined by

$$(n_v, x) > 0, \qquad \forall v, \qquad x \in D. \qquad (5.30)$$

The intersection of D with a sphere centred at 0 is a spherical simplex, or convex spherical polygon limited by arcs of grand circles crossing at right angles or acute angles $\pi/2, \pi/3, \ldots$

The matrix $p_{\mu v}$ is ordinarily represented by a *graph* with r vertices with an unmarked line linking μ and v if $p_{\mu v} = 3$, a line marked $4, 5, \ldots$ is $p_{\mu v} = 4, 5, \ldots$, etc. Such a diagram characterizes the simplex, and thus the group. We shall return to this in Subsection 5.2.3.

Carrying on with our analogy after this definition of the fundamental region, the elementary wavefunctions that we have already encountered appear as a finite sum over the operations of G:

$$\psi_{\{k\}}(x) = \sum_{g \in G} A(g) e^{i(x, gk)} \qquad (5.31)$$

with

$$k \in D, \qquad x \in D.$$

A restrictive hypothesis is made, namely that the function $\psi(x)$ is *symmetric* in $\mathbb{R}_N$, in other words that its data in D suffices to determine it in the whole of $\mathbb{R}_N$ using the properties (5.28):

$$\psi_{\{k\}}(gx) = \psi_{\{k\}}(x), \qquad x \in D. \qquad (5.32)$$

According to (5.31), ψ is a solution to the wave equation in D:

$$-\Delta_N \psi = (k, k)\psi, \qquad x \in D. \qquad (5.33)$$

In contrast, ψ will be determined by conditions at the boundary of D, which are not arbitrary. In the examples treated thus far, they are of the type of a linear relation between ψ and its normal derivative on the limit hyperplanes of D:

(a) disc. $\left(\dfrac{d\psi}{dn_v}\right) - 2c_v \psi|_{x \in H_v} = 0, \qquad \dfrac{d}{dn_v} = (n_v, \nabla); \qquad (5.34)$

(b) $\psi = 0$, $x \in$ certain H_ν, which is equivalent to $c_\nu = \infty$ on these hyperplanes. The constants c_ν remain to be determined.

Using the symmetry (5.32), one easily sees that condition (5.34) will be verified if the coefficients $A(g)$ satisfy

$$\frac{A(g_\nu g)}{A(g)} = \frac{(n_\nu, gk) + ic_\nu}{(n_\nu, gk) - ic_\nu} \tag{5.35}$$

irrespective of $g \in G$ and irrespective of which generator g_ν is.

5.2.2 Wavefunctions associated with semi-simple groups

We now make use of the known properties of certain finite-order reflection groups to construct a class of elementary solutions of the boundary problem (5.33)–(5.34) by solving equations (5.35) for certain choices of the c_ν coefficients. The spherical simplexes are enumerated by Coxeter and up to two exceptions can be simply defined using root diagrams of simple Lie groups, despite the fact that the correspondence is not bijective and that there might be two different vector diagrams, for example B_N and C_N (relative to $SO(2N+1)$ and $Sp(2N)$), having an identical reflection group (Weyl group), which is precisely the order $2^N N!$ group of Example 5.1.1. It is precisely with the root diagram of a semi-simple Lie group Γ that we shall associate a solution to the aforementioned problem.

Let $\nu, \mu, \ldots$ be a system of r fundamental roots of Γ; the spherical simplex is defined together with the fundamental domain in $\mathbb{R}_N$:

$$n_\nu = \nu(\nu, \nu)^{-1/2}, \quad D : (\nu, x) > 0.$$

Let $\alpha, \beta, \ldots$ be the positive roots, whose components on the fundamental roots are thus non-negative, which leads to $(\alpha, x) > 0$, $x \in D$. We have the classic result

$$\alpha > 0 \ (\alpha \neq \nu) \Rightarrow g_\nu \alpha > 0, \tag{5.36}$$

where g_ν is the reflection with respect to the hyperplane H_ν normal to the fundamental root ν. Such a result can be verified on each diagram; for A_{N-1}, B_N, C_N and D_N, it is very straightforward; the verification for F_4 is made for illustrative purposes in Appendix E.

From (5.36), we deduce the following solution to relations (5.35) by choosing

$$c_\nu = c(\nu, \nu)^{-1/2}, \quad c > 0 \tag{5.37}$$

$$A(g) = \prod_{\alpha > 0} \left(1 - \frac{ic}{(\alpha, gk)} \right). \tag{5.38}$$

Indeed, according to (5.38),

$$A(g_v g) = \prod_{\alpha > 0} \left(1 - \frac{ic}{(g_v \alpha, gk)} \right) = \left(1 + \frac{ic}{(v, gk)} \right) \prod_{\substack{\alpha > 0 \\ \alpha \neq v}} \left(1 - \frac{ic}{(g_v \alpha, gk)} \right).$$

(5.39)

But the $g_v \alpha$ are positive according to (5.36), and distinct since

$$g_v(\alpha - \beta) = 0 \Rightarrow \alpha - \beta = 0.$$

We thus obtain from (5.39)

$$A(g_v g) = \left(1 + \frac{ic}{(v, gk)} \right) \left(1 - \frac{ic}{(v, gk)} \right)^{-1} A(g),$$

(5.40)

in other words (5.35), with the choice (5.37) for c_v. We can thus enunciate the following

Proposition: The elementary symmetric wavefunction

$$\psi_{\{k\}}(x) = \sum_{g \in G} \prod_{\alpha > 0} \left(1 - \frac{ic}{(\alpha, gk)} \right) e^{i(x, gk)}, \qquad k, x \in D$$

(5.41)

is a solution to the Schrödinger equation

$$- \Delta_N \psi + 2c \sum_{\alpha > 0} \delta((x, \alpha)) \psi = (k, k) \psi$$

(5.42)

in which α are the positive roots of a semi-simple Lie group Γ, and G the corresponding Weyl group. It is straightforward to see, by applying Stokes' theorem, that the conditions at the boundary of D (5.34) indeed arise from (5.42).

5.2.3 Specialization and extension

(a) A_{N-1}, order $N!$

Let e_i, $i = [1, N]$ be an orthonormalized basis of $\mathbb{R}_N$. The fundamental roots of A_{N-1} are $v = e_{i+1} - e_i$ and the positive roots are $\alpha = e_j - e_i$, $j > i$. The Weyl group G is the permutation group π_N, whose generators are the $N - 1$ transpositions $g_{e_{i+1}-e_i} = P_{i,i+1}$. The $N - 1$ vertex graph is the following:

$$(x, e_i) = x_i,$$

$$g_{e_j - e_i} e_i = e_j,$$

$$g_{e_j - e_i} k = P_{ij} k, \qquad g \leftrightarrow P \in \pi_N$$

$$(gk, \alpha) \Rightarrow (Pk, e_j - e_i) = (Pk)_j - (Pk)_i.$$

Fundamental region in $\mathbb{R}_N$:

$$(x_{i+1} - x_i) > 0 \Rightarrow D : x_1 < x_2 < \ldots < x_N.$$

Potential (in the Schrödinger equation (5.42)):

$$\sum_{\alpha > 0} \delta((x, \alpha)) = \sum_{j > i} \delta(x_j - x_i). \tag{5.43}$$

The specialization of (5.41) to A_{N-1} thus gives the results obtained above in (4.5)–(4.8).

(b) B_N and C_N; D_N

The fundamental roots are, for example,

$$\left. \begin{matrix} B_N & e_1 \\ C_N & 2e_1 \\ D_N & e_1 + e_2 \end{matrix} \right\}, \qquad e_2 - e_1, e_3 - e_2, \ldots, e_N - e_{N-1} \tag{5.44}$$

and the positive roots are

$$e_j - e_i, \qquad e_j + e_i \quad (j > i);$$

with e_i for B_N and $2e_i$ for C_N. \tag{5.45}

The Weyl group common to B_N and C_N is the hypercubic group encountered previously, of order $2^N N!$. The operations g of this group are such that

$$k \rightarrow gk \qquad \text{or} \qquad k_j \rightarrow \varepsilon_{Pj} k_{Pj} \tag{5.46}$$

in which P is a permutation of π_N and $\varepsilon_i = \pm 1$, with the 2^N possible choices $\{\varepsilon\}$. The graph has N vertices:

The group D_N is a subgroup of the preceding one, such that $\varepsilon_1 \varepsilon_2 \ldots \varepsilon_N = 1$, with order $2^{N-1} N!$. The graph is:

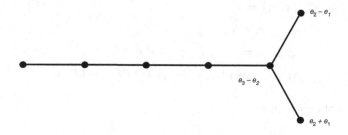

The fundamental domains in $\mathbb{R}_N$ are

$$D^+ : 0 < x_1 < x_2 < \ldots < x_N, \quad \text{for } B_N, C_N$$

and

$$x_1 < x_2 < \ldots < x_N, \quad x_1 + x_2 > 0, \quad \text{for } D_N.$$

In the three cases, the potential is written as

$$V(x) = 2b \sum_i \delta(x_i) + 2c \sum_{i<j} \left(\delta(x_i - x_j) + \delta(x_i + x_j) \right), \tag{5.47}$$

with $b/c = 0, 1/2, 1$ for D_N, C_N and B_N, respectively.

Extension

The Schrödinger equation can be resolved with potential (5.47) for an arbitrary value of b and c. The solution is as follows:

$$\psi_{\{k\}}(x) = \sum_{g \in G} A(g) e^{i(x,gk)}, \quad x \in D^+ \tag{5.48}$$

in which G is the hypercubic group of order $2^N N!$ and $A(g)$ is defined as follows:

$$A(g) = \prod_{j=1}^N \left(1 - \frac{ib}{(e_j, gk)} \right) \prod_{j<l} \left(1 - \frac{ic}{(e_l - e_j, gk)} \right) \left(1 - \frac{ic}{(e_l + e_j, gk)} \right). \tag{5.49}$$

One can easily verify relations (5.35) associated with the fundamental reflections $g_{e_1}, g_{e_2-e_1}, \ldots, g_{e_N-e_{N-1}}$. For example, g_{e_1} leaves the double product (5.49) invariant, and one finds

$$\frac{A(g_{e_1}g)}{A(g)} = \frac{(e_1, gk) + ib}{(e_1, gk) - ib}. \tag{5.50}$$

Let us finally note that the elementary wavefunction (5.11) of the system of bosons on the semi-infinite line is exactly of the limiting form (5.48) for $b = \infty$

(after multiplying by $\prod_j |k_j|/(ib)^N$). Indeed, the potential $b\delta(x)$, having infinite intensity, forces the vanishing of the wavefunction at the origin. On the positive semi-axis, the $\sum \delta(x_i + x_j)$ in potential (5.47) is ineffective.

(c) We shall not discuss the *exceptional Lie groups* since they do not lead to an analogy of pairwise interacting particles. We shall moreover not mention anything of a more interesting problem consisting of constructing the wavefunctions belonging to an irreducible representation of G of dimension higher than 1:

$$\psi(gx) = T(g)\psi(x),$$

in other words having an arbitrary 'type of symmetry', except for algebra A_{N-1} corresponding to the permutation group π_N whose problem is treated completely in Chapters 11 and 12.

5.3 The open chain

The superposition method described in Subsection 5.1.1 for finding the solution of the problem of bosons with wavefunctions vanishing at the ends of a given interval can be applied to the problem of the spin-1/2 chain in which the first and last atom do not interact with each other. The system's Hamiltonian is thus the following:

$$H = \sum_{n=1}^{N-1} S_n^x S_{n+1}^x + S_n^y S_{n+1}^y + \Delta\left(S_n^z S_{n+1}^z - \frac{1}{4}\right). \tag{5.51}$$

Following the method described in Section 1.1, one finds the equation for the amplitudes and the energy levels E:

$$\sum_{\{n'\}} [a\{n'\} - \Delta a\{n\}] = 2Ea\{n\}, \tag{5.52}$$

in which the sum is performed over all allowed transitions $\{n\} \to \{n'\}$. We extend the definition of $a\{n\}$ to the cases in which the n_i can be equal and, additionally, to the interval $[0, N+1]$.

It is then possible to decouple system (5.52) into the system of differences (1.20) and the system analogous to (1.21) which cancels all the forbidden terms by (5.52). We however have two additional forbidden terms, which when put to zero constitute the boundary conditions

$$a(0, n_2, n_3, \ldots) - \Delta a(1, n_2, n_3, \ldots) = 0,$$
$$a(n_1, \ldots, n_{M-1}, N+1) - \Delta a(n_1, \ldots, n_{M-1}, N) = 0. \tag{5.53}$$

We look for the solution in the following form, analogous to (5.7):

$$a(n_1 n_2 \ldots n_M) = \sum_{\{\varepsilon\}} C(\varepsilon_1 \ldots \varepsilon_M) \sum_P \exp \left\{ \frac{i}{2} \sum_{\alpha < \beta} \psi(\varphi_{P\alpha} - \varphi_{P\beta}) + i \sum_\alpha k_{P\alpha} n_\alpha \right\},$$

(5.54)

with $k_\alpha = \varepsilon_\alpha |k_\alpha|$, $\varphi_\alpha = \varepsilon_\alpha |\varphi_\alpha|$, with the M quantities $|\varphi_\alpha|$ are a set of distinct complex numbers with real parts situated between 0 and π. Equations (1.20) are satisfied, if the energy of the corresponding state is given by the formula

$$E = \sum_{\alpha=1}^{M} (\cos k_\alpha - \Delta). \tag{5.55}$$

In contrast, equations (1.21) are also satisfied when the antisymmetric phases and the momenta are related by equation (1.25), or explicitly

$$\cot \frac{k_\alpha}{2} = \coth \frac{\Phi}{2} \tan \frac{\varphi_\alpha}{2}, \qquad \Delta = \cosh \Phi \tag{5.56}$$

$$\cot \frac{1}{2} \psi(\varphi_\alpha - \varphi_\beta) = \coth \Phi \, \tan \frac{\varphi_\alpha - \varphi_\beta}{2}. \tag{5.57}$$

There remain the boundary conditions. The first equation (5.53) gives

$$\prod_{\alpha \neq 1} e^{\frac{i}{2} \psi(\varphi_{P1} - \varphi_{P\alpha})} C(\varepsilon_1 \ldots \varepsilon_{P1} \ldots \varepsilon_M)(1 - \Delta e^{i k_{P1}})$$

$$+ \prod_{\alpha \neq 1} e^{-\frac{i}{2} \psi(\varphi_{P1} + \varphi_{P\alpha})} C(\varepsilon_1 \ldots - \varepsilon_{P1} \ldots \varepsilon_M)(1 - \Delta e^{-i k_{P1}}) = 0. \tag{5.58}$$

With the help of (1.34) we write

$$\frac{1 - \Delta e^{-i k_\alpha}}{1 - \Delta e^{i k_\alpha}} = -e^{-i \psi(2\varphi_\alpha)}, \tag{5.59}$$

which allows us to put relations (5.58) in the form

$$\frac{C(\varepsilon_1 \ldots - \varepsilon_\alpha \ldots)}{C(\varepsilon_1 \ldots \varepsilon_\alpha \ldots)} = \exp \left\{ \frac{i}{2} \sum_{\beta(\neq \alpha)} (\psi(\varphi_\alpha - \varphi_\beta) + \psi(\varphi_\alpha + \varphi_\beta)) + i \psi(2\varphi_\alpha) \right\}. \tag{5.60}$$

Their unique solution is

$$C(\varepsilon_1 \ldots \varepsilon_M) = \exp \left\{ -\frac{i}{2} \sum_{\alpha \leq \beta} \psi(\varphi_\alpha + \varphi_\beta) \right\}. \tag{5.61}$$

The amplitude can thus be written in the rather elegant form

$$a\{n\} = \sum_P \sum_{\{\varepsilon\}} \exp \left\{ \frac{i}{2} \sum_{\alpha < \beta} (\psi(\varphi_{P\alpha} - \varphi_{P\beta}) - \psi(\varphi_{P\alpha} + \varphi_{P\beta})) \right.$$

$$\left. - \frac{i}{2} \sum_\alpha \psi(2\varphi_\alpha) + i \sum_\alpha n_\alpha k_{P\alpha} \right\}. \tag{5.62}$$

We must now satisfy the second boundary condition (5.53), which reads

$$e^{iNk_{PM}} (e^{ik_{PM}} - \Delta) \exp \left\{ \frac{i}{2} \sum_{\alpha \leq M} (\psi(\varphi_{P\alpha} - \varphi_{PM}) - \psi(\varphi_{P\alpha} + \varphi_{PM})) \right\}$$

$$+ e^{-iNk_{PM}} (e^{-ik_{PM}} - \Delta) \exp \left\{ \frac{i}{2} \sum_{\alpha \leq M} (\psi(\varphi_{P\alpha} + \varphi_{PM}) - \psi(\varphi_{P\alpha} - \varphi_{PM})) \right\} = 0 \tag{5.63}$$

and must be verified for any P and any sign system $\{\varepsilon\}$. This is the case when the quantities φ_α and k_α are related by the equations

$$(N + 1)k_\alpha = \pi \lambda_\alpha + \psi(2\varphi_\alpha) + \frac{1}{2} \sum_{\beta(\neq\alpha)} (\psi(\varphi_\alpha - \varphi_\beta) + \psi(\varphi_\alpha + \varphi_\beta)) \tag{5.64}$$

in which $\{\lambda\}$ is a set of integer numbers. We have not determined the allowed sets of such numbers. It is likely that Griffiths' method described in Section 1.4 could allow us to demonstrate the existence of real solutions for certain classes of quantum numbers. One can also proceed using continuity in the Φ parameter (or Θ, since equations (5.64) can be demonstrated in the same way for $|\Delta| < 1$).

Let us consider, for example, the case of an odd number of sites $N = 2M + 1$ with $S^z = \frac{1}{2}$. There exists a single state for which all k are real on the whole interval $\Delta > 1$. The λ are easily determined by looking at the $\Delta = \infty$ limit; one finds

$$\{\lambda\} = \{2, 4, 6, \ldots, 2M\}. \tag{5.65}$$

It is curious to note that in the $\Delta = 1$ limit, the system (5.64) for the spin chain of length N is identical to a class of systems (1.22) for the ring of length $2N$. In the case of the isotropic interaction, we thus have the analogous situation to that mentioned in Subsection 5.1.2 for the system of bosons in a box L and with period $2L$ (formula (5.18)).

Appendix E

Proposition: Let $\alpha > 0$ be the positive roots of the diagram of a Lie algebra Γ; ν the fundamental roots, g_ν the reflections with respect to the H_ν hyperplane normal to ν and passing through the origin of the diagram. We have

$$g_\nu\alpha > 0, \qquad \forall \nu \neq \alpha. \tag{E.1}$$

We shall not give the proof of this classic theorem, but only verify it on the exceptional group F_4.

· By hypothesis, α is a form with non-negative coefficients on the fundamental roots

$$\alpha = \sum_\mu c_\mu\mu, \qquad c_\mu \geq 0 \tag{E.2}$$

and we have

$$g_\nu\alpha = \alpha - 2\frac{(\nu, \alpha)}{(\nu, \nu)}\nu = \left(c_\nu - 2\frac{(\nu, \alpha)}{(\nu, \nu)}\right)\nu + \sum_{\mu\neq\nu} c_\mu\mu. \tag{E.3}$$

It therefore suffices to show that the coefficient of ν is non-negative

$$c_\nu \leq -\sum_{\mu(\neq\nu)} 2\frac{(\nu, \mu)}{(\nu, \nu)}c_\mu, \tag{E.4}$$

or equivalently, after introducing the angles of the simplex,

$$(\nu, \nu)^{1/2}c_\nu \leq \sum_{\mu(\neq\nu)} 2\cos\frac{\pi}{p_{\mu\nu}}(\mu, \mu)^{1/2}c_\mu. \tag{E.5}$$

Let us take the example of F_4. Let e_j be the basis vectors of $\mathbb{R}_4$. The chosen fundamental roots are

$$\left\{\frac{1}{2}(e_4 - e_1 - e_2 - e_3), e_1, e_2 - e_1, e_3 - e_2\right\} \tag{E.6}$$

and the corresponding fundamental domain $(\nu, x) > 0$:

$$D : 0 < x_1 < x_2 < x_3; \qquad x_1 + x_2 + x_3 < x_4, \tag{E.7}$$

with the Dynkin diagram whose vertices are associated with the four fundamental roots:

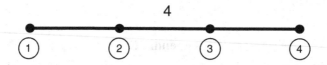

which means, $p_{13} = p_{14} = p_{24} = 2$; $p_{12} = p_{34} = 3$; $p_{23} = 4$.

The 24 positive roots of F_4 have the following components on the fundamental (underlined) ones:

$$
\begin{aligned}
& & e_2 - e_1 &= \underline{(0\,0\,1\,0)} & e_2 + e_1 &= (0\,2\,1\,0) \\
e_1 &= \underline{(0\,1\,0\,0)} & e_3 - e_1 &= (0\,0\,1\,1) & e_3 + e_1 &= (0\,2\,1\,1) \\
e_2 &= (0\,1\,1\,0) & e_4 - e_1 &= (2\,2\,2\,1) & e_4 + e_1 &= (2\,4\,2\,1) \\
e_3 &= (0\,1\,1\,1) & e_3 - e_2 &= \underline{(0\,0\,0\,1)} & e_3 + e_2 &= (0\,2\,2\,1) \\
e_4 &= (2\,3\,2\,1) & e_4 - e_2 &= (2\,2\,1\,1) & e_4 + e_2 &= (2\,4\,3\,1) \\
& & e_4 - e_3 &= (2\,2\,1\,0) & e_4 + e_3 &= (2\,4\,3\,2)
\end{aligned}
\tag{E.8}
$$

and

$$
\frac{1}{2}(e_4 + e_1 + e_2 + e_3) = (1\,3\,2\,1)
$$

$$
\frac{1}{2}(e_4 - e_1 + e_2 + e_3) = (1\,2\,2\,1)
$$

$$
\frac{1}{2}(e_4 + e_1 - e_2 + e_3) = (1\,2\,1\,1)
$$

$$
\frac{1}{2}(e_4 + e_1 + e_2 - e_3) = (1\,2\,1\,0)
$$

$$
\frac{1}{2}(e_4 - e_1 - e_2 - e_3) = (1\,1\,1\,1)
$$

$$
\frac{1}{2}(e_4 - e_1 + e_2 - e_3) = (1\,1\,1\,0)
$$

$$
\frac{1}{2}(e_4 + e_1 - e_2 - e_3) = (1\,1\,0\,0)
$$

$$
\frac{1}{2}(e_4 - e_1 - e_2 - e_3) = \underline{(1\,0\,0\,0)}.
\tag{E.9}
$$

Taking (E.8) into account, the relations (E.5) which have to be verified are written

$$
\begin{aligned}
c_1 &\le c_2, \\
c_2 &\le c_1 + 2c_3, \\
c_3 &\le c_2 + c_4, \\
c_4 &\le c_3,
\end{aligned}
\tag{E.10}
$$

which one can rapidly do by using the above table.

6

Continuum limit of the spin chain

6.1 δ-Interacting bosons and the Heisenberg–Ising chain

6.1.1 Amplitude and energy

Knowing the expressions for the wavefunctions, amplitudes and energy levels of
the system of repulsive bosons studied in Chapter 4, and of the spin chain studied
in Chapter 1, it is straightforward to show that the first system can be viewed as
a limit of the second when the lattice spacing $a = L/N$ tends to zero, the length
of the chain remaining finite and the anisotropy parameter Δ tending to unity such
that the product $N(\Delta - 1)$ remains finite.

Referring to Chapter 1, Section 1.3, we examine the limit of the chain in which
$N \to \infty$, $\Delta = \cos \Theta \to 1$, such that we have

$$\lim N\Theta^2 = cL, \qquad 1 - \Delta \approx \frac{cL}{2N}. \tag{6.1}$$

If the phases $\psi_{\alpha\beta}$ are supposed finite in the considered limit, we have $\Im\theta_\alpha = \pi$
while $\Re\theta_\alpha$ and k_α tend to zero, which leads to the equivalences

$$\cot \frac{\psi_{\alpha\beta}}{2} = \frac{\theta_\alpha - \theta_\beta}{2\Theta}, \qquad k_\alpha = \frac{\Theta}{2}\Re\theta_\alpha. \tag{6.2}$$

We shall set

$$Lq_\alpha = \lim Nk_\alpha. \tag{6.3}$$

We deduce from (6.1) and (6.2)

$$\cot \frac{\psi_{\alpha\beta}}{2} = \frac{1}{c}(q_\alpha - q_\beta). \tag{6.4}$$

The coupled equations (1.32) take the limiting form

$$Lq_\alpha = 2\pi\lambda_\alpha + {\sum_{\beta}}' \psi_{\alpha\beta}, \tag{6.5}$$

which are precisely the equations for the momenta q_α of the system of repulsive bosons in a volume L (4.60).

If we now consider the amplitudes $a\{n\}$ for arguments n_α such that

$$\frac{x_\alpha}{L} = \lim \frac{n_\alpha}{N}, \qquad k_\alpha n_\alpha = q_\alpha x_\alpha, \tag{6.6}$$

we have

$$\lim a(n_1 n_2 \ldots n_M) = \psi_B(x_1 x_2 \ldots x_M) \equiv \sum_P A(P) \exp(i q_{P1} x_1 + \ldots + i q_{PM} x_M) \tag{6.7}$$

in which ψ_B is the wavefunction (4.5) of the boson system.

As far as the energy is concerned, we obtain to order N^{-2}

$$E = \sum_{\alpha=1}^{M} (k_\alpha^2 - \Theta^2) = \frac{L^2}{N^2} \sum_\alpha q_\alpha^2 - \frac{MLc}{N} \tag{6.8}$$

and thus

$$E = a^2 E_B - acM + O(a^3), \tag{6.9}$$

with $a = L/N$ and $E_B = \sum q_\alpha^2$ the energy of state ψ_B. One can also write

$$E_B = \lim_{a \to 0} \frac{E}{a^2} + \frac{cM}{a}. \tag{6.10}$$

6.1.2 Boson Hamiltonian equivalent to that of Heisenberg–Ising

It is possible to show directly, without invoking wavefunctions, that the spin chain Hamiltonian is equivalent, after a convenient rescaling, to the boson Hamiltonian H_B. We have already highlighted in Chapter 1 the equivalence with a system of spinless fermions after Jordan–Wigner transformation. The equivalence with bosons is a consequence of a representation used by Dyson (1956) for the study of ferromagnetic systems, relating a non-Hermitian bosonic Hamiltonian to a spin wave system. Defining the Fourier modes

$$S_k^+ = \sum_{n=1}^{N} e^{ikn} S_n^+, \qquad k = \frac{2\pi}{N} p, \quad p \text{ integer}, \quad -\pi < k \le \pi$$

$$S_k^- = \sum_n e^{-ikn} S_n^-,$$

$$S_k^z = \sum_n e^{ikn} S_n^z, \tag{6.11}$$

we get the commutation relations

$$[S_k^+, S_{k'}^-] = 2S_{k-k'}^z,$$
$$[S_k^z, S_{k'}^+] = S_{k+k'}^+,$$
$$[S_k^-, S_{k'}^z] = S_{k-k'}^- \tag{6.12}$$

and Hamiltonian (1.3) can be written ($J \equiv 1$)

$$H_\Delta = \frac{N\Delta}{2} - \frac{2}{N}\sum_k \cos k(S_k^+ S_k^- + \Delta S_k^z S_{-k}^z). \tag{6.13}$$

We shall expand the eigenstates $|\psi\rangle$ on the non-orthogonal and overcomplete basis

$$|k_1 \ldots k_M\rangle = N^{-\frac{M}{2}} S_{k_1}^- \ldots S_{k_M}^- |S^z = N/2\rangle. \tag{6.14}$$

Calculating $H|\psi\rangle$ requires operating with $S_k^+ S_k^-$ on vector $|k_1 k_2 \ldots k_M\rangle$. Commutating the operator through the set of $S_{k_j}^-$, one obtains

$$S_k^- S_k^+ |k_1 \ldots k_M\rangle \equiv N \left(\sum_{j=1}^M \delta(k - k_j)\right) |k_1 \ldots k_M\rangle$$
$$-2\sum_{\substack{j<l \\ \uparrow \\ j}} |k_1 \ldots, \underset{\uparrow}{k}, \ldots, \underset{\uparrow}{k_j + k_l - k}, \ldots k_M\rangle. \tag{6.15}$$

In the same way, we find for $k \neq 0$,

$$S_k^z S_{-k}^z |k_1 \ldots k_M\rangle \equiv M|k_1 \ldots k_M\rangle$$
$$+2\sum_{\substack{j\neq l}} |k_1 \ldots, \underset{\uparrow}{k_j - k}, \ldots, \underset{\uparrow}{k_l + k}, \ldots k_M\rangle. \tag{6.16}$$

We then associate with each vector $|k_1 \ldots k_M\rangle$ the state of a system of bosons

$$|k_1 \ldots k_M\rangle \rightarrow a_{k_1}^\dagger \ldots a_{k_M}^\dagger |0\rangle \equiv |k_1 \ldots k_M\rangle_B \tag{6.17}$$

where $|0\rangle$ is the vacuum for the algebra $a, a^\dagger$,

$$\left[a_k, a_{k'}^\dagger\right] = \delta(k - k'). \tag{6.18}$$

We notice that the operation of $S_k^- S_k^+$ on $|k_1 \ldots k_M\rangle$ is equivalent to that of a certain bosonic operator on state $|k_1 \ldots k_M\rangle_B$; more specifically, we have

$$S_k^- S_k^+ \rightarrow N a_k^\dagger a_k - \sum_{p,p'} a_k^\dagger a_{p+p'-k}^\dagger a_p a_{p'}, \tag{6.19}$$

and also, if $k \neq 0$,

$$S_k^z S_{-k}^z \rightarrow M + \sum_{p,p'} a_{p-k}^\dagger a_{p'+k}^\dagger a_p a_{p'}. \tag{6.20}$$

If we make use of expression (6.20) for all k, even zero, it is necessary to correct the sum by the value

$$(N/2 - M)^2 - M - M(M - 1) = \frac{N^2}{4} - MN,$$

with

$$M = \sum_k a_k^\dagger a_k. \tag{6.21}$$

We arrive at the conclusion that operating with Hamiltonian H in the basis $|k\rangle$ is equivalent to operating with $\tilde{H}$ in the basis of bosonic states, with the following expression for $\tilde{H}$:

$$\tilde{H} = 2\Delta M - 2\sum_k \cos k\, a_k^\dagger a_k$$

$$+ \frac{2}{N}\sum_k \cos k \sum_{p,p'} \left(a_k^\dagger a_{p+p'-k}^\dagger a_p a_{p'} - \Delta a_{p'+k}^\dagger a_{p-k}^\dagger a_p a_{p'} \right). \tag{6.22}$$

$\tilde{H}$ is non-Hermitian and, by construction, its spectrum contains that of H.

6.1.3 Weak limit of $\tilde{H}$

We write $\tilde{H}$ in the form

$$\tilde{H} = 2M(\Delta - 1) + \sum_k 2(1 - \cos k)a_k^\dagger a_k$$

$$+ \frac{2}{N}\sum_{k,p,p'} \left((1 - \Delta)\cos k + \cos(p' + k) - \cos k \right) a_{p'+k}^\dagger a_{p-k}^\dagger a_p a_{p'}. \tag{6.23}$$

We change the discrete indices by taking

$$kN = qL, \quad q = 2\pi\nu/L, \quad -\frac{N}{2} < \nu < \frac{N}{2}, \tag{6.24}$$

and we have

$$\tilde{H} + \frac{MLc}{N} = \sum_q 4\sin^2\frac{qL}{2N}a_q^\dagger a_q + \frac{cL}{N^2}\sum_{q,r,r'}\cos\frac{qL}{2N}a_{r'+q}^\dagger a_{r-q}^\dagger a_r a_{r'}$$

$$+ \frac{4}{N}\sum_{q,r,r'} \left(\sin^2\frac{(r'+q)L}{2N} - \sin^2\frac{qL}{2N} \right) a_{r'+q}^\dagger a_{r-q}^\dagger a_r a_{r'}. \tag{6.25}$$

Multiplying each side of (6.25) by $(N/L)^2$ and taking the limit of each matrix element of $\tilde{H}$ associated with finite momenta, we obtain the weak limit

$$\lim\left(\frac{N^2}{L^2}\tilde{H} + \frac{NMc}{L}\right) = \sum_q q^2 a_q^\dagger a_q + \frac{c}{L}\sum_{q,r,r'} a_{r'+q}^\dagger a_{r-q}^\dagger a_r a_{r'}. \tag{6.26}$$

The sums run over momenta of the form $q = 2\pi\nu/L$, $\nu \in \mathbb{Z}$. The first negligible term of (6.25), of order N^{-1}, is

$$\frac{1}{N}\sum_{q,r,r'}(r'^2 + 2qr')a_{r'+q}^\dagger a_{r-q}^\dagger a_r a_{r'}, \tag{6.27}$$

which is non-Hermitian. We recognize on the right-hand side of (6.26) the Hamiltonian of a system of bosons with local interactions in a volume of periodicity L, the coupling constant being $2c$. If we look at the relation between the eigenvalues E of $\tilde{H}$ and E_B of H_B, we fall back on (6.10). The link between the quantum 'nonlinear Schrödinger' model and the anisotropic Heisenberg–Ising chain is thus established.

6.1.4 Extension to the XYZ model

We would like to extend the preceding limiting process to the XYZ Hamiltonian (1.5). The supplementary term that we need to add to Hamiltonian H_Δ is

$$H_r = 2\Gamma\sum_n\left(S_n^x S_{n+1}^x - S_n^y S_{n+1}^y\right) = \frac{\Gamma}{N}\sum_k \cos k\left(S_k^+ S_{-k}^+ + S_k^- S_{-k}^-\right). \tag{6.28}$$

Using the correspondence described in Subsection 6.1.2, we have

$$S_k^- S_{-k}^- \to N a_k^\dagger a_{-k}^\dagger, \tag{6.29}$$

in which the factor N comes from the normalization chosen in (6.14). We also have the equivalence

$$S_k^+ S_{-k}^+ \to N(a_k a_{-k} + O(N^{-1})) \tag{6.30}$$

since the exact correspondence is

$$S_k^+ \to \sqrt{N}\left(a_k - \frac{2}{N}\sum_{p,p'} a_{p+p'-k}^\dagger a_p a_{p'}\right). \tag{6.31}$$

The associated Hamiltonian thus contains the supplementary term

$$\tilde{H}_\Gamma = \Gamma\sum_k \cos k\left(a_k^\dagger a_{-k}^\dagger + a_{-k} a_k\right). \tag{6.32}$$

For this term to be of the same order as $\tilde{H}$, in other words $O(N^{-2})$, we must choose the scale

$$\Gamma = \gamma\frac{L^2}{N^2} = \gamma a^2, \tag{6.33}$$

to be compared with $(1 - \Delta) = O(N^{-1})$. We thus obtain the limit

$$\lim_{a \to 0} \left(\frac{1}{a^2} (\tilde{H} + \tilde{H}_\Gamma) + \frac{Mc}{a} \right) = \sum_q q^2 a_q^\dagger a_q + \sum_q \gamma (a_q^\dagger a_{-q}^\dagger + a_{-q} a_q)$$

$$+ \frac{c}{L} \sum_{q,r,r'} a_{r'+q}^\dagger a_{r-q}^\dagger a_r a_{r'}, \tag{6.34}$$

in which M appearing on the left-hand side must be understood as $\dfrac{N}{2} - S^z$. Indeed, the boson Hamiltonian defined by the right-hand side of (6.34) does not conserve particle number. Its spectrum is thus not *a priori* simply related to that of the XYZ Hamiltonian solved by Baxter (1971a). The problem to be resolved would be that of the anisotropic chain in the presence of a magnetic field.

6.2 Luttinger and Thirring models

6.2.1 Equivalent fermionic system

The Jordan–Wigner transformation applied to the XYZ Hamiltonian

$$H_{\Delta,\Gamma} = \sum_{n=1}^{N} \left(S_n^+ S_{n+1}^- + S_n^- S_{n+1}^+ \right) + \Gamma \left(S_n^+ S_{n+1}^+ + S_n^- S_{n+1}^- \right) + 2\Delta \left(S_n^z S_{n+1}^z - \frac{1}{4} \right) \tag{6.35}$$

gives us the equivalent fermionic operator Hamiltonian

$$H = \sum_{n=1}^{N} \left(a_{n+1}^\dagger a_n + a_n^\dagger a_{n+1} \right) + \Gamma \left(a_n a_{n+1} + a_{n+1}^\dagger a_n^\dagger \right) + 2\Delta \left(a_n^\dagger a_n a_{n+1}^\dagger a_{n+1} - a_n^\dagger a_n \right). \tag{6.36}$$

This is only exact up to a minor restriction: the transformation indeed induces a boundary effect, already mentioned in Subsection 1.3.4, which could in any case be treated in a general fashion. However, to preserve simplicity, we shall limit ourselves to states for which the cyclic form (6.36) is exact. We have seen that, for $\Gamma = 0$, it sufficed to restrict ourselves to M odd. For Γ not equal to zero, M is not conserved anymore, but its parity still is since fermions are created or absorbed in pairs. The form (6.36) is thus appropriate for the class of states belonging to quantum number $(-1)^M = -1$.

In order to exploit the cyclic invariance of the chain, we shall use the momentum basis defined by the canonical transformation

$$a_k^\dagger = N^{-\frac{1}{2}} \sum_{n=1}^{N} a_n^\dagger e^{ikn},$$

$$kN = 2\pi \times \text{integer}; \quad -\pi < k \le \pi. \tag{6.37}$$

One thus obtains

$$H = \sum_k 2\cos k\, a_k^\dagger a_k - i\Gamma \sin k (a_k^\dagger a_{-k}^\dagger + a_k a_{-k}) - \frac{4\Delta}{N} \sum_p \sin^2 \frac{p}{2} \rho_{-p}\rho_p, \quad (6.38)$$

with the expression for the densities

$$\rho_p = \sum_k a_{k+p}^\dagger a_k. \quad (6.39)$$

The sums run over the discrete values of k or p (mod 2π).

The operator $S^z = N/2 - M$ is determined from the expression

$$M = \sum_n a_n^\dagger a_n = \sum_k a_k^\dagger a_k = \frac{1}{N} \sum_p \rho_{-p}\rho_p, \quad (6.40)$$

which is a constant of motion only if $\Gamma = 0$.

The interaction term of H is proportional to Δ. The free fermion system corresponding to $\Delta = 0$, or the XY model, is easily diagonalized. The quasiparticle energy is $E(k) = (\cos^2 k + \Gamma^2 \sin^2 k)^{1/2}$. The correlation functions are expressed in terms of determinants, which are amenable to elaborate calculations in the thermodynamic limit. The link between the XY model and the two-dimensional Ising model is mentioned in Chapter 7 and has been exploited for the calculation of critical exponents, etc. (McCoy and Wu, 1973).

6.2.2 Luttinger model. Hamiltonian

Before addressing the definition of the Thirring model as a certain continuum limit of the XYZ model, let us examine a weak limit of the Heisenberg–Ising Hamiltonian as a fermionic system, under conditions in which the spectrum of the limit Hamiltonian coincides with the limit of the discrete spectrum of the spin chain. Such was the behaviour observed for the equivalent bosonic system in Subsection 6.1.3.

Dividing the Brillouin zone into two parts according to the sign of the velocity ($d\varepsilon/dk = -\sin k$) of the unperturbed excitations, we perform the canonical transformation

$$\begin{aligned} a_{-\frac{\pi}{2}+k} &= b_{k,1}, \\ a_{\frac{\pi}{2}+k} &= b_{k,2}, \end{aligned} \qquad -\frac{\pi}{2} \le k < \frac{\pi}{2}, \quad (6.41)$$

defining two fermion species $b_{k,\sigma}$ ($\sigma = 1, 2$) within a half-zone. According to (6.38) and (6.41) we can write H_Δ in terms of the new b operators

$$H_\Delta + 2M\Delta = \sum_{|k|<\frac{\pi}{2}} 2\sin k (b_{k1}^\dagger b_{k1} - b_{k2}^\dagger b_{k2}) + \frac{2\Delta}{N} \sum_{|p|<\pi} \cos p\, \rho_{-p}\rho_p, \quad (6.42)$$

in which we have decomposed the densities in the following manner:

$$\rho_p = \rho_p^0 + \rho_{\pi-p}^\pi, \qquad 0 \le p \le \pi \tag{6.43}$$

$$\rho_p^0 = \sum_{-\frac{\pi}{2} < k < \frac{\pi}{2} - p} b_{k+p,1}^\dagger b_{k,1} + b_{k+p,2}^\dagger b_{k,2},$$

$$\rho_p^\pi = \sum_{p-\frac{\pi}{2} < k < \frac{\pi}{2}} b_{k-p,2}^\dagger b_{k,1} + b_{k-p,1}^\dagger b_{k,2}. \tag{6.44}$$

We also have the constant

$$M = \sum_{|k| < \frac{\pi}{2}} b_{k,1}^\dagger b_{k,1} + b_{k,2}^\dagger b_{k,2}. \tag{6.45}$$

Let us examine the weak limit of operator $(H + 2M\Delta)/a$ when the lattice spacing a tends to zero, such that the length of the chain remains finite. More precisely, we look for the limit of the energy matrix defined between states with a finite number M of excitations above the vacuum (ferromagnetic state $|S^z = N/2\rangle$) associated with momenta:

$$k = aq, \qquad q = \frac{2\pi}{L} n; \qquad n \in \mathbb{Z}. \tag{6.46}$$

One thus obtains

$$H_\alpha = \lim_{a \to 0} \frac{1}{2a}(H + 2M\Delta) = \sum_q q(b_{q1}^\dagger b_{q1} - b_{q2}^\dagger b_{q2}) + \frac{\Delta}{L}(\rho_{-q}^0 \rho_q^0 - \rho_{-q}^\pi \rho_q^\pi), \tag{6.47}$$

with the definitions of the limit densities

$$\rho_q^0 = (\rho_{-q}^0)^\dagger = \sum_r b_{r+q,1}^\dagger b_{r,1} + b_{r+q,2}^\dagger b_{r,2},$$

$$\rho_q^\pi = (\rho_{-q}^\pi)^\dagger = \sum_r b_{r-q,2}^\dagger b_{r,1} + b_{r-q,1}^\dagger b_{r,2}, \tag{6.48}$$

in which the sums run over all momenta $q, r = (2\pi/L) \times$ integer, without restriction.

It is useful to express the Hamiltonian in terms of local fields

$$b_\sigma(x) = L^{-1/2} \sum_q b_{q,\sigma} e^{iqx}, \tag{6.49}$$

verifying the usual anticommutation relations

$$[b_\sigma^\dagger(x), b_{\sigma'}(x')] = \delta_{\sigma\sigma'}\delta(x - x'), \dots \tag{6.50}$$

The interaction potential is expressed in terms of the local densities

$$b^\dagger(x)b(x) = \frac{1}{L}\sum_q \rho_q^0 e^{-iqx}, \qquad b^\dagger(x)\gamma b(x) = \frac{1}{L}\sum_q \rho_q^\pi e^{-iqx}, \tag{6.51}$$

in which γ is the matrix

$$\gamma = \begin{pmatrix} 0 & 1 \\ 1 & 0 \end{pmatrix}. \tag{6.52}$$

The corresponding Hamiltonian density is

$$\Delta \left[(b^\dagger(x)b(x))^2 - (b^\dagger(x)\gamma b(x))^2 \right]. \tag{6.53}$$

The result is unambiguous if we define

$$b^\dagger(x)b(x) = \lim_{x' \to x} b^\dagger(x)b(x').$$

One then finds, in a shortened notation,

$$\begin{aligned}(b^\dagger b)^2 - (b^\dagger \gamma b)^2 &= b_1^\dagger b_1 b_2^\dagger b_2 + b_2^\dagger b_2 b_1^\dagger b_1 + b_1^\dagger b_1 (b_1^\dagger b_1 - b_2 b_2^\dagger) \\ &\quad + b_2^\dagger b_2 (b_2^\dagger b_2 - b_1 b_1^\dagger) = 4 b_1^\dagger(x) b_1(x') b_2^\dagger(x) b_2(x'),\end{aligned} \tag{6.54}$$

in which we used $[b_1^\dagger(x), b_1(x')] = [b_2^\dagger(x), b_2(x')]$, $b^2(x) = 0$. Introducing the matrix

$$\alpha = \begin{pmatrix} 1 & 0 \\ 0 & -1 \end{pmatrix},$$

the limit Hamiltonian is thus written

$$H_\alpha = \int_0^L dx \left\{ -i b^\dagger(x) \alpha \frac{db(x)}{dx} + 4\Delta b_1^\dagger(x) b_1(x) b_2^\dagger(x) b_2(x) \right\}, \tag{6.55}$$

or alternately

$$H_\alpha = \sum_q \left(q b_q^\dagger \alpha b_q + \frac{4\Delta}{L} \rho_{-q1} \rho_{q2} \right). \tag{6.56}$$

The particle number of each species is conserved:

$$M = \sum_q b_q^\dagger b_q, \qquad R = \sum_q b_q^\dagger \alpha b_q. \tag{6.57}$$

6.2.3 Luttinger model. Spectrum and wavefunctions

According to (6.55), the Schrödinger equation for M particles having coordinates x_j and internal variables α_j subjected to Hamiltonian H_α is the following:

$$\sum_{j=1}^M -i\alpha_j \frac{\partial \Psi}{\partial x_j} + 2\Delta \sum_{j<l} (1 - \alpha_j \alpha_l)\delta(x_j - x_l)\Psi = E\Psi, \tag{6.58}$$

where Ψ is completely antisymmetric under particle exchange. The α_j are constants of motion.

It is important to note here that equation (6.58) is undefined, in contrast to the second-order equation for the δ-interacting system of particles. Indeed, if the potential $\delta(x_1 - x_2)$ nontrivially acts between particles 1 and 2, in other words if $\alpha_1 \neq \alpha_2$, the wavefunction cannot be continuous in $x_1 = x_2$, and the product $\delta\Psi$ is ill-defined as a distribution.

However, in view of the linear character of the derivatives in the Schrödinger equation, an arbitrary regularization of the δ distribution gives a unique solution corresponding to that of Luttinger. We can illustrate this in the following way:

$$\frac{d\Psi}{dx} + 2\Delta\delta(x)\Psi = 0 \Rightarrow \Psi(x) = e^{-\Delta\varepsilon(x)}, \tag{6.59}$$

provided we have the relation $2\delta(x) = \dfrac{d\varepsilon(x)}{dx}$ between the regularized functions.

Under this hypothesis, the gluing conditions on the interacting planes are then

$$
\begin{array}{lll}
\text{(a)} \; \alpha_1 = \alpha_2, & \quad \Psi_{\alpha_1\alpha_2}(x_1, x_1) = 0, & \\
\text{(b)} \; \alpha_1 = -\alpha_2, & \quad -i\alpha_1 \text{ disc. } \ln \Psi = 2\Delta, & \tag{6.60}
\end{array}
$$

in other words

$$\Psi_{\alpha_1\alpha_2}(x_1, x_1 + 0) = e^{i(\alpha_1 - \alpha_2)\Delta}\Psi_{\alpha_1\alpha_2}(x_1, x_1 - 0). \tag{6.61}$$

From there, we immediately deduce the Luttinger wavefunction, associated with a set of M real 'energies' $\{q_j\}$ and 'velocities' $\{\alpha_j\}$:

$$
\begin{aligned}
\Psi_{\{\alpha\}}(x_1 \ldots x_M) &= \left| e^{iq\alpha_1 x_1}, e^{iq\alpha_2 x_2}, \ldots, e^{iq\alpha_M x_M} \right| \\
&\quad \times \exp\left\{ -\frac{i\Delta}{2} \sum_{j<l} (\alpha_j - \alpha_l)\varepsilon(x_j - x_l) \right\}
\end{aligned}
\tag{6.62}
$$

belonging to total energy $E = \displaystyle\sum_{j=1}^{M} q_j$.

We notice, however, that form (6.62) is not general enough for us to impose the periodicity conditions. In fact, we obtain yet another solution by multiplying Ψ with a phase factor $\exp\{-i(R\Delta/L)\sum_j x_j\}$. The energy of state Ψ becomes

$$E = \sum_j q_j - \frac{\Delta}{L}R^2. \tag{6.63}$$

This new form is compatible with the periodicity conditions and the q_j then verify the quantization conditions

$$q_j = \frac{2\pi}{L}n_j + \frac{\Delta}{L}M \tag{6.64}$$

from which we get the discrete (but still massively degenerate) levels

$$E = \frac{2\pi}{L} \sum_j n_j - \frac{\Delta}{L}(R^2 - M^2) \quad (|R| \le M). \tag{6.65}$$

The energy displacement due to the interaction is proportional to each particle species' number, $\frac{1}{2}(M + R)$ and $\frac{1}{2}(M - R)$.

6.2.4 Pathology of the δ-interaction

We quickly notice that the limit of the spectrum and of the states of the Heisenberg spin chain do not coincide with the spectrum and states of the limit Hamiltonian if we adopt the Luttinger boundary conditions formulated in (6.61). The limit of the spectrum is obtained from the coupled equations (1.32), by looking for a solution of the form

$$k_j = \alpha_j \left(\frac{\pi}{2} - a\hat{q}_j\right).$$

According to (6.47), the limit energy is

$$E_\alpha = \lim \frac{1}{2a} \sum_j 2\cos k_j = \sum_j \hat{q}_j. \tag{6.66}$$

The limit phases are given by (1.26):

$$\lim \psi_{jl} = \pm\pi - (\alpha_j - \alpha_l)\mu, \tag{6.67}$$

in which we have set

$$\Delta = \tan\mu, \quad -\frac{\pi}{2} < \mu < \frac{\pi}{2}. \tag{6.68}$$

When $(N/2)$ is even, equations (1.32) take the limit form

$$-L\alpha_j\hat{q}_j = 2\pi\lambda'_j + (R - \alpha_j M)\mu, \quad \lambda' \in \mathbb{Z} \tag{6.69}$$

from which we get the energy

$$E_\alpha = \frac{2\pi}{L} \sum_j n'_j + \frac{\mu}{L}(M^2 - R^2), \quad n' \in \mathbb{Z}. \tag{6.70}$$

Owing to the fact that M and R necessarily have the same parity, we note that the whole limit spectrum has periodicity π in the μ parameter, which is consistent with relation (6.68), $\Delta = \tan\mu$.

Comparing formulas (6.70) and (6.65) on the one hand, and (6.69) and (6.64) on the other hand, shows that it is sufficient to replace Δ by μ in the Luttinger model expressions in order to obtain the limit states of the discrete chain. The boundary conditions appropriate for Hamiltonian (6.55) or (6.58) considered as a continuum limit of the spin chain are thus not (6.61), but rather the following:

$$-i\alpha_1 \text{ disc. } \Psi = 2\Delta \cdot \frac{1}{2}(\Psi(x_1, x_1 + 0) + \Psi(x_1, x_1 - 0)), \qquad (6.71)$$

which gives exactly formula (6.61) in which we will have replaced Δ by μ. Formula (6.71) is equivalent to defining $\Psi(x_1, x_1)$ as the half-sum of left and right values. This prescription can be justified by noticing that the product $\delta\Psi$ is defined directly by (6.56), in other words by the Fourier transform. Indeed, at a simple discontinuity point, the Fourier series converges to the half-sum (Titchmarsh, 1939), which we can illustrate analogously to (6.59):

$$iq\Psi(q) + 2\Delta \cdot \frac{1}{2\pi} \sum_{q'} \Psi(q') = 0 \Rightarrow \Psi(q) = \delta_q + \frac{i\Delta}{\pi q}, \qquad (6.72)$$

which expresses the prescription $\varepsilon(x)\delta(x) = 0$.

In conclusion, for the fermion model as well as for the boson one, the weak limit of Hamiltonian H_Δ so defined admits the limit states as eigenstates, provided that the interaction is correctly defined and that the number of particles involved remains finite. There is no reason to expect that such a property should remain were the number of interacting particles to become infinite (existence of a Fermi sea), as is the case for example in the Thirring field theory. The solution of Mattis and Lieb (1965) for the Luttinger field theory (massless Thirring model), based on a correct treatment of the density operator algebra, would not *a priori* reproduce the dispersion curve of the antiferromagnetic spin wave over its linear part. Formula (2.7) (des Cloizeaux and Gaudin, 1966) gives us the excitation curve

$$\eta_1(q) = \frac{\pi}{2} \frac{\sin \Theta}{\Theta} |\sin q|, \qquad \Delta = \cos \Theta.$$

Mattis and Lieb obtain

$$\eta_\alpha(q) = \left(1 + \frac{4\Delta}{\pi}\right)^{1/2} q.$$

The two formulas coincide only to linear order in the coupling.

6.3 Massive Thirring model

6.3.1 Continuum limit

Our task here is simply to extend to the Hamiltonian $H_{\Delta,\Gamma}$ (6.38) of the XYZ model the weak limit applied in the case of $\Gamma = 0$ in Section 6.2. However, in order that the supplementary term in Γ be interpreted as a mass term, it is necessary to perform a canonical transformation which, in contrast to (6.41), redefines the fermion

vacuum (corresponding to the ferromagnetic states $M = 0$). The new vacuum is
defined by the filling of the negative velocity states in the $\cos k (\sin k < 0)$ band.
This vacuum has a very artificial character, but it is from this that we can still obtain
a solvable limit model with a finite number of particles. This vacuum plays an
intermediate role between the initial vacuum and the vacuum of the particle–hole
symmetric theory, corresponding to the antiferromagnetic state (half-filled $\cos k$
band). We thus perform the canonical transformation

$$a_{-\frac{\pi}{2}+k} = \psi_{k,1}, \qquad a_{\frac{\pi}{2}-k}^{\dagger} = \psi_{k,2}, \tag{6.73}$$

which is equivalent to performing the substitution

$$b_{k,2} = \psi_{-k,2}^{\dagger}, \qquad b_{k,1} = \psi_{k,1}. \tag{6.74}$$

We thus obtain, from (6.38),

$$H_{\Delta,\Gamma} + 2M\Delta = \sum_{|k|<\frac{\pi}{2}} 2\sin k (\psi_{k1}^{\dagger}\psi_{k1} - \psi_{k2}^{\dagger}\psi_{k2})$$

$$+ \sum_{|k|<\frac{\pi}{2}} i\Gamma \cos k (\psi_{k2}^{\dagger}\psi_{k1} - \psi_{k1}^{\dagger}\psi_{k2}) + \frac{2\Delta}{N} \sum_{|p|<\pi} \cos p \, \rho_{-p}\rho_p, \tag{6.75}$$

in which the definition of ρ_p results from (6.44) and (6.74).

We shall examine a weak limit of Hamiltonian (6.75) in conditions such that we
have

$$\lim_{a\to 0} \frac{\Gamma}{2a} = m, \qquad \langle S^z \rangle \text{ finite.} \tag{6.76}$$

We thus obtain

$$H_{\mathscr{T}} = \lim_{a\to 0} \frac{1}{2a}(H_{\Delta,\Gamma} + 2M\Delta) = \sum_q \psi_q^{\dagger}(\alpha q + \beta m)\psi_q + \frac{4\Delta}{L}\rho_{-q1}\rho_{q2} + \text{cst} \tag{6.77}$$

with the following matrix representation:

$$\alpha = \begin{pmatrix} 1 & 0 \\ 0 & -1 \end{pmatrix}, \quad \beta = \begin{pmatrix} 0 & -i \\ i & 0 \end{pmatrix}, \quad \alpha^2 = \beta^2 = 1, \quad \alpha\beta + \beta\alpha = 0. \tag{6.78}$$

We note the existence of two limit constants of motion

$$-S^z = R = \sum_q \psi_q^{\dagger}\alpha\psi_q, \tag{6.79}$$

$$Q = \sum_q \psi_q^{\dagger}\psi_q, \tag{6.80}$$

in other words the particle number of each species is constant.

In terms of the local fields

$$\psi_\sigma(x) = L^{-1/2} \sum_q \psi_{q,\sigma} e^{iqx}, \tag{6.81}$$

the Hamiltonian

$$H_{\mathscr{J}} = \int_0^L dx\, \psi^\dagger(x) \left(-i\alpha \frac{\partial}{\partial x} + \beta m\right) \psi(x) + 4\Delta \int_0^L dx\, \psi_1^\dagger(x) \psi_1(x) \psi_2^\dagger(x) \psi_2(x) \tag{6.82}$$

describes relativistic fermions of positive and negative energy in $1+1$ dimensions, the bare mass being m and the coupling 4Δ. The charge quantum number Q is the limit form of the operator $\sum_k \sin ka_k^\dagger a_k$, transform of $\sum_n (S_n^+ S_{n+1}^- - S_{n+1}^+ S_n^-)$, which commutes with the XY Hamiltonian, but not with the XYZ one, except in the continuum limit.

If one wants to construct a particle–hole symmetric field theory starting from (6.82), one must perform the usual canonical transformation or perform the filling of all negative energy states, which contradicts our weak limit hypothesis in which R and Q remain finite. It is not our objective to study the Thirring field theory; we only examine the finite problem posed by (6.82).

6.4 Diagonalization of H𝒢

Hamiltonian (6.82) describes a system of Dirac particles with two internal states having a mutual δ function interaction potential, with periodicity conditions over L. The number of particles being constant and taken equal to Q, the wave equation for Q particles of coordinates x_j and internal variables α_j, β_j is naturally written

$$\sum_{j=1}^{Q} \left(-i\alpha_j \frac{\partial}{\partial x_j} \Psi + \beta_j m \Psi\right) + 2\Delta \sum_{j<l} (1 - \alpha_j \alpha_l) \delta(x_j - x_l) \Psi = E\Psi. \tag{6.83}$$

The interaction is of current–current type

$$4\Delta \rho_1 \rho_2 = (\psi^\dagger \psi)^2 - (\psi^\dagger \alpha \psi)^2,$$

from which one gets the operator $1 - \alpha_1 \alpha_2$ in front of $\delta(x_1 - x_2)$.

We shall write the matrix equation (6.83) using the representation in which the velocities α_j are diagonal, in such a way that the wavefunction has 2^Q components which are completely antisymmetric in the particle indices:

$$\Psi_{\alpha_1 \alpha_2 \alpha_3 \ldots}(x_1 x_2 x_3 \ldots) = -\Psi_{\alpha_2 \alpha_1 \alpha_3 \ldots}(x_2 x_1 x_3 \ldots).$$

Similarly to the massless case (6.58), the δ function interaction makes sense only in the weak limit process starting from the discrete case, which leads to the gluing conditions (6.71) on the interaction planes

$$\frac{i}{2}(\alpha_1 - \alpha_2)(\Psi_{\alpha_1 \alpha_2}(x, x+0) - \Psi_{\alpha_1 \alpha_2}(x, x-0)) = \Delta(\Psi_{\alpha_1 \alpha_2}(x, x+0) + \Psi_{\alpha_1 \alpha_2}(x, x-0)).$$
$$(6.84)$$

We note that this condition, specialized to the case $\alpha_1 = \alpha_2$, does not mean anything other than antisymmetry and does not necessarily imply the result $\Psi_{\alpha, \alpha}(x, x) = 0$; one would require a continuity hypothesis which we are not obliged to make following the original hypothesis leading us to (6.84), a condition that we can rewrite in the form

$$\Psi_{\alpha_1 \alpha_2}(x_1, x_1 + 0) = -\exp(i(\alpha_1 - \alpha_2)\mu)\Psi_{\alpha_2 \alpha_1}(x_1, x_1 + 0), \qquad (6.85)$$

valid only for $\alpha_1 \neq \alpha_2$.

We must underline again that this possible discontinuity (we shall see that it exists) of the wavefunction in the absence of interaction ($\alpha_1 = \alpha_2$) is a sure sign of the pathologic character of the Hamiltonian. The latter is not defined. It becomes so only after a Luttinger-like regularization, or as the limit of a discrete system. The same phenomenon occurs in the recent solution of the Kondo Hamiltonian restricted to the neighbourhood of the Fermi surface (Wiegmann, 1981), a limit Hamiltonian which is as pathological as that of Thirring in its form (6.83). The Kondo problem with the usual Hamiltonian is in fact a diffractive problem which was exactly solved for two electrons in the presence of the magnetic impurity (Gaudin, 1978).

6.4.1 Solution

Away from the interaction planes, the wave equation (6.83) admits a family of factorized solutions made up of products of individual functions of momentum q and energy ε:

$$\psi(x) = \chi_\alpha(x) \exp(iqx), \qquad (6.86)$$

verifying

$$-i\alpha \frac{\partial}{\partial x}\psi + \beta m \psi = \varepsilon \psi, \qquad (6.87)$$

in other words

$$(\alpha q - \varepsilon)\chi_\alpha + m\chi_{-\alpha} = 0, \qquad \varepsilon = \pm(q^2 + m^2)^{1/2}. \qquad (6.88)$$

Bethe's method leads one to look for a wavefunction whose restriction to sector

$$D_I: \quad x_1 < x_2 < \ldots < x_Q$$

takes the form of a superposition

$$\Psi_{\alpha_1\alpha_2\ldots\alpha_Q}(x_1x_2\ldots x_Q) = \sum_{P\in\pi_Q} B(P)I(P)\chi_{\alpha_1}(\varepsilon_{P1})\chi_{\alpha_2}(\varepsilon_{P2})\ldots\chi_{\alpha_Q}(\varepsilon_{PQ})$$
$$\times \exp(ix_1q_{P1}+\ldots+ix_Qq_{PQ}), \tag{6.89}$$

associated with given sets $\{q\}$ and $\{\varepsilon\}$ specifying the signs of the energies. Once $\Psi_{\{\alpha\}}$ is determined in sector D_I, the antisymmetry determines $\Psi_{\{\alpha\}}$ in the other sectors.

The gluing conditions (6.85) give us

$$B(P)\chi_{\alpha_3}(\varepsilon_{P3})\chi_{\alpha_4}(\varepsilon_{P4}) - B(P(34))\chi_{\alpha_3}(\varepsilon_{P4})\chi_{\alpha_4}(\varepsilon_{P3})$$
$$= -\exp(i(\alpha_3-\alpha_4)\mu)\left\{B(P)\chi_{\alpha_4}(\varepsilon_{P3})\chi_{\alpha_3}(\varepsilon_{P4}) - B(P(34))\chi_{\alpha_4}(\varepsilon_{P4})\chi_{\alpha_3}(\varepsilon_{P3})\right\},$$

in other words

$$\frac{B(P(34))e^{2i\alpha\mu}-B(P)}{B(P)e^{2i\alpha\mu}-B(P(34))} = \frac{\chi_\alpha(\varepsilon_{P4})\chi_{-\alpha}(\varepsilon_{P3})}{\chi_{-\alpha}(\varepsilon_{P4})\chi_\alpha(\varepsilon_{P3})} = \frac{\alpha q_{P3}-\varepsilon_{P3}}{\alpha q_{P4}-\varepsilon_{P4}}, \tag{6.90}$$

which is manifestly independent of $\alpha = (\alpha_1-\alpha_2)/2 = \pm 1$. Note that this formula is not valid for $\alpha = 0$, according to (6.85).

Relation (6.90) must be satisfied for all adjacent pairs $(j, j+1)$ (we have written (34)). We can thus set

$$B(P) = \exp\left(\frac{i}{2}\sum_{j<l}\psi_{Pj,Pl}\right), \tag{6.91}$$

in which the ψ_{jl} are antisymmetric phases determined by the relation originating from (6.90):

$$\frac{\exp(-i\psi_{jl}+2i\mu)-1}{\exp(2i\mu)-\exp(-i\psi_{jl})} = \frac{q_j-\varepsilon_j}{q_l-\varepsilon_l}, \quad \forall j, l. \tag{6.92}$$

One defines the rapidity parameters λ_j such that one has

$$q_j = m\sinh\lambda_j, \qquad \varepsilon_j = m\cosh\lambda_j \tag{6.93}$$

with $\Im\lambda_j = 0$ or π, according to the sign of the real part of the energy. Relation (6.92) is then written

$$\tan\frac{\psi_{jl}}{2} = \tan\mu\tanh\frac{\lambda_j-\lambda_l}{2}. \tag{6.94}$$

In this problem the rapidities λ play an analogous role to the θ parameters introduced in (1.38) for the XXZ model in the $|\Delta| < 1$ region, and offer the same advantages. In the massless limit, one has

$$\varepsilon_j = q_j\,\mathrm{sgn}\,\lambda_j, \qquad \psi_{lj} = \mu(\,\mathrm{sgn}\,\lambda_l - \mathrm{sgn}\,\lambda_j), \tag{6.95}$$

which allows us to find wavefunction (6.62) again after the replacement $\Delta \rightarrow \mu$ by identifying $\alpha_j = \text{sgn}\, \lambda_j = $ velocity of particle j.

Finally, the periodicity conditions in L are written

$$Lm \sinh \lambda_j = 2\pi n_j + \sum_{l=1}^{Q} \psi_{jl}, \qquad j = 1, 2, \ldots, Q \tag{6.96}$$

where n is integer or half-odd integer according to Q being odd or even. The total energy is

$$E = \sum_j \varepsilon_j = m \sum_j \cosh \lambda_j. \tag{6.97}$$

The theory for the complex roots and bound states corresponding to the particles of this finite model could be performed in parallel fashion to that of the Heisenberg–Ising system in Chapter 1. The roots of system (6.96) are grouped into strings $C_n(\lambda_0)$ of order n:

$$C_n(\lambda_0) : \quad \lambda_v = \lambda_0 + i v \left(\frac{\pi}{2} - \mu \right), \tag{6.98}$$

with $v = -(n - 1), -(n - 3), \ldots, (n - 3), (n - 1)$.

According to the reasoning of Subsection 1.5.3, the particles are bound provided we have the inequality

$$(n - 1) \left(\frac{\pi}{2} - \mu \right) < \pi. \tag{6.99}$$

The mass of the order-n bound states is

$$m_n = m \frac{\sin n(\pi/2 - \mu)}{\cos \mu} \tag{6.100}$$

for real rapidity λ_0, i.e., positive energy. One must note the exception of the maximal order string such that

$$(n - 1) < \pi \left(\frac{\pi}{2} - \mu \right)^{-1} < n,$$

a string for which the energy is negative, i.e., the rapidity $\lambda_0 + i\pi$.

We shall end here this introduction to the Thirring model, in other words at the threshold of the true field theory, since the equations cannot be exploited in this direction without going beyond what our hypotheses allow and without exiting the finite domain to which we restricted ourselves. The construction of the physical vacuum by filling negative energy states requires the introduction of a cutoff energy (or rapidity), and defining excitations becomes possible only after renormalization of masses and couplings. We are thus in a certain sense thrown back onto the discrete model of the beginning. Using the results of Johnson *et al.* (1973) for the excitations of the chain with three couplings (results which extend those of Subsection 2.1.3 to the $\Gamma \neq 0$ case), we can evaluate the renormalized mass as a function of the discrete lattice spacing. One can also notice as in Subsection 6.2.4 that, in the case of zero mass, the field theory is distinct in its results from the continuum limit of the magnetic chain. It is then obvious that all the richness of the spectrum of the Thirring Hamiltonian originates from that of the Heisenberg–Ising one, or more precisely of the vicinity of the critical point ($\Gamma = am \rightarrow 0$) of the three-constant chain. The exact themodynamics of the chain (Gaudin, 1971d; Takahashi and Suzuki, 1972) clearly reveals the structure of the antiferromagnetic ground state as the condensate of an infinite number of bound states. The vacuum structure of the massive Thirring model is analogous. All these questions are the subject of more recent developments (Bergknoff and Thacker, 1979a,b; Korepin, 1980).

7

The six-vertex model

7.1 The ice model

7.1.1 Introduction

The *six-vertex model*, which is the object of this chapter, is a special case of the eight-vertex model on a two-dimensional square lattice introduced by Fan and Wu (1970) in order to summarize a class of exactly solvable models in classical statistical mechanics. The thermodynamics of the six-vertex model is by now known in its full generality (Yang and Yang, 1966a–d; Lieb, 1967a) which is not the case for the eight-vertex model, but only the self-conjugate one (Baxter, 1971a).

The general model can be considered as a two-dimensional idealization of a crystalline system in which pairs of adjacent atoms or radicals on the network are linked by 'hydrogen bonds'. Of ionic type, this link between two neighbouring electronegative atoms is realized by a proton H^+ which is located closer to one of the atoms than to the other. On each link of the network there thus exist two equilibrium positions for H^+ (Pauling, 1960). If the coordination number at each site equals 4, as is the case for the oxygen atom in hexagonal ice, we indeed have eight possible proton configurations on a given site, which give rise to the eight vertices.

In contrast, for the physical systems for which this model could be viewed as a valid idealization – ice H_2O, the ferroelectric PO_4H_2K, the antiferroelectric $PO_4H_2NH_4$ – there exist at most two H^+ next to each site. This is the 'ice rule'; in ice each oxygen atom is the originator of four links according to the heights of a tetrahedron with the neighbouring oxygen atoms, which gives six configurations or six positions for the H_2O molecule (Onsager and Dupuis, 1960) with respect to the network.

In the case of ice, and within a sufficiently low temperature window such that the preceding description remains valid, but not too low so that the residual dipolar interactions remain negligible, the six vertex configurations have more or less the same energy in the absence of external electric field. One of the problems consists of enumerating the configurations of the system which are compatible with the

Pauling condition, in order to calculate the observed 'residual' entropy. This notion does not contradict Nernst's theorem since at a temperature in the vicinity of zero one must take into account the mean electric fields which lift the degeneracy.

In ferroelectric substances the internal field and the applied external field polarize the H^+R^- dipoles and lift the degeneracy of the orientational states of the molecule, favouring at low temperature the ferroelectric or antiferroelectric ordered states depending on the internal structure. The problem consists of studying the order–disorder transition on such models (Nagle, 1966; Lieb, 1967b–d; Sutherland, Yang and Yang, 1967; C. P. Yang, 1967).

7.1.2 Definition of the two-dimensional model

Let us consider a square lattice (with P lines and N columns) with toric topology in order to avoid boundary effects ($N + 1 = 1$, $P + 1 = 1$). On each link, an arrow oriented towards one of the adjacent summits (or sites) or the other indicates which atom the proton of the hydrogen bond is closest to. At each summit two ingoing and two outgoing arrows, defining a vertex configuration, realize the ice rule, in other words electric neutrality. The vertices numbered from 1 to 6 (Figure 7.1) represent configurations of energy e_i with which are associated statistical weights $\omega_i = e^{-\beta e_i}$, for temperature β^{-1}.

The first four vertices have a residual dipole moment and we can write the associated energies as a function of the horizontal and vertical electric fields. If the minimum energy vertex is of type 1, 2, 3, 4 the system will be called ferroelectric.

We undertake the calculation of the partition function

$$Z = \sum_{\Gamma} \prod_{S} \omega_{i(S)}, \qquad V : S \to i(S), \tag{7.1}$$

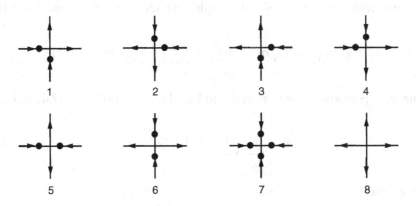

Figure 7.1 The eight vertices of the two-dimensional model.

in which the sum runs over all configurations Γ of PN compatible vertices; the product is that of the weights associated with vertices V of Γ and runs over the summits S of the lattice. The free energy is the limit

$$-\beta f = \lim_{P \to \infty} \lim_{N \to \infty} \frac{1}{PN} \ln Z_{P,N}. \tag{7.2}$$

7.2 The transfer matrix

7.2.1 Definition and method

The coordinates of the sites of the toric lattice are denoted by the two integers (n, p), $n = [1, N]$, $p = [1, P]$ (with $O \equiv N$, etc.). With each 'vertical' arrow $(n, p) \leftrightarrow (n, p+1)$ we associate a spin $\uparrow$ or $\downarrow$, $\sigma_n^p = \pm 1$, and with each horizontal arrow $(n, p) \leftrightarrow (n+1, p)$ a spin $\uparrow$ or $\downarrow$, $\tau_n^p = \pm 1$. The statistical weight of vertex $V_{(n,p)}$ is defined entirely by $\{\sigma_n^{p-1}, \sigma_n^p, \tau_{n-1}^p, \tau_n^p\}$ and does not depend explicitly on (n, p) because of the presupposed homogeneity of the lattice. We can thus define the 4×4 matrix A such that

$$\omega_{i(V_{n,p})} = \langle \sigma_n^{p-1} \tau_{n-1}^p | A | \sigma_n^p \tau_n^p \rangle. \tag{7.3}$$

We shall give its explicit form later on. The partition function is thus written according to (7.1):

$$Z = \sum_{\{\sigma\}} \sum_{\{\tau\}} \prod_{p=1}^{P} \prod_{n=1}^{N} \langle \sigma_n^{p-1} \tau_{n-1}^p | A | \sigma_n^p \tau_n^p \rangle. \tag{7.4}$$

For each value of p, we are free to perform the sums over the τ_n^p of line p, $n = [1, N]$; we then obtain a matrix defined between the configurations $\sigma^{(p-1)}$ and $\sigma^{(p)}$ of two consecutive rows of vertical spins, which we call the transfer matrix T:

$$\langle \sigma_1^{p-1} \sigma_2^{p-1} \ldots \sigma_N^{p-1} | T | \sigma_1^p \sigma_2^p \ldots \sigma_N^p \rangle = \sum_{\{\tau_n\}} \prod_{n=1}^{N} \langle \sigma_n^{p-1} \tau_{n-1}^p | A | \sigma_n^p \tau_n^p \rangle. \tag{7.5}$$

It remains to perform the sums over the $\{\sigma\}$ to obtain, according to (7.4) and (7.5),

$$Z = \sum_{\{\sigma_n^p\}} \prod_{p=1}^{P} \langle \sigma^{(p-1)} | T | \sigma^{(p)} \rangle, \tag{7.6}$$

in other words

$$Z = \text{Tr } T^P \tag{7.7}$$

in which T is a $2^N \times 2^N$ matrix between states $|\sigma_1 \sigma_2 \dots \sigma_N\rangle$. In contrast, we define 2×2 matrices A_n whose elements depend on σ_n and σ_n':

$$\langle \tau' | A_n | \tau \rangle = \langle \sigma_n' \tau' | A | \sigma_n \tau \rangle, \tag{7.8}$$

in order to write the transfer matrix element in the form (Fan and Wu, 1970; Sutherland, 1970)

$$\langle \sigma_1' \sigma_2' \dots \sigma_N' | T | \sigma_1 \sigma_2 \dots \sigma_N \rangle = \sum_{\{\tau\}} \prod_n \langle \tau_{n-1} | A | \tau_n \rangle = \operatorname*{Tr}_\tau A_1 A_2 \dots A_N. \tag{7.9}$$

In order to avoid writing the matrix elements, we express the A_n matrices in terms of spin operators S_n with $S_n^z = \frac{1}{2}\sigma_n$ and T_n with $T^z = \frac{1}{2}\tau$.

By simple inspection of Figure 7.1 one can write

$$A_n = \omega_1 \left(\frac{1}{2} + S_n^z\right) \left(\frac{1}{2} + T^z\right) + \omega_2 \left(\frac{1}{2} - S_n^z\right) \left(\frac{1}{2} - T^z\right)$$

$$+ \omega_3 \left(\frac{1}{2} + S_n^z\right) \left(\frac{1}{2} - T^z\right) + \omega_4 \left(\frac{1}{2} - S_n^z\right) \left(\frac{1}{2} + T^z\right)$$

$$+ \omega_5 S_n^- T^+ + \omega_6 S_n^+ T^- + \omega_7 S_n^+ T^+ + \omega_8 S_n^- T^-. \tag{7.10}$$

In the τ representation, matrix A_n of the six-vertex model is thus written

$$A_n = \begin{pmatrix} \omega_1 \left(\frac{1}{2} + S_n^z\right) + \omega_4 \left(\frac{1}{2} - S_n^z\right) & \omega_5 S_n^- \\ \omega_6 S_n^+ & \omega_3 \left(\frac{1}{2} + S_n^z\right) + \omega_2 \left(\frac{1}{2} - S_n^z\right) \end{pmatrix} \tag{7.11}$$

and the transfer matrix, function of the S_n spins, is

$$T = \operatorname*{Tr}_\tau A_1 A_2 \dots A_N. \tag{7.12}$$

In the self-conjugate eight-vertex model, we have the restrictions

$$\omega_1 = \omega_2 = a, \quad \omega_3 = \omega_4 = b, \quad \omega_5 = \omega_6 = c, \quad \omega_7 = \omega_8 = d \tag{7.13}$$

$$A_n = \frac{1}{2}(a+b) + 2\left\{(c+d)S_n^x T^x + (c-d)S_n^y T^y + (a-b)S_n^z T^z\right\}, \tag{7.14}$$

and the corresponding transfer matrix will be studied in Chapter 8.

Finally, the calculation of $\lim P^{-1} \ln Z$ leads us to diagonalize the matrix T. Its elements are positive or zero and the greatest eigenvalue t is thus positive. From inequalities

$$t^P \leq \operatorname{Tr} T^P \leq N t^P \tag{7.15}$$

we deduce from (7.2)–(7.7)

$$-\beta f = \lim_{N\to\infty} N^{-1} \ln t. \qquad (7.16)$$

7.2.2 Transfer matrix elements in the spin-conserving basis

The six-vertex model follows the ice rule. This is represented by the conservation of the number of vertical arrows from one line of the lattice to the next, in other words by the conservation of the spin component

$$S^z = \sum_n S_n^z, \quad [S^z, T] = 0. \qquad (7.17)$$

This is easily seen by noting that the vertices which do not conserve S^z are of type 5 or 6. But along a given horizontal line vertices 5 and 6 necessarily alternate as displayed in Figure 7.2, and can be associated pairwise due to the boundary condition $N + 1 \equiv 1$. However each 5, 6 pair globally conserves the vertical component of spin, which proves (7.17). We shall put

$$S^z = \frac{N}{2} - M, \quad M = 0, 1, 2, \ldots, \frac{N}{2} \text{ or } \frac{N-1}{2} \qquad (7.18)$$

and expand the eigenstate $|M\rangle$ of T on the basis already used in Chapter 1 for the spin-1/2 chain:

$$|M\rangle = \sum C(n_1 n_2 \ldots n_M)|n_1 n_2 \ldots n_M\rangle. \qquad (7.19)$$

The sum runs over all allowed configurations

$$1 \le n_1 < n_2 < n_3 \ldots < n_M \le N. \qquad (7.20)$$

We must now determine the matrix elements. With the help of (7.12), one obtains

$$\langle m_1 m_2 \ldots | T | n_1 n_2 \ldots \rangle \equiv \langle \{m\} | T | \{n\} \rangle$$
$$= \operatorname*{Tr}_\tau \langle 0 | S_{m_1}^+ \ldots S_{m_M}^+ A_1 A_2 \ldots A_N S_{n_1}^- \ldots S_{n_M}^- | 0 \rangle. \qquad (7.21)$$

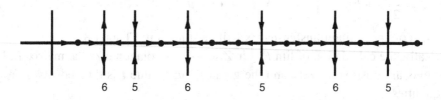

Figure 7.2 Alternating 5 and 6 vertices.

Lemma:

$$T = T_+ + T_-,$$ (7.22)

$$\langle\{m\}|T_+|\{n\}\rangle \neq 0 \text{ iff } 1 \leq m_1 \leq n_1 \leq m_2 \leq n_2 \leq \ldots \leq m_M \leq n_M \leq N,$$ (7.23)

$$\langle\{m\}|T_-|\{n\}\rangle \neq 0 \text{ iff } 1 \leq n_1 \leq m_1 \leq n_2 \leq m_2 \leq \ldots \leq n_M \leq m_M \leq N.$$ (7.24)

Proof: On the right-hand side of (7.21) we can commute each S_n^- through the A of higher index so as to position it immediately to the right of A_n; similarly towards the left for the S_n^+. In order to establish the lemma, it is sufficient to show that we cannot have more than one S^+ strictly between two consecutive S^-, in other words we must show that

$$\langle 0|\ldots A_{n_1}S_{n_1}^- \ldots S_m^+ A_m A_{m+1} \ldots A_{p-1}S_p^+ A_p \ldots A_{n_2}S_{n_2}^- \ldots |0\rangle = 0$$ (7.25)

with

$$n_1 < m < p < n_2.$$

But product (7.25) contains the factor

$$\langle \uparrow | S_m^+ A_m \ldots A_{p-1}S_p^+ A_p | \uparrow \rangle$$ (7.26)

which, according to (7.11), is equal to

$$\begin{pmatrix} 0 & \omega_5 \\ 0 & 0 \end{pmatrix} \cdot \begin{pmatrix} \omega_1 & 0 \\ 0 & \omega_3 \end{pmatrix}^{p-m-1} \cdot \begin{pmatrix} 0 & \omega_5 \\ 0 & 0 \end{pmatrix} \equiv 0$$ (7.27)

which proves (7.25). We thus have the two possible alternations (7.23) and (7.24) of which the only common element corresponds to the diagonal term $\{m\} = \{n\}$, which will be split between T_+ and T_- as we will see later.

Figure 7.3 illustrates the structure of the T_- matrix elements. This in particular shows how vertices of type 5 and 6 alternate in this representation.

Before determining the explicit form of the matrix elements, let us note that these are homogeneous of degree N in the set of ω_i, $i = 1, \ldots, 6$. In addition, following the remark illustrated in Figure 7.2, vertices 5 and 6 always coming in pairs, the

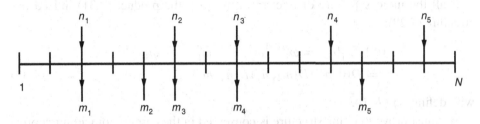

Figure 7.3 Structure of $\langle\{m\}|T_-|\{n\}\rangle$ ($N = 15$, $M = 5$).

T matrix depends on ω_5 and ω_6 only via the product $\omega_5\omega_6 = c^2$. In the following calculations we shall thus set $\omega_5 = \omega_6 = 1$, at the cost of having to re-establish the inhomogeneity later on.

Proposition:

(a) $\langle\{m\}|T_-|\{n\}\rangle = \bar{D}(0, m_1)D(m_1, n_1)\bar{D}(n_1, m_2)\ldots\bar{D}(n_{M-1}, m_M)$

$\quad D(m_M, n_M)\bar{D}(n_M, N+1), 1 \leq m_1 \leq n_1 \leq \ldots \leq n_N \leq N.$ (7.28)

(b) $\langle\{m\}|T_+|\{n\}\rangle = D(0, n_1)\bar{D}(n_1, m_1)D(m_1, n_2)\ldots D(m_{M-1}, n_M)$

$\quad \bar{D}(n_M, m_M)D(m_M, N+1), 1 \leq n_1 \leq m_1 \leq \ldots \leq m_M \leq N.$ (7.29)

(c) $\quad \bar{D}(n, m) = \begin{cases} \omega_1^{m-n-1} & n < m \\ \omega_2 & n = m \end{cases}$ (7.30)

$\quad D(n, m) = \begin{cases} \omega_3^{m-n-1} & n < m \\ \omega_4 & n - m. \end{cases}$

Proof: Starting from (7.21) and commuting the S through the A such that each S_m^+ is immediately to the left of A_m and each S_n^- to the right of A_n, we obtain the contribution to T_-, by distinguishing the case of coinciding indices, for example the first, m_1, n_1.

- $m_1 < n_1$

$$\langle\{m\}|T_-|\{n\}\rangle = \text{Tr } \langle 0|A_1 A_2 \ldots S_{m_1}^+ A_{m_1} \ldots A_{n_1} S_{n_1}^- \ldots |0\rangle$$

$$= \text{Tr } \begin{pmatrix} \omega_0 & 0 \\ 0 & \omega_3 \end{pmatrix}^{m_1-1} \cdot \begin{pmatrix} 0 & 1 \\ 0 & 0 \end{pmatrix} \cdot \begin{pmatrix} \omega_1 & 0 \\ 0 & \omega_3 \end{pmatrix}^{n_1-m_1-1} \cdot \begin{pmatrix} 0 & 0 \\ 1 & 0 \end{pmatrix} \ldots \text{ (7.31)}$$

- $m_1 = n_1$

$$\langle\{m\}|T_-|\{n\}\rangle = \text{Tr } \langle 0|A_1 \ldots S_{n_1}^+ A_{n_1} S_{n_1}^- \ldots |0\rangle$$

$$= \text{Tr } \begin{pmatrix} \omega_1 & 0 \\ 0 & \omega_3 \end{pmatrix}^{n_1-1} \cdot \begin{pmatrix} \omega_4 & 0 \\ 0 & \omega_2 \end{pmatrix} \ldots \begin{pmatrix} 0 & 1 \\ 0 & 0 \end{pmatrix} \ldots \begin{pmatrix} 0 & 0 \\ 1 & 0 \end{pmatrix} \ldots \text{ (7.32)}$$

The structure of the product of the matrices which haven't been written out in (7.31) or (7.32) is identical to those which have been written: alternation of the matrices T_+ and T_- interspersed with diagonal matrices.

If all the indices $\{n\}$ are distinct from the $\{m\}$, the product (7.31) indeed has structure (7.28):

$$\langle\{m\}|T_-|\{n\}\rangle = \omega_1^{m_1-1}\omega_3^{n_1-m_1-1}\omega_1^{m_2-n_1-1}\omega_3^{n_2-m_2-1}\ldots$$

$$= \bar{D}(0, m_1)D(m_1, n_1)\bar{D}(n_1, m_2)\ldots \text{ (7.33)}$$

with definition (7.30).

We must prove that this structure is conserved in the case of coinciding coordinates n and m. Let us take, for example, $n_1 = m_1$. The matrix element is given by the right-hand side of (7.32); since we suppose $\{m\} \neq \{n\}$ for the moment, there exists in the matrix product at least one pair $\ldots \begin{vmatrix} 0 & 1 \\ 0 & 0 \end{vmatrix} \begin{vmatrix} \cdots \end{vmatrix} \begin{vmatrix} 0 & 0 \\ 1 & 0 \end{vmatrix} \ldots$ which

suffices to project the product on the type $\begin{vmatrix} * & 0 \\ 0 & 0 \end{vmatrix}$. We then obtain from (7.32)

$$\langle\{m\}|T_-|\{n\}\rangle = \omega_1^{n_1-1}\omega_4\omega_1^{m_2-n_1-1}\ldots = \bar{D}(0,n_1)D(n_1,n_1)\bar{D}(n_1,m_2)\ldots \quad (7.34)$$

Formula (7.28) is thus general. Let us, however, note that it is only defined for the allowed sets $\{n\}$ (and $\{m\}$) for which the coordinates are *distinct*. The method for obtaining T_+ is analogous. Matrices T_- and T_+ alternate in this order, which projects on the second diagonal element. Finally, the $\{m\} = \{n\}$ case is obvious:

$$\langle\{n\}|T|\{n\}\rangle = \text{Tr}\begin{pmatrix} \omega_1 & 0 \\ 0 & \omega_3 \end{pmatrix}^{n_1-1} \cdot \begin{pmatrix} \omega_4 & 0 \\ 0 & \omega_2 \end{pmatrix} \cdot \begin{pmatrix} \omega_1 & 0 \\ 0 & \omega_3 \end{pmatrix}^{n_2-n_1-1} \cdot \begin{pmatrix} \omega_4 & 0 \\ 0 & \omega_2 \end{pmatrix}\ldots$$
$$= \langle\{n\}|T_+|\{n\}\rangle + \langle\{n\}|T_-|\{n\}\rangle, \quad (7.35)$$

in which the second equality is valid due to (7.28), (7.29) and (7.30). The proposition is thus proven.

7.3 Diagonalization

7.3.1 Cyclic invariance

The equation for the amplitudes $c\{n\}$ which diagonalize the transfer matrix is written

$$\sum_{\{n\}} c(n_1, n_2, \ldots, n_M)\left(\langle\{m\}|T_+|\{n\}\rangle + \langle\{m\}|T_-|\{n\}\rangle\right) = tc(m_1, m_2, \ldots, m_M).$$
$$(7.36)$$

With the help of results (7.28) and (7.29) the matrix elements can be written in the condensed form

$$\langle\{m\}|T_+|\{n\}\rangle = \prod_{i=1}^{M} D(m_{i-1}n_i)\bar{D}(n_i m_i),$$

$$\langle\{m\}|T_-|\{n\}\rangle = \prod_{i=1}^{M} D(m_i n_i)\bar{D}(n_i m_{i+1}) \quad (7.37)$$

under the convention that

$$m_0 = m_M - N, \qquad m_{M+1} = m_1 + N \quad (7.38)$$

since one easily verifies from (7.30) that

$$D(0, n_1)D(m_M, N+1) = D(m_0, n_1),$$
$$\bar{D}(0, m_1)\bar{D}(n_M, N+1) = \bar{D}(n_M, m_{M+1}). \tag{7.39}$$

Let us now make the *hypothesis* that the coefficients $c\{m\}$ solving (7.36) can be extended outside the interval of definition $[1, N]$ so as to verify the cyclicity condition

$$c(m_1 m_2 \ldots m_M) \equiv c(m_2 m_3 \ldots m_{M+1}) \equiv c(m_0 m_1 \ldots m_{M-1}). \tag{7.40}$$

Let us consider the T_+ term on the left-hand side of (7.36) and perform the change of summation variables

$$n_1 \to n_0, \ n_2 \to n_1, \ \ldots, \ n_M \to n_{M-1}. \tag{7.41}$$

The corresponding sum is written

$$\sum_{1 \le n_0 < n_1 < \ldots \le N} c(n_0 n_1 \ldots n_{M-1}) \prod_{i=1}^{M} D(m_i n_i) \bar{D}(n_i m_{i+1}) \tag{7.42}$$

in which we used the relation

$$D(m_0 n_0)\bar{D}(n_0 m_1) = D(m_M n_M)\bar{D}(n_M m_{M+1}), \quad n_M \equiv n_0 + N. \tag{7.43}$$

Using the cyclicity hypothesis (7.40), we see that the contribution (7.42) coming from T_+ has the same form as the contribution coming from T_-. Only the variation domains differ. In both cases we still have

$$m_i \le n_i \le m_{i+1}, \quad i = [1, M], \tag{7.44}$$

but for T_- we have

$$m_1 \le n_1 < n_2 < \ldots < n_M \le N \tag{7.45}$$

and for T_+, by virtue of (7.43),

$$m_1 \le n_1 < n_2 < \ldots < n_{M-1} \le N < n_M \le m_{M+1}. \tag{7.46}$$

The union of both summation domains thus leaves us with only the restriction (7.44) and the condition $n_1 < n_2 < \ldots < n_{M+1}$. The eigenvalue equation is thus written

$$\sum_{\{n\}} c(n) \prod_{i=1}^{M} D(m_i n_i) \bar{D}(n_i m_{i+1}) = tc(m) \tag{7.47}$$

with the understanding that $m_{M+1} - m_1 = N$. Equation (7.47) re-establishes the formal cyclic invariance of the problem.

7.3.2 Lieb's method

Equations (7.47) would be factorizable in the absence of the condition requiring the n to be distinct. This leads to the idea of extending the definition of the $c\{n\}$ coefficients to the domain $n_1 \le n_2 \le n_3 \ldots$ and looking for the solution in the form of a superposition of factorized solutions. Given the convolutive form of the equation, we look for a Bethe-type solution

$$c(n_1 n_2 \ldots n_M) = \sum_{Q \in \pi_M} B(Q) q_{Q1}^{n_1} q_{Q2}^{n_2} \cdots q_{QM}^{n_M} \tag{7.48}$$

in which the sum runs over the $M!$ permutations Q and in which the quantities $q_1, q_2, \ldots, q_M$ are distinct. The cyclic conditions (7.40) will be fulfilled if the coefficients B verify

$$B(Q) = q_{Q1}^{N} B(QC), \tag{7.49}$$

in which C is the cyclic permutation $(12 \ldots M)$.

Let us now perform the sums over $n_1, n_2, \ldots$ as if the conditions $n_1 \ne n_2, \ldots$ did not exist, but underlining the forbidden terms which are of the form

$$n_i = m_{i+1} = n_{i+1}, \quad i = 1, \ldots, M. \tag{7.50}$$

Substituting (7.48) into the left-hand side of (7.47), one obtains

$$tc(m) = \sum_{Q} \sum_{\{n\}} B(Q) \prod_{i=1}^{M} q_{Qi}^{n_i} D(m_i n_i) \bar{D}(n_i m_{i+1}). \tag{7.51}$$

Let us perform the sum independently over each n_i while underlining the *dangerous terms* which are $n_i = m_i$ and $n_i = m_{i+1}$, for example for $i = 2$

$$\sum_{n_2 = m_2}^{m_3} q_{Q2}^{n_2} D(m_2 n_2) \bar{D}(n_2 m_3)$$

$$= q_{Q2}^{m_2} \omega_4 \omega_1^{m_3 - m_2 - 1} + \sum_{n = m_2 + 1}^{m_3 - 1} q_{Q2}^{n} \omega_3^{n - m_2 - 1} \omega_1^{m_3 - n - 1} + q_{Q2}^{m_3} \omega_2 \omega_3^{m_3 - m_2 - 1} \tag{7.52}$$

$$\equiv \left(\omega_4 + \frac{q_{Q2}}{\omega_1 - \omega_3 q_{Q2}} \right) q_{Q2}^{m_2} \omega_1^{m_3 - m_2 - 1} + \left(\omega_2 - \frac{1}{\omega_1 - \omega_3 q_{Q2}} \right) q_{Q2}^{m_3} \omega_3^{m_3 - m_2 - 1}. \tag{7.53}$$

We shall write this sum over two terms in the symbolic form $X_2 + Y_3$, each containing a dangerous term. The sum over distinct n of the product $\prod_i$ on the right-hand side of (7.51) thus has the following structure:

$$\prod = (X_1 + Y_2)(X_2 + Y_3)(X_3 + Y_4) \ldots (X_M + Y_{M+1}) \tag{7.54}$$

in which the indicated couplings between Y_2X_2, Y_3X_3, etc. indicate that the product of the two dangerous terms appearing in Y_2X_2, etc. must be omitted. The quantity $\prod$ can be expanded in the following way:

$$\prod = \prod_0 + \prod_1 + \prod_2 + \cdots \tag{7.55}$$

with

$$\prod_0 = X_1X_2\ldots X_M + Y_2Y_3\ldots Y_{M+1},$$

$$\prod_1 = \sum_i \underline{Y_iX_i}\cdots,$$

$$\prod_2 = \sum_{i-j>1} \underline{Y_jX_j}\,\underline{Y_iX_i}\cdots, \text{ etc.} \tag{7.56}$$

The expansion shows precisely the distributivity of the 'product' $\prod$ with respect to the operation $|\underline{}|$. Besides $\prod_0$, all terms contain a pair $\underline{Y_iX_i}$ as a factor. These will give a vanishing contribution to (7.51) if it is possible to choose the $B(Q)$ such that

$$\sum_Q \cdots \underline{Y_jX_j}\cdots \underline{Y_iX_i}\cdots B(Q) = 0, \qquad i > j+1. \tag{7.57}$$

However by (7.53)

$$Y_j \equiv Y_j(Q(j-1)), \qquad X_j \equiv X_j(Qj),$$
$$Y_i \equiv Y_i(Q(i-1)), \qquad X_i \equiv X_i(Qi)$$

and we thus have $Q(i-1) \neq Qj$, etc. It then suffices to take, for $j > 1$,

$$\underline{Y_j(Q(j-1))X_j(Qj)}B(Q) + \underline{Y_j(Qj)X_j(Q(j-1))}B(QQ_{j-1,j}) = 0. \tag{7.58}$$

This gives us

$$\left(\omega_2 - \underline{\frac{1}{\omega_1 - \omega_3 q_{Qj-1}}}\right)\left(\omega_4 + \frac{q_{Qj}}{\omega_1 - \omega_3 q_{Qj}}\right)B(Q)$$

$$+\left(\omega_2 - \underline{\frac{1}{\omega_1 - \omega_3 q_{Qj}}}\right)\left(\omega_4 + \frac{q_{Qj-1}}{\omega_1 - \omega_3 q_{Qj-1}}\right)B(QQ_{j-1,j}) = 0. \tag{7.59}$$

As far as the X_1Y_{M+1} terms are concerned, it is sufficient to notice that

$$\sum_Q \underline{X_1(Q1)Y_{M+1}(QM)}B(Q) \equiv \sum_Q \underline{X_1(Q2)Y_{M+1}(Q1)}B(QC). \tag{7.60}$$

With the help of the cyclicity condition (7.49), the right-hand side is written

$$\sum_Q \underbrace{X_1(Q2)Y_{M+1}(Q1)q_{Q1}^{-N}B(Q)} \equiv \sum_Q \underbrace{X_1(Q2)Y_1(Q1)B(Q)}, \qquad (7.61)$$

in which we used (7.53). But the right-hand side of (7.61) vanishes by virtue of relations (7.59) for $j = 2$. Relations (7.59) and (7.49) will determine the B and the q.

7.3.3 Coupled equations

We must now show that the right-hand side of (7.51) is of the form $tc(m)$. The contribution only comes from terms of the type $\prod_0$ (7.56). One finds

$$tc(m) = \sum_Q B(Q) \left\{ \prod_{i=1}^M \left(\omega_4 + \frac{q_i}{\omega_1 - \omega_3 q_i} \right) \omega_1^{N-M} q_{Q1}^{m_1} \cdots q_{QM}^{m_M} \right.$$
$$\left. + \prod_{i=1}^M \left(\omega_2 - \frac{1}{\omega_1 - \omega_3 q_i} \right) \omega_3^{N-M} q_{Q1}^{m_2} \cdots q_{QM}^{m_{M+1}} \right\}, \qquad (7.62)$$

in other words we verify that state $|M\rangle$ whose coefficients are $c(m)$ is an eigenstate with eigenvalue

$$t = \omega_1^{N-M} \prod_{i=1}^M \left(\omega_4 + \frac{\omega_5 \omega_6 q_i}{\omega_1 - \omega_3 q_i} \right) + \omega_3^{N-M} \prod_{i=1}^M \left(\omega_2 - \frac{\omega_5 \omega_6}{\omega_1 - \omega_3 q_i} \right) \qquad (7.63)$$

in which we have re-established homogeneity in the six parameters ω_j.

Relations (7.59) for the $B(Q)$ coefficients are written

$$\frac{B(Q)}{B(QQ_{12})} = -\frac{\omega_1\omega_4 + \omega_2\omega_3 q_{Q1} q_{Q2} - (\omega_1\omega_2 + \omega_3\omega_4 - \omega_5\omega_6)q_{Q1}}{\omega_1\omega_4 + \omega_2\omega_3 q_{Q1} q_{Q2} - (\omega_1\omega_2 + \omega_3\omega_4 - \omega_5\omega_6)q_{Q2}} = e^{i\psi_{Q1,Q2}} \qquad (7.64)$$

in which $\psi_{\alpha\beta}$ is an antisymmetric quantity in its indices.

Let us now introduce the notation

$$\Delta = \frac{\omega_1\omega_2 + \omega_3\omega_4 - \omega_5\omega_6}{2(\omega_1\omega_2\omega_3\omega_4)^{1/2}}, \qquad (7.65)$$

$$e^{2\beta h} = \left(\frac{\omega_1\omega_4}{\omega_2\omega_3} \right)^{1/2}, \qquad e^{2\beta v} = \left(\frac{\omega_1\omega_3}{\omega_2\omega_4} \right)^{1/2} \qquad (7.66)$$

in which h and v are respectively the horizontal and vertical components of the external electric field. We also introduce the quantities k_j linked to q_j by the relation

$$q_j = e^{ik_j + 2\beta h}. \qquad (7.67)$$

In these conditions, equations (7.64) are formally identical to those obtained for the $A(Q)$ coefficients in the anisotropic spin chain problem of Chapter 1, formula (1.27). The solution for (7.64) is thus

$$B(Q) = \exp\left\{\frac{i}{2}\sum_{j<l}\psi_{Qj,Ql}\right\} \tag{7.68}$$

in which the $\psi_{j,l}$ and the k_j are related by (1.16). In contrast, the coupled equations for the k_j, which come from the periodicity condition (7.49), do not form a real system anymore, except for zero horizontal field. They are written

$$Nk_j = 2\pi\lambda_j + 2iN\beta h + \sum_l{}' \psi_{jl}, \qquad j = [1, M]. \tag{7.69}$$

If the condition $\omega_1\omega_4 = \omega_2\omega_3$ is fulfilled, the transfer matrix T and the Hamiltonian H of the closed Heisenberg–Ising chain have identical eigenvectors. One could check directly that H commutes with T. The proof (Sutherland, 1970) will be given in Chapter 8 for the (self-adjoint) eight-vertex model (see Section 8.3).

7.4 The free energy

7.4.1 Determination of the largest eigenvalue

Before passing to the thermodynamic limit, the problem consists of choosing the eigenvector of T, or if one prefers the set $\{k\}$, such that, for given M, the corresponding eigenvalue t is positive and maximal. In zero horizontal field, the last remark in the preceding subsection and the Perron–Frobenius theorem give us the answer.

In each subspace specified by value of S^z, the Hamiltonians $H_\Delta + N\Delta/2$ for $\Delta > 0$ or H_Δ for $\Delta < 0$ have positive or vanishing matrix elements. Their highest eigenvalue is thus associated with an eigenstate whose components are real and all of the same sign. But this state is also an eigenstate of T and, according with the stated property of the components, it is associated with the highest eigenvalue of T. For each value of M we are thus led back to the already solved problem of specifying the highest-energy state of H_Δ or the ground state of $H_{-\Delta}$. One should not forget that the corresponding $\{k\}$ sets are related in the following manner:

$$\{k\}_\Delta = \{k - \pi\}_{-\Delta}.$$

We are thus led to distinguish three cases:

F.	$\Delta > 1$.	$\Delta = \cosh\tilde{\Phi}$,	$\tilde{\Phi} > 0$	
D.	$-1 < \Delta < 1$.	$\Delta = \cos\tilde{\Theta} = -\cos\Theta$,	$\tilde{\Theta} = \pi - \Theta$,	$0 < \Theta < \pi$
AF.	$\Delta < -1$.	$\Delta = -\cosh\Theta$,	$\Theta > 0$	(7.70)

and use the results given in Chapter 1.

F. $\Delta > 1$. The maximal energy state of H_Δ is the ferromagnetic state $|0\rangle$ (or $|N\rangle$). When M is fixed, it is the state $|M\rangle$ characterized by the presence of a single 'string' (see Chapter 1, Subsection 1.5.1) of variable $\varphi = \pi$; in the $N \gg 1$ limit, we have seen that the φ_ν parameters were given by

$$\varphi_\nu = \pi + i\nu\tilde{\Phi}, \quad \nu = [-(M-1), M-1].$$

The solution will be generalized later to the case $h \neq 0$.

D. $-1 < \Delta < 1$. With Orbach's parametrization already introduced for the anisotropic chain, in the notation of Section 1.5, formulas (1.100) and (1.101)

$$k_1 = f_\Theta(\theta_1), \quad \left(\tan \frac{1}{2} f_{P\Theta}(\theta) = \cot \frac{P\Theta}{2} \tanh \frac{\theta}{2}\right)$$

$$\psi_{12} = \pm\pi + f_{2\Theta}(\theta_1 - \theta_2) \tag{7.71}$$

the system (7.69) is written

$$Nf_\Theta(\theta_J) = 2\pi J + 2iN\beta h + \sum_{J'} f_{2\Theta}(\theta_J - \theta_{J'}), \quad J = -\frac{M-1}{2}, \ldots, \frac{M-1}{2}.$$

AF. $\Delta < 1$.

$$f_1 = f_\Phi(\varphi_1), \text{ etc.} \quad Nf_\Phi(\varphi_J) = 2\pi J + 2iN\beta h + \sum_{J'} f_{2\Phi}(\varphi_J - \varphi_{J'}).$$

We have supposed that the choice of the ground state quantum numbers remains correct when h is sufficiently close to zero.

7.4.2 Expression for the t eigenvalue

It is natural to express t with the help of Orbach's parameters (φ or θ, according to the interval of Δ):

$$-1 < \Delta < 1. \quad e^{ik} = e^{if_\Theta(\theta)} = \frac{1 - e^{\theta + i\Theta}}{e^\theta - e^{i\Theta}}, \quad \Delta = -\cos\Theta.$$

$$\Delta < -1. \quad i\theta \to \varphi, \quad i\Theta \to \Phi, \quad f_\Theta(\theta) \to f_\Phi(\varphi). \tag{7.72}$$

Let us introduce the notation that will be used for the eight-vertex model:

$$a = (\omega_1\omega_2)^{1/2}, \quad b = (\omega_3\omega_4)^{1/2}, \quad c = (\omega_5\omega_6)^{1/2}, \tag{7.73}$$

and the notation of Sutherland and Yang:

$$\eta = \frac{a}{b} = e^{\beta\delta}, \quad \xi = \frac{c^2}{ab}. \tag{7.74}$$

According to (7.65), we have

$$c^2 = a^2 + b^2 - 2\Delta ab, \qquad 2\Delta = \eta + \eta^{-1} - \xi. \tag{7.75}$$

The symmetry in a and b allows us to choose $a \geq b$ or $\Delta \geq 0$. Homogeneity also allows the choice $ab = 1$. Taking (7.66) into account, the expression for the ω_i in terms of the field is

$$\omega_1 = \exp\left(\frac{\delta}{2} + h + v\right)\beta, \qquad \omega_3 = \exp\left(\frac{-\delta}{2} - h + v\right)\beta,$$

$$\omega_2 = \exp\left(\frac{\delta}{2} - h - v\right)\beta, \qquad \omega_4 = \exp\left(\frac{-\delta}{2} - h - v\right)\beta. \tag{7.76}$$

We shall finally define the vertical and horizontal polarizations x and y, $-1 < x, y < +1$:

$$1 - y = \frac{2M}{N}, \qquad x = \frac{\partial(\beta f)}{\partial v}. \tag{7.77}$$

Substituting the expressions (7.67) for q_j into the formula for t, we obtain

$$t = P_+ e^{\beta N(\delta/2 + h + vy)} + P_- e^{\beta N(-\delta/2 - h + vy)} \tag{7.78}$$

with

$$P_\pm = \prod_{j=1}^{M} \frac{e^{\pm ik_j}(2\Delta - \eta^{\pm 1}) - 1}{e^{\pm ik_j} - \eta^{\pm 1}}. \tag{7.79}$$

Let us define the parameters τ and χ by the relations

$$e^{i\tau} = \frac{\eta e^{i\Theta} + 1}{\eta + e^{i\Theta}}, \qquad e^{\chi} = \frac{\eta e^{\Phi} + 1}{e^{\Phi} + \eta}, \qquad 0 \leq \tau \leq \Theta, \quad 0 \leq \chi \leq \Phi. \tag{7.80}$$

We obtain

$$P_\pm = (-1)^M e^{\pm M\Phi} \prod_{j=1}^{M} \frac{e^{i\varphi_j + \chi \pm 2\Phi} - 1}{e^{i\varphi_j + \chi} - 1}, \tag{7.81}$$

in the region $\Delta < -1$. The above formula suggests considering t as a function of the parameter $\varphi = i\chi$ and defining the trigonometric polynomial

$$Q(\varphi) = \prod_{j=1}^{M} \sin\frac{1}{2}(\varphi - \varphi_j) \tag{7.82}$$

to obtain

$$P_\pm = (-1)^M \frac{Q(\varphi \mp 2i\Phi)}{Q(\varphi)}. \tag{7.83}$$

If we define

$$g(\varphi) = \left(\sin\frac{\varphi}{2}\right)^N, \tag{7.84}$$

the original relation (7.63) is equivalent to the following one:

$$t(\varphi)Q(\varphi) = g(\varphi + i\Phi)Q(\varphi - 2i\Phi)e^{\beta N(h+vy)} + g(\varphi - i\Phi)Q(\varphi + 2i\Phi)e^{\beta N(-h+vy)}, \tag{7.85}$$

provided we choose the correct normalization for $t(\varphi)$, by re-establishing homogeneity:

$$c = \sinh\Phi, \quad a = \sinh\frac{1}{2}(\Phi - i\varphi), \quad b = \sinh\frac{1}{2}(\Phi + i\varphi).$$

7.4.3 *Thermodynamic limit* ($\Delta > 1$)

According to the conclusions of Subsection 7.4.1, the highest eigenvalue of T in zero horizontal field is associated with the maximal energy state of H which is thus the ferromagnetic state $S = \dfrac{N}{2}$, $S^z = \dfrac{N}{2} - M$, in the domain $\Delta = \cosh\tilde{\Phi} > 1$. We have seen how the roots φ_ν of the coupled equations were grouped into a single exceptional string of length M and abscissa π. When h is non-vanishing, but in a certain neighbourhood of 0, we can think that the string structure is not modified. In fact we find that a solution of the form

$$\varphi_\nu = \pi + i(\bar{\varphi} + \nu\tilde{\Phi} + \delta_\nu), \quad \nu = -(M-1),\dots,M-1$$

is still valid, in which δ_ν are quantities which are negligible in the exponential approximation already defined for the spin chain. To determine $\bar{\varphi}$, it is sufficient to perform the sum of equations (7.69), which eliminates all the antisymmetric phases and leaves us with an equation of the form

$$e^{-2\beta hM} = \prod_\nu \frac{\sin\frac{1}{2}(\varphi_\nu + i\tilde{\Phi})}{\sin\frac{1}{2}(\varphi_\nu - i\tilde{\Phi})} = \frac{\cosh\frac{1}{2}(\bar{\varphi} + M\tilde{\Phi})}{\cosh\frac{1}{2}(\bar{\varphi} - M\tilde{\Phi})}. \tag{7.86}$$

We deduce that

$$\bar{\varphi} = -2\beta hM, \tag{7.87}$$

provided h is not so large and verifies the inequality

$$|2\beta h| < \tilde{\Phi}. \tag{7.87'}$$

Under these assumptions, it is possible to show as in Chapter 1 that the δ_ν are in fact negligible, and thus to prove the consistency of solution (7.85). For the values of h which do not verify inequality (7.87'), the structure (7.85) is not valid and a subset of the roots φ_J condenses into a continuous distribution of type $\varphi_j = \pi + iu_j$ with real u_j. The coupled equations are written

$$\prod_{j \neq i} \frac{\sinh \frac{1}{2}(u_i - u_j + 2\tilde{\Phi})}{\sinh \frac{1}{2}(u_i - u_j - 2\tilde{\Phi})} = e^{-2N\beta h} \left(\frac{\cosh \frac{1}{2}(u_i + \tilde{\Phi})}{\cosh \frac{1}{2}(u_i - \tilde{\Phi})} \right)^N \tag{7.88}$$

and before writing an integral equation for the density of roots u_j, it would be necessary to study this system in more detail, which we have not done.

If we limit ourselves to the region $2\beta|h| < \tilde{\Phi}$, after the transformations

$$e^{ik} = \frac{e^{-i\varphi + \tilde{\Phi}} - 1}{e^{-i\varphi} - e^{\tilde{\Phi}}}, \qquad e^{\tilde{\chi}} = \frac{\eta e^{\tilde{\Phi}} - 1}{\eta - e^{\tilde{\Phi}}}, \tag{7.89}$$

we obtain

$$P_\pm = \prod_\nu \frac{\cosh \frac{1}{2}(\tilde{\varphi} - \tilde{\chi} + (\nu \pm 2)\tilde{\Phi})}{\cosh \frac{1}{2}(\tilde{\varphi} - \tilde{\chi} + \nu\tilde{\Phi})} \tag{7.90}$$

or in other words

$$\ln P_+ = -\ln P_- = -2M\beta h. \tag{7.91}$$

By virtue of relations (7.16) and (7.76), the free energy per site as a function of h and v is expressed as

$$-\beta f(h, v) = \operatorname*{Max}_y \begin{cases} \beta \left(\dfrac{\delta}{2} + h + vy \right) - \beta h(1 - y) \\[2mm] -\beta \left(\dfrac{\delta}{2} + h - vy \right) + \beta h(1 - y) \end{cases}$$

or

$$f(h, v) = \operatorname{Min} \{e_1, e_2\} = -\frac{\delta}{2} - |h + v|, \qquad |h| < \frac{1}{2\beta}\tilde{\Phi}, \ \forall v. \tag{7.92}$$

A more detailed study (Sutherland *et al.*, 1967) shows that expression (7.92) for the free energy is valid in a wider domain than that obtained with the help of the partial solution described here. The completely ordered domain (magnetization equal to unity) is defined by the following inequalities:

$$\begin{cases} (\eta e^{2\beta h} - 1)(\eta e^{2\beta v} - 1) \geq \eta\xi, & h + v \geq 0 \\ (\eta e^{-2\beta h} - 1)(\eta e^{-2\beta v} - 1) \geq \eta\xi, & h + v \leq 0. \end{cases} \tag{7.93}$$

This domain contains the one obtained in (7.92).

We can further rewrite inequalities (7.93) in the form

$$\begin{cases} (\omega_1 - \omega_3)(\omega_1 - \omega_4) \geq \omega_5\omega_6, & \omega_1 > \omega_2 \\ (\omega_3 - \omega_1)(\omega_3 - \omega_2) \geq \omega_5\omega_6, & \omega_1 < \omega_2. \end{cases} \tag{7.94}$$

Outside this domain, a detailed study of the free energy remains to be done.

$$-f = \underset{y}{\text{Max}} \left\{ \mp \left(\frac{\delta}{2} + h \right) + vy \mp \frac{1}{\beta} \sum_j \ln \frac{\cosh \frac{1}{2}(\tilde{\chi} - u_j \pm 2\tilde{\Phi})}{\cosh \frac{1}{2}(\tilde{\chi} - u_j)} \right\}. \tag{7.95}$$

7.4.4 Thermodynamic limit ($\Delta < 1$)

Let us consider the cases $-1 < \Delta < 1$ and $\Delta < -1$ simultaneously, in the $N \to \infty$ limit, $\dfrac{2M}{N} = 1 - y$ in which y ($-1 \le y \le 1$) is the vertical magnetization. Extrapolating the classic results on the ground state of Hamiltonian $H_{-\Delta}$, we suppose that the roots θ_j (or φ_j) of the coupled equations (7.72) or (7.73) condense into a continuous distribution on an arc of an analytic curve Γ, symmetric with respect to the imaginary axis in the θ (or φ) plane, with end points $-\zeta^*$ and ζ (using the same notation in both cases). If we define a complex density $\rho(\theta)$ (or $\rho(\varphi)$) such that along arc Γ we have arg $(\rho(\theta)d\theta) = 0$, with $N\rho(\theta)d\theta =$ number of θ_j on Γ between θ and $\theta + d\theta$, the equations (7.83) for example take the limit form

$$f_\Phi(\varphi) = -R_\zeta(\varphi) + 2i\beta h + \int_{-\zeta^*}^{\zeta} f_{2\Phi}(\varphi - \varphi')\rho_\zeta(\varphi')d\varphi' \tag{7.96}$$

in which function $f_\Phi(\varphi)$ is defined in (7.71). By derivation, we obtain

$$\frac{\sinh \Phi}{\cosh \Phi - \cos \varphi} = 2\pi \rho_\zeta(\varphi) + \int_{-\zeta^*}^{\zeta} \frac{\sinh 2\Phi}{\cosh 2\Phi - \cos(\varphi - \varphi')} \rho_\zeta(\varphi')d\varphi' \tag{7.96'}$$

with

$$\frac{dR_\zeta}{d\varphi}(\varphi) = 2\pi \rho_\zeta(\varphi), \qquad R_\zeta(\zeta) = -R_\zeta(-\zeta^*) \tag{7.97}$$

and the expression for the magnetization

$$1 - y = 2 \int_{-\zeta^*}^{\zeta} \rho_\zeta(\varphi)d\varphi, \tag{7.98}$$

and for the horizontal component of the field

$$\beta h(1 - y) = -i \int_{-\zeta^*}^{\zeta} f_\Phi(\varphi)\rho_\zeta(\varphi)d\varphi. \tag{7.99}$$

The preceding formulas are written for $\Delta < -1$, but we would find analogous expressions for $-1 < \Delta < 1$ using the substitution $\varphi \to i\theta$:

$$f(h, v) = \underset{y}{\text{Min}} : \left\{ -\left(\frac{\delta}{2} + h \right) - vy, -\beta^{-1} \int_{-\zeta^*}^{\zeta} if_\Phi(\varphi + i\Phi - i\chi)\rho_\zeta(\varphi)d\varphi \right\}, \tag{7.100}$$

in which we have chosen the smallest of the two terms.

The integral equation (7.96') shows us that $\rho_\zeta(\varphi)$ is an analytic function of φ in the strip $|\operatorname{Im}\varphi| < \Phi$. In this strip, we shall thus be allowed to deform the contour Γ, its end points remaining fixed, in order to compute ρ_ζ most conveniently. Setting $\zeta = b + i\gamma$, we can write the integral equation in the form

$$f_\Phi(\varphi + i\gamma) = -R_\zeta(\varphi + i\gamma) + 2i\beta h + \int_{-b}^{+b} f_{2\Phi}(\varphi - \varphi')\rho_\zeta(\varphi' + i\gamma)d\varphi', \quad (7.101)$$

in which the integral is performed along the segment $[-b, +b]$ of the real axis. Once the density $\rho_\zeta(\varphi)$ has been obtained as a function of φ and ζ, we can hope that relations (7.98) and (7.99) determine ζ uniquely for given h and y.

The original contour Γ, where roots condense, will be defined by the differential equation

$$-\text{Argument of } \rho(\varphi) = \text{Angle between the real axis and the tangent to } \Gamma \text{ in } \varphi.$$

It now remains to give explicitly the minimum with respect to y condition indicated in expression (7.100), in order to calculate the horizontal magnetization x. The calculation is done in Appendix F and gives the following results:

(a)

$$F(h, v) = -\left(\frac{\delta}{2} + h + v\right) - \beta^{-1}Z(h, v), \quad (7.102)$$

in which $Z(h, v)$ is a symmetric function of h and v.

(b) If we define the complex number $\zeta_1 = b + i\gamma_1$ as a function of $\zeta = b + i\gamma$ via the relation

$$\gamma_1 + \gamma = \chi - \Phi(\text{or } \tau - \Theta), \quad (7.103)$$

we have the symmetric result

$$Z(h, v) = \int_{-\zeta^*}^{\zeta} (if_\Phi(\varphi + i\Phi - i\chi) - 2\beta v)\rho_\zeta(\varphi)d\varphi$$

$$\equiv \int_{-\zeta_1^*}^{\zeta_1} (if_\Phi(\varphi + i\Phi - i\chi) - 2\beta h)\rho_{\zeta_1}(\varphi)d\varphi. \quad (7.104)$$

(c) The quantities $\zeta(\zeta_1)$ are defined by the relations

$$\begin{cases} 1 - y = 2 \int_{-\zeta^*}^{\zeta} \rho_\zeta(\varphi)d\varphi, \\[2mm] 1 - x = 2 \int_{-\zeta_1^*}^{\zeta_1} \rho_{\zeta_1}(\varphi)d\varphi, \end{cases} \quad (7.105)$$

$$
\begin{cases}
\beta h(1-y) = -i \displaystyle\int_{-\zeta^*}^{\zeta} f_\Phi(\varphi)\rho_\zeta(\varphi)d\varphi, \\[2mm]
\beta v(1-x) = -i \displaystyle\int_{-\zeta_1^*}^{\zeta_1} f_\Phi(\varphi)\rho_{\zeta_1}(\varphi)d\varphi,
\end{cases}
\tag{7.106}
$$

which allow in principle the calculation of $\zeta(h, v)$, and thereafter the magnetizations. This system is perfectly symmetric under the exchange

$$
(\zeta, h, y) \leftrightarrow (\zeta_1, v, x).
$$

The preceding formulas can be applied immediately to the case $\Delta = 0$ ($\Theta = \pi/2$), in which the kernel $\sin 2\Theta(\cosh(\theta - \theta') - \cos 2\Theta)^{-1}$ of integral equation (7.96′) vanishes identically. The density $\rho(\theta)$ is then independent of the ζ parameter,

$$
\rho(\theta) = \frac{1}{2\pi}\frac{1}{\cosh\theta}.
\tag{7.107}
$$

From equations (7.105), we deduce

$$
\tan\frac{1}{2}\pi y = \frac{\cos\gamma}{\sinh b}, \qquad \tan\frac{1}{2}\pi x = \frac{\cos\gamma_1}{\sinh b}
\tag{7.108}
$$

with

$$
\gamma_1 + \gamma = \tau - \frac{\pi}{2}, \qquad \eta = \tan\left(\frac{\pi}{4} + \frac{\tau}{2}\right) > 1, \qquad 0 < \tau < \pi/2.
$$

Equations (7.106) give

$$
\tanh 2\beta h = \frac{\sin\gamma}{\cosh b}, \qquad \tanh 2\beta v = \frac{\sin\gamma_1}{\cosh b},
\tag{7.109}
$$

which gives

$$
\sinh 2\beta h = \tan\tau \sin\frac{1}{2}\pi y - \frac{1}{\cos\tau}\cos\frac{1}{2}\pi y \tan\frac{1}{2}\pi x
\tag{7.110}
$$

or

$$
\sinh 2\beta h = \frac{1}{2}(\eta - \eta^{-1})\sin\frac{1}{2}\pi y - \frac{1}{2}(\eta + \eta^{-1})\cos\frac{1}{2}\pi y \tan\frac{1}{2}\pi x
\tag{7.111}
$$

and the similar equation for $\sinh 2\beta v$, by exchange $x \leftrightarrow y$.

7.4.5 Nature of the transitions in zero field

Despite the simplicity of the kernels appearing in the integral equations (7.96) and (7.99), these can in general not be solved simply for arbitrary values of ζ. However, the values of ζ defined by $\Re\zeta = \pi$ or $\Re\zeta = \infty$, respectively, constitute a notable exception in which the solution can be found by Fourier transform. These particular values correspond to the case in which the magnetizations x and y are zero, which as we shall see does not necessarily imply that the fields h and v vanish. In the vicinity of these values, in other words for small magnetizations x and y, perturbative solutions to equations (7.96) can be developed.

Let us put ourselves initially in zero field. For the whole region $\Delta < 1$, inequalities (7.81) allow us to choose the appropriate sign in expression (7.100) for the free energy in zero field $F(h = 0, v = 0)$, to obtain

$$\Delta \leq 1, \qquad F(0,0) = -\frac{\delta}{2} - \frac{1}{2\beta} \int_{-\pi}^{+\pi} d\varphi \rho(\varphi) \ln \left(\frac{\cosh(2\Phi - \chi) - \cos \varphi}{\cosh \chi - \cos \varphi} \right), \quad (7.113)$$

$$-1 \leq \Delta \leq 1, \quad F(0,0) = -\frac{\delta}{2} - \frac{1}{2\beta} \int_{-\infty}^{+\infty} d\theta \rho(\theta) \ln \left(\frac{\cosh \theta - \cos(2\Theta - \tau)}{\cosh \theta - \cos \tau} \right), \quad (7.114)$$

in which the densities ρ are the functions already encountered in Chapter 2 for the anisotropic Heisenberg chain

$$\begin{cases} 2\pi \rho(\varphi) = \displaystyle\sum_{k=-\infty}^{+\infty} \frac{e^{ik\varphi}}{2 \cosh k\Phi} = \frac{K}{\pi} \operatorname{dn}\left(\frac{K\varphi}{\pi}; k \right), \\[4mm] 2\pi \rho(\theta) = \displaystyle\int_{-\infty}^{+\infty} \frac{e^{i\omega\theta}}{2 \cosh \omega\Theta} d\theta = \frac{\pi}{2\Theta} \cosh \frac{\pi\theta}{2\Theta} \end{cases} \quad (7.115)$$

and which indeed give us $y = 0$, $h = 0$ by virtue of (7.97) and (7.98). We can then transform expressions (7.101) and (7.102) into the following series of integrals, which give us the zero-field free energy in the whole domain of Δ:

$$\Delta \leq -1, \qquad \Delta = -\cosh \Theta, \qquad F(0,0) = -\frac{\delta}{2} - \frac{1}{\beta} Z_1(\Delta)$$

with

$$Z_1(\Delta) = \frac{\Phi - \chi}{2} + \sum_{k=1}^{\infty} \frac{1}{k} e^{-k\Phi} \frac{\sinh k(\Phi - \chi)}{\cosh k\Phi}. \quad (7.116)$$

$$-1 \leq \Delta \leq 1, \qquad \Delta = -\cos \Theta,$$

$$Z_2(\Delta) = \int_0^{\infty} \frac{d\omega}{\omega} \frac{\sinh \omega(\pi - \Theta)}{\sinh \omega\pi} \frac{\sinh \omega(\Theta - \tau)}{\cosh \omega\Theta}. \quad (7.117)$$

$$\Delta \geq 1, \qquad F(0,0) = -\frac{\delta}{2}. \quad (7.118)$$

The expressions for Z_1 and Z_2 are sufficiently simple to allow the study of their analytical properties in Δ, via the parameters Φ or Θ, the quantities χ (or τ) being themselves functions of Φ (or Θ) by (7.80). The dependence of F on the temperature $T = \beta^{-1}$ is thus obtained through Δ and the η parameter (7.72).

According to the definitions, Z_1 is analytical in Φ in the region $|\Re \chi| < \Re\Phi$, which means, by virtue of $\eta > 1$, in the region $\Re\Phi > 0$. Similarly, Z_2 is analytical in

$$-\Re\Theta < \Re\tau < \Re\Theta < \pi,$$

which simply means for $0 < \Re\Theta < \pi$.

The only possible physical singularities of the free energy are thus for the following Δ values:

$$\begin{cases} \Delta = -1, & \text{sing. } Z_1 \text{ at } \Phi = 0 \\ & \text{sing. } Z_2 \text{ at } \Theta = 0 \\ \Delta = 1, & \text{sing. } Z_2 \text{ at } \Theta = \pi. \end{cases}$$

The functions Z_1 and Z_2 have the same limit at $\Delta = -1$ as seen from their definition, and since Z_2 vanishes at $\Theta = \pi$, the free energy is continuous in the whole Δ domain.

Setting $\Phi = i\Theta$, $\chi = i\tau$, their analytic continuation to the half-strip

$$\Re\Phi > 0, \qquad 0 < \Im\Phi < \pi$$

is linked in the following manner:

$$Z_2 = Z_1 + i \sum_{l=0}^{\infty} \frac{(-1)^l}{2l+1} \frac{\cos(2l+1)\frac{\pi \chi}{2\Phi}}{e^{(2l+1)\pi^2/\Phi} - 1}. \tag{7.119}$$

The line $\Re\Phi = 0$ is a dense singularity line forming a natural boundary. However, around $\Phi = 0$, one can expand Z_1 and Z_2 in powers of Φ or $i\Theta$, and these asymptotic series coincide. In the Δ variable, Z_1 is cut along $[-1, +1]$, Z_2 is cut along $]-\infty, -1]$ and $[+1, +\infty[$. The singularity is of type

$$\exp\left(-\frac{\pi^2}{\Phi}\right) \sim \exp\left\{-\pi^2 \left(\frac{2}{-\Delta - 1}\right)^{1/2}\right\}. \tag{7.120}$$

The condition $\Delta = -1$, or $c = a + b$, admits a unique solution for the temperature when the vertex energies verify

$$e_3 + e_4 > e_5 + e_6 \quad \text{or} \quad e_1 + e_2 > e_5 + e_6, \tag{7.121}$$

thus favouring an antiferroelectric ordered state at low temperature (F model). The high-temperature phase corresponds to

$$-1 < \Delta < \frac{1}{2},$$

and the antiferroelectric phase to $\Delta < -1$.

Given the continuity of all derivatives of $F(\beta)$ at the critical temperature and the existence of a singularity of type mentioned in (7.120), the transition can be labelled as being of *infinite order*.

If the vertex energies verify the following inequality:

$$e_1 + e_2 < e_5 + e_6, \tag{7.122}$$

which favours an ordered ferroelectric state at low temperatures (KDP model[1]), then there exists a single critical temperature corresponding to $\Delta = 1$, or $c = a - b$. The phase corresponding to the high-temperature region is defined in the interval $1/2 < \Delta < 1$. The transition is manifest and is of first order. Indeed, in the vicinity of $\Delta = 1$ (or $\Theta = \pi$), $F(0, 0)$ admits according to (7.105) the limited expansion in powers of $\tilde{\Theta} = \pi - \Theta$:

$$F(0, 0) = -\frac{\delta}{2} + \frac{4\tilde{\Theta}^2}{\beta(\eta - 1)} \int_0^\infty \frac{\omega d\omega}{\sinh 2\omega\pi} + \frac{2\tilde{\Theta}^3}{\beta(\eta - 1)} \int_0^\infty \frac{\omega^2 d\omega}{\cosh^2 \omega\pi} + O(\tilde{\Theta}^4),$$

in other words

$$F(0, 0) = -\frac{\delta}{2} + \frac{1 - \Delta}{2\beta(\eta - 1)} + \frac{\sqrt{2}}{3\pi} \frac{(1 - \Delta)^{3/2}}{\beta(\eta - 1)} + \ldots \tag{7.123}$$

The latent heat of the transition thus takes the following value for the KDP model ($b = c$, $\eta_c = 2$):

$$U_c = \frac{\partial \beta F}{\partial \beta} \Big|_{\beta_c} = \frac{1}{2} \frac{1}{\eta_c - 1} \frac{\partial \Delta}{\partial \beta_c} = \frac{\delta}{2}$$

and the specific heat above the critical temperature diverges with exponent $-1/2$:

$$C = \frac{\partial U}{\partial \beta^{-1}} \propto \frac{\sqrt{2}}{4\pi} (\beta_c \delta)^{3/2} \left(1 - \frac{\beta}{\beta_c}\right)^{-1/2}, \qquad \beta < \beta_c. \tag{7.124}$$

7.4.6 Singularities of the free energy in presence of a field

(Lieb, 1967a; Sutherland *et al.*, 1967)

It would be necessary, for each of the ferroelectric $\left(\Delta > \dfrac{1}{2}\right)$ and antiferroelectric $\left(\Delta < \dfrac{1}{2}\right)$ models, to explore the analytical properties of the free energy function in the whole domain of variation of the external field (h, v). The complexity of the integral equations to study forces us to limit ourselves to the vicinity of certain

---

[1] KDP stands for potassium dihydrogen phosphate, KH_2PO_4.

solvable cases, where the free energy is in fact singular. We shall again follow Sutherland *et al.* (1967) in their description of the singularities.

$\Delta > \dfrac{1}{2}$. *Ferroelectric case*

In the presence of a field, the free energy of the ordered phase (Δ necessarily greater than 1) is written $F(h, v) = -\dfrac{\delta}{2} - |h+v|$ according to (7.92), an expression which is valid in the region of sufficiently large field defined by (7.39). The magnetization is complete, and parallel to the symmetry axis of the network ($x = y$). Above the critical temperature defined by inequalities (7.93), the free energy is given by the limit form of (7.95) for $\Delta > 1$ and by the analogous form of (7.100) for $\dfrac{1}{2} < \Delta < 1$. The point $\Delta = 1$ is not singular in the presence of a field and the first-order transition becomes second order. An explicit study, however, remains to be performed.

$\Delta < \dfrac{1}{2}$. *Antiferroelectric case*

Let us show that in the low-temperature phase (Δ necessarily less than -1), the magnetization remains zero in the presence of a sufficiently small electric field.

Let us start from the zero-field point $h = v = 0$. We know that in this case $\zeta = \pi$ and equation (7.96′) can be solved by Fourier transformation, giving $\rho_n(\varphi) = \mathrm{Dn}\,\varphi$. Let us now vary ζ while maintaining $\Re\zeta = \pi$, $\Im\zeta = \gamma \neq 0$. The integral equation (7.96′) remains a convolution. The arc Γ can be moved and deformed in such a way as to coincide with the segment $[-\pi, +\pi]$ without crossing the singularities present on $\Im\varphi = 2\Phi$, provided we have $|\gamma| < 2\Phi$. The $\rho_\gamma(\varphi)$ function is thus independent of γ in the region $|\gamma| < 2\Phi$. We conclude thus from (7.98) that $y = 0$. There exists, however, a nonzero value of h given by

$$\beta h = -\int_{-\pi+i\gamma}^{\pi+i\gamma} i f_\Phi(\varphi)\rho_\gamma(\varphi)d\varphi. \tag{7.125}$$

Indeed $f(\varphi)$ is not periodic, but

$$f_\Phi(\varphi) = -\varphi + \text{ odd periodic function},$$

from which we deduce

$$\beta h = \frac{1}{2\pi i}\int_{-\pi+i\gamma}^{\pi+i\gamma} \varphi\,\mathrm{Dn}\,\varphi d\varphi = H(\gamma). \tag{7.126}$$

Substituting the series (7.115) in the integrand, we obtain

$$H(\gamma) = \frac{\gamma}{2} + \sum_{k=1}^{\infty} \frac{(-1)^k}{k}\frac{\sinh k\gamma}{\cosh k\Phi}, \tag{7.127}$$

or, in terms of the *elliptic amplitude* of periods $2K$ and $4i\,K'$ ($\Phi = \pi K'/K$),

$$H(\gamma) = \text{iam}\left(K\left(1+\frac{i\gamma}{\pi}\right); k\right).\tag{7.128}$$

Starting from the zero-field point $h = v = 0$, we have thus managed to increase the value of the horizontal field up to the value given by (7.125), the magnetization y remaining zero. This occurs as long as the γ parameter remains lower than 2Φ. Invoking the symmetry lemma (equation (7.104)), we can perform the same steps for the vertical field v, starting from the point $v = 0, h \neq 0, x = y = 0$; the v field can be increased while maintaining a zero magnetization, up to the point where the associated γ_1 parameter reaches the value $\chi - \Phi - \gamma$, $|\gamma_1| < 2\Phi$, with boundary

$$-\beta v = \int_{-\pi+i\gamma_1}^{\pi+i\gamma_1} if_\Phi(\varphi)\rho_{\gamma_1}(\varphi)d\varphi.\tag{7.129}$$

As a consequence, in the interior of the closed curve parametrically defined by the relations

$$\begin{cases} \beta h = H(\gamma), & -2\Phi \le \gamma \le 2\Phi \\ \beta v = H(\gamma + \Phi - \chi), \end{cases}\tag{7.130}$$

the magnetization remains zero, $x = y = 0$.

This boundary curve for the electric field is an algebraic curve of type 1 in the variables

$$\cosh \beta h = \text{sn}\left(K + K\frac{i\gamma}{\pi}; k\right) = \text{dn}^{-1}\left(K'\frac{\gamma}{\Phi}; k'\right),$$

$$\cosh \beta v = \text{sn}\left(K + iK' + iK\frac{\gamma - \chi}{\pi}; k\right) = k^{-1}\,\text{dn}\left(K'\frac{\gamma - \chi}{\Phi}; k'\right),\tag{7.131}$$

or further still

$$\tanh \beta h = k'\,\text{sn}\left(K'\frac{\gamma}{\Phi}; k'\right),$$

$$\tanh \beta v = k'\,\text{cd}\left(K'\frac{\gamma - \chi}{\Phi}; k'\right).\tag{7.132}$$

Elimination by use of an addition formula gives, if one prefers, the algebraic form

$$\frac{\tanh \beta v \left(\left(\frac{k'}{k}\right)^2 - \sinh^2 \beta h\right)^{1/2} - \tanh \beta h \left(\left(\frac{k'}{k}\right)^2 - \sinh^2 \beta v\right)^{1/2}}{1 - \tanh \beta v \tanh \beta h \left(\left(\frac{k'}{k}\right)^2 - \sinh^2 \beta h\right)^{1/2}\left(\left(\frac{k'}{k}\right)^2 - \sinh^2 \beta v\right)^{1/2}}$$

$$= k'\,\text{cd}\left(K'\frac{\chi}{\Phi}; k'\right).\tag{7.133}$$

Equations (7.130) or (7.133) can be considered as defining the critical temperature in the presence of a field. Above this temperature, the susceptibility is nonzero and the expression for $F(h, v)$ is given by (7.104) or the analogous form in Θ depending on whether Δ is lower or higher than -1. The point $\Delta = -1$ is, however, not singular and the transition, which was of infinite order in zero field, gives way to a second-order transition. The detailed study of these questions is not our aim here.

The singularities of the free energy find a handy representation is terms of magnetization. The thermodynamic function (Legendre transform)

$$F_0(x, y) = F(h, v) - hx - vy \tag{7.134}$$

is defined in the square of magnetizations

$$-1 < x < 1, \quad -1 < y < 1.$$

We distinguish two cases.

(a) $-1 < \Delta < 1$.

F_0 possesses an absolute minimum at the origin, around which we prove that

$$F_0(x, y) = F(0, 0) + \frac{(\pi - \Theta)}{4 \cos \frac{\pi \tau}{2\Theta}} \left(x^2 + y^2 - 2xy \sin \frac{\pi \tau}{2\Theta} \right) + \dots \tag{7.135}$$

(b) $\Delta < -1$.

The origin has a conical singularity, since in this point there exists a one-parameter family of normals to the surface F_0 whose components

$$h = -\frac{\partial F_0}{\partial x}, \quad v = -\frac{\partial F_0}{\partial y}$$

are linked from (7.130).

The other singularities of the free energy are located at $x = \pm 1$ ($y = \pm 1$), where F_0 is in general infinite. We deduce in Appendix G (formula (G.10)) the behaviour of the vertical field in the vicinity of the unit vertical magnetization

$$\beta v = \ln \frac{2\xi}{\pi} \frac{\cos \frac{\pi x}{2}}{1 - y} + \dots \quad (|x| \neq 1). \tag{7.136}$$

Appendix F

Working again with the notation of Subsection 7.4.4, we have defined the function $Z(h, v)$ in conformity with (7.100) and (7.102):

$$Z(h, v) = \text{Max}_\zeta \int_{-\zeta^*}^{\zeta} (i f_\Phi(\varphi - i\chi + i\Phi) - 2\beta v) \rho_\zeta(\varphi) d\varphi. \tag{F.2}$$

The maximum over ζ is taken with the constraint $h = $ constant:

$$\frac{\partial}{\partial y} \int_{-\zeta^*}^{\zeta} (i f_\Phi(\varphi) + 2\beta h) \rho_\zeta(\varphi) d\varphi = 0. \qquad \text{(F.3)}$$

Let us first show the following

Lemma: Let $\rho_{\zeta_1}(\varphi)$ be a solution to the integral equation (7.96) for values ζ_1 and h_1 of the boundary and field such that

$$\zeta = b + i\gamma, \qquad \zeta_1 = b + i\gamma_1, \qquad \gamma_1 + \gamma = \chi - \Phi. \qquad \text{(F.4)}$$

We have the identity

$$\int_{-\zeta^*}^{\zeta} (i f_\Phi(\varphi - i\chi + i\Phi) - 2\beta h_1) \rho_\zeta(\varphi) d\varphi = \int_{-\zeta_1^*}^{\zeta_1} (i f_\Phi(\varphi - i\chi + i\Phi) - 2\beta h) \rho_{\zeta_1}(\varphi) d\varphi. \qquad \text{(F.5)}$$

Proof: The equation (7.101) can be written for the pair h_1, ζ_1 and the argument $-\varphi$:

$$f_\Phi(-\varphi + i\gamma_1) = -R_{\zeta_1}(-\varphi + i\gamma_1) + 2i\beta h_1 - \int_{-b}^{+b} f_{2\Phi}(\varphi - \varphi') \rho_{\zeta_1}(-\varphi' + i\gamma_1) d\varphi'. \qquad \text{(F.6)}$$

Multiplying equation (7.101) by $\rho_{\zeta_1}(-\varphi + i\gamma_1)$ and equation (F.6) by $\rho_\zeta(\varphi + i\gamma)$, subtracting term by term and integrating over $[-b, +b]$, we have

$$\int_{-b}^{+b} (f_\Phi(-\varphi + i\gamma_1) - 2i\beta h_1) \rho_\zeta(\varphi + i\gamma) d\varphi$$

$$- \int_{-b}^{+b} (f_\Phi(\varphi + i\gamma) - 2i\beta h) \rho_{\zeta_1}(-\varphi + i\gamma_1) d\varphi$$

$$= -\frac{1}{2\pi} \int_{-b}^{+b} d\varphi \frac{d}{d\varphi} (R_\zeta(\varphi + i\gamma) R_{\zeta_1}(-\varphi + i\gamma_1)). \qquad \text{(F.7)}$$

The right-hand side gives us the integrated term

$$R_\zeta(\zeta) R_{\zeta_1}(-\zeta_1^*) - R_\zeta(-\zeta^*) R_{\zeta_1}(\zeta_1),$$

which vanishes by virtue of (7.97): $R_\zeta(\zeta) = -R_\zeta(-\zeta^*)$. After appropriate translations $\varphi \to \varphi - i\gamma$ in the first integral, $\varphi \to \varphi + i\gamma_1$ in the second and change of the sign of φ, the left-hand side of (F.7) reduces to (F.5), which is therefore proved. □

Let us now address the proof of relations (7.105) and (7.106), by going back to the maximum condition (F.2). In order to lighten the derivation, we shall put in $\rho_\zeta(\varphi)$ the ζ dependence coming from the integral boundaries. This will allow us to avoid writing boundary terms. We thus write the condition of the maximum of y as

$$\int_{(\zeta)} d\varphi (i f_\Phi(\varphi + i\Phi - i\chi) - 2\beta v) \frac{\partial \rho_\zeta}{\partial y}(\varphi) = 0 \qquad \text{(F.8)}$$

and the condition (F.3)

$$\int_{(\zeta)} d\varphi (i f_\Phi(\varphi) + 2\beta h)\frac{\partial \rho_\zeta}{\partial y}(\varphi) = 0 \tag{F.9}$$

with

$$\int_{(\zeta)} 2\frac{\partial \rho_\zeta}{\partial y}(\varphi)d\varphi = -1. \tag{F.10}$$

We can further write (F.8) in the form

$$\beta v = -\int_{-b}^{+b} i f_\Phi(\varphi - i\gamma_1)\frac{\partial \rho_\zeta}{\partial y}(\varphi + i\gamma)d\varphi. \tag{F.11}$$

The variation $\dfrac{\partial \rho_\zeta}{\partial y}$ is obtained by differentiating (7.96):

$$\frac{\partial R_\zeta}{\partial y}(\varphi + i\gamma) = \int_{-b}^{+b} d\varphi' f_{2\Phi}(\varphi - \varphi')\frac{\partial \rho_\zeta}{\partial y}(\varphi' + i\gamma). \tag{F.12}$$

Multiplying the two sides of (F.12) by $\rho_{\zeta_1}(-\varphi + i\gamma_1)$ and integrating over $[-b, +b]$, taking (F.6) into account, we have

$$\int_{-b}^{+b} \left(\rho_{\zeta_1}(-\varphi + i\gamma_1)\frac{\partial R_\zeta}{\partial y}(\varphi + i\gamma) - R_{\zeta_1}(-\varphi + i\gamma_1)\frac{\partial \rho_\zeta}{\partial y}(\varphi + i\gamma) \right) d\varphi$$

$$= \int_{-b}^{+b} (f_\Phi(-\varphi + i\gamma_1) - 2i\beta h_1)\frac{\partial \rho_\zeta}{\partial y}(\varphi + i\gamma)d\varphi, \tag{F.13}$$

or in other words, after translation in the integral on the right-hand side of (F.13):

$$\int_{-\zeta^*}^{\zeta} (i f_\Phi(\varphi - i\chi + i\Phi) - 2\beta h_1)\frac{\partial \rho_\zeta}{\partial y}d\varphi$$

$$= i\left\{ \frac{\partial R_\zeta}{\partial y}(\zeta)R_{\zeta_1}(-\zeta_1^*) - \frac{\partial R_\zeta}{\partial y}(-\zeta^*)R_{\zeta_1}(\zeta_1) \right\}$$

$$= -i R_{\zeta_1}(\zeta_1)\left[\frac{\partial R_\zeta}{\partial y}(-\zeta^*) + \frac{\partial R_\zeta}{\partial y}(\zeta) \right]. \tag{F.14}$$

Let us now make use of condition (F.9). Multiplying the two sides of equality

$$\frac{\partial R_\zeta}{\partial y}(\varphi) = \int_{(\zeta)} f_{2\Phi}(\varphi - \varphi')\frac{\partial \rho_\zeta}{\partial y}(\varphi')d\varphi' \tag{F.15}$$

by $\rho_\zeta(\varphi)$ and integrating over $[-b, b]$, we obtain with the use of (7.96)

$$-\int_{-\zeta^*}^{\zeta} (f_\Phi(\varphi) - 2i\beta h)\frac{\partial \rho_\zeta}{\partial y}(\varphi)d\varphi = \int_{-\zeta^*}^{\zeta} \left(\frac{\partial R_\zeta}{\partial y}\rho_\zeta - R_\zeta \frac{\partial \rho_\zeta}{\partial y} \right)(\varphi)d\varphi. \tag{F.16}$$

The right-hand side can be integrated to give, with (7.97),

$$R_\zeta(\varphi)\frac{\partial R_\zeta}{\partial y}(\varphi)\big|_{-\zeta^*}^{\zeta} = R_\zeta(\zeta)\left[\frac{\partial R_\zeta}{\partial y}(\zeta) + \frac{\partial R_\zeta}{\partial y}(-\zeta^*)\right].$$

Since the left-hand side vanishes by virtue of (F.9), we deduce that the right-hand side of (F.14) is zero, and thus also the left-hand side. Comparing with (F.8) gives us the identification $h_1 = v$.

From the symmetry lemma (F.5) we thus deduce

$$Z(h, v) = \underset{\zeta_1}{\text{Max}} \int_{-\zeta_1^*}^{\zeta_1} (i f_\Phi(\varphi - i\chi + i\Phi) - 2\beta h)\rho_{\zeta_1} d\varphi, \tag{F.17}$$

in which ζ and ζ_1 are related by (F.4).

As a consequence, the horizontal magnetization x is given by the relation

$$1 - x = -\frac{1}{\beta}\frac{\partial Z}{\partial h} = 2\int_{-\zeta_1^*}^{\zeta_1} \rho_{\zeta_1}(\varphi)d\varphi. \tag{F.18}$$

The vertical field is obtained by integrating (7.96) relative to the pair $(h_1 = v, \zeta_1)$, after multiplication by $\rho_{\zeta_1}(\varphi)$:

$$\beta v(1 - x) = \int_{-\zeta_1^*}^{\zeta_1} (-i) f_\Phi(\varphi)\rho_{\zeta_1}(\varphi)d\varphi. \tag{F.19}$$

This completes the proof of relations (7.105) and (7.106).

Appendix G

We are concerned here with the extraction of the leading behaviour of the field (h, v) as a function of x, y in the vicinity of $y = 1$, starting from equations (7.105) and (7.96).

According to the first equation (7.105), $y \approx 1$ leads to small $b = \Re\zeta$. We however have $\Re\zeta_1 = b$, and in order to have x not necessarily close to 1, it is necessary that ζ_1 be close to a singularity of the integrand $\pm i\Phi$. This implies

$$\gamma_1 \approx -\Phi, \qquad \gamma \approx \chi, \qquad b \approx 0. \tag{G.1}$$

The integral equation (7.96) can be solved by perturbation. Setting

$$u = \varphi + i\Phi \approx 0, \qquad \zeta_1 = b + i(\gamma_1 + \Phi), \qquad \rho_{\zeta_1}(\varphi) = \rho_1(u), \tag{G.2}$$

we obtain

$$2\pi\rho_1(u) + \int_{-\zeta_1^*}^{\zeta_1} du' \frac{\sinh 2\Phi}{\cosh 2\Phi - \cos(u - u')}\rho_1(u') = \frac{\sinh \Phi}{\cosh \Phi - \cosh(\Phi + iu)}, \tag{G.3}$$

or in other words, taking the definition of $1 - x$ into account,

$$2\pi\rho_1(u) = \frac{i}{u} + \coth\Phi\frac{1 - x}{2} \qquad (G.4)$$

with

$$\tan(\pi x/2) = (\gamma_1 + \Phi)/b. \qquad (G.5)$$

Similarly, equation (7.96) for $\rho_\zeta(\varphi)$, $\zeta = b + i\gamma$ gives us for $\varphi \approx \chi$:

$$2\pi\rho_\zeta(\chi) = \frac{\sinh\Phi}{\cosh\Phi - \cosh\chi},$$

from which we deduce from (7.105)

$$1 - y = \frac{2b}{\pi}\frac{\sinh\Phi}{\cosh\Phi - \cosh\chi}, \qquad (G.6)$$

and from (7.106)

$$2\beta h = if_\Phi(i\chi) = -\beta\delta, \qquad (G.7)$$

or further still $h = -\delta/2$. This is the dominant term in h.

To calculate v, we have

$$
\begin{aligned}
\beta v(1 - x) &= \int_{-\xi_1^*}^{\xi_1} (-i)f_\Phi(u - i\Phi)\rho_1(u)du \\
&= \frac{1}{2\pi}\int_{-\xi_1^*}^{\xi_1}\frac{du}{u}\ln\left(\frac{-iu}{2\sinh\Phi}\right) = \frac{1 - x}{2}\ln\left(\frac{2\sinh\Phi}{b}\cos\frac{\pi x}{2}\right) \quad (G.8)
\end{aligned}
$$

in which we have used (G.2) and (G.5). Substituting the value of b given by (G.6) into the right-hand side of (G.7), we obtain

$$2\beta v = \ln\left(\frac{4\sinh^2\Phi}{\pi(\cosh\Phi - \cosh\chi)}\frac{\cos(\pi x/2)}{1 - y}\right) \qquad (G.9)$$

or further still

$$2\beta v = \ln\left(\frac{2\xi}{\pi}\frac{\cos(\pi x/2)}{1 - y}\right), \qquad (G.10)$$

with definition (7.72) of ξ: $\xi/2 = \cosh\Phi + \cosh\beta\delta$.

8

The eight-vertex model

8.1 Definition and equivalences

8.1.1 Definition

The models of ferro- or antiferroelectric systems encountered in Chapter 7 are characterized by what is known as the 'ice rule', in other words the condition of electrical neutrality on each site of the lattice. The ice model, the F and KDP models in the presence of an external field constitute, in their two-dimensional reduction, special cases of the six-vertex model depending on six arbitrary energies, of which the general solution is known. The thermodynamic properties have been worked out and are exposed in the synthesis article of Lieb and Wu (1972).

The thermodynamic problem boils down to the evaluation of the partition function of a system on a torus of dimension $N \times M$, as a function of the statistical weights $\omega_j = e^{-\beta e_j}$, $j = 1, \ldots, 8$, associated with each vertex:

$$Z(\omega_1, \ldots, \omega_8) = \sum_{\Gamma} \prod_{V} \omega_{i(V)}. \tag{8.1}$$

The product runs over the MN sites V, and the sum over all compatible vertex configurations $\Gamma: V \to i(V)$. The periodic boundary conditions (our lattice is a torus $\mathbb{Z}_N \oplus \mathbb{Z}_M$) are a technical commodity. The free energy per site exists and can be calculated as the double limit

$$-\beta f = \lim_{M \to \infty} \lim_{N \to \infty} \frac{1}{MN} \ln Z. \tag{8.2}$$

The general eight-vertex model, which depends on six inhomogeneous parameters, is not solved. It was considered by Fan and Wu (1970) as a model encompassing the previously solved ferroelectric cases, but also equivalent under certain parameter choices to Ising or dimer-type models. Let us give a few examples of these equivalences without intending to go into details.

If we abandon the ice rule, again within the hypothesis of the square lattice, the number of configurations of the four hydrogen bonds equals 16. For the 16-vertex model, nothing is known besides certain equivalences with the Ising model in a field (Gaaff and Hijmans, 1975). If we restrict ourselves to even-degree entrant vertices, we obtain the eight-vertex model.

8.1.2 Equivalence with an Ising model

With each vertex configuration Γ on the square lattice, we associate two configurations of 'spins' s' and s'' on the sites of the dual lattice, that is to say a system of signs $+$ and $-$, or spins ± 1, on each plaquette of the original lattice. The rule for constructing these is as follows: given Γ, and thus a system of arrows on the links, the North$\rightarrow$South and East$\rightarrow$West arrows must separate opposite signs on two adjacent plaquettes (Figure 8.1). The construction is possible since the incident degree on each site is even. We obtain for each Γ two configurations of spins s' and s'', symmetric under the global exchange $+ \leftrightarrow -$.

Let us then suppose that the spins of the dual lattice interact with their nearest neighbours with an energy $K_\alpha \sigma \sigma'$ ($\alpha = 1, 2, 3, 4$) according to whether the link $\sigma \sigma'$ is vertical, horizontal, diagonal or antidiagonal. The energy of a configuration s' is the sum of the energies of the partial configurations of four spins around each site under the condition of splitting the energy of the vertical and horizontal links in two. In terms of the K parameters, the statistical weights of each vertex are

$$\omega_1 = \exp(K_1 + K_2 + K_3 + K_4),$$
$$\omega_2 = \exp(-K_1 - K_2 + K_3 + K_4),$$
$$\omega_3 = \exp(-K_1 + K_2 - K_3 - K_4),$$
$$\omega_4 = \exp(K_1 - K_2 - K_3 - K_4),$$

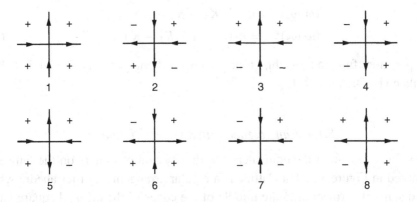

Figure 8.1 The eight vertices, numbered from 1 to 8.

$$\omega_5 = \omega_6 = \exp(K_3 - K_4),$$
$$\omega_7 = \omega_8 = \exp(-K_3 + K_4). \tag{8.3}$$

We shall see later that the equalities $\omega_5 = \omega_6$ and $\omega_7 = \omega_8$ do not lead to restrictions. The model above is a four-parameter one with the only effective relation between the ω being

$$\omega_1\omega_2\omega_3\omega_4 = \omega_5\omega_6\omega_7\omega_8. \tag{8.4}$$

The solution is not known, except for $K_1 = K_2 = 0$ which reduces to the ordinary Ising model on two independent lattices: $Z = Z_{\text{Ising}}^2$.

8.1.3 Equivalence of the self-dual model with an Ising model with 4-spin interaction

The self-dual eight-vertex model, whose weights are invariant under a global reversal of the arrow directions on a vertex, is equivalent to an Ising model with 4-spin interaction between two independent sublattices (Kadanoff and Wegner, 1971). The partition function is written

$$Z = \sum_{\{\sigma\}} \exp \left\{ \sum_{j=1}^{M} \sum_{k=1}^{N} K_3 \sigma_{j,k} \sigma_{j+1,k+1} + K_4 \sigma_{j+1,k} \sigma_{j,k+1} \right.$$
$$\left. + K \sigma_{j,k} \sigma_{j+1,k+1} \sigma_{j,k+1} \sigma_{j+1,k} \right\}; \tag{8.5}$$

we have the correspondence between the K and the ω:

$$\omega_1 = \omega_2 = \exp(K_3 + K_4 + K) = a,$$
$$\omega_3 = \omega_4 = \exp(-K_3 - K_4 + K) = b,$$
$$(\omega_5\omega_6)^{1/2} = \exp(K_3 - K_4 - K) = c,$$
$$(\omega_7\omega_8)^{1/2} = \exp(-K_3 + K_4 - K) = d. \tag{8.6}$$

The partition function of this four-parameter homogeneous model has been calculated by Baxter (1971a).

8.1.4 Equivalence with a system of dimers

Let us finally mention the equivalence with a system of dimers on the lattice R_1 illustrated in Figure 8.2. This lattice is a regular arrangement of tetrahedra whose joined summits project onto the middle of the edges of the original square lattice supporting the system of arrows. The dimer configurations on the tetrahedra are

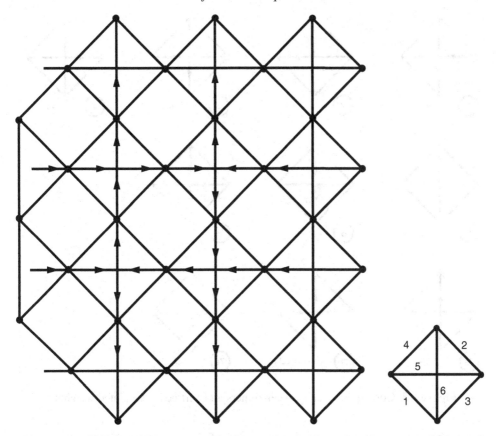

Figure 8.2 Indices of the statistical weights of each link of a tetrahedron.

represented in Figure 8.3, in which we indicate the rules for the correspondence between dimers and the arrow system.

By inspection of Figure 8.3, we notice that the partition function of the dimer system on the non-planar graph R_1 is equal to $Z(\omega_1, \ldots, \omega_8)$ when the constraint $\omega_7\omega_8 = \omega_1\omega_2 + \omega_3\omega_4 + \omega_5\omega_6$ is obeyed. The particular case $\omega_5 = \omega_6 = 0$ studied by Baxter gives a system of dimers on the diagonal square lattice of Figure 8.2. The problem can then be solved using the Pfaffian method (Kasteleyn, 1967). Kasteleyn's method can, however, be extended to the non-planar graph R_1 provided we modify the constraint to

$$\omega_1\omega_2 + \omega_3\omega_4 - \omega_5\omega_6 = \omega_7\omega_8. \tag{8.7}$$

We recognize in (8.7) the condition of Fan and Wu (1970) for the free fermion model defined as a case of the eight-vertex model, solvable by the Pfaffian method (the (7c) configuration appears with a sign opposite to that of (7a) or (7b), since

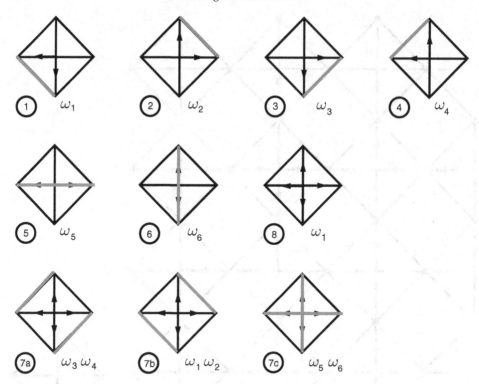

Figure 8.3 Correspondence between dimers and vertices. Associated weights.

it modifies by one unit the number of cycles of the determinant associated with Kasteleyn's method).

One can easily show that this problem is equivalent to that of the Ising model in zero field. Indeed, it is sufficient to define a lattice R_2 by doubling the junctions of summits of tetrahedra of R_1 in order to create a new link between each tetrahedra. If we exchange by conjugation ($1 \leftrightarrow 2$, $3 \leftrightarrow 4$, $5 \leftrightarrow 6$, $7 \leftrightarrow 8$) the vertices belonging to one of the two diagonal sublattices, we can faithfully associate with a dimer configuration on R_1 another configuration of dimers on R_2. By reduction of the tetrahedra, we notice that the remaining links support a system of closed polygons of Ising type for which we could easily calculate the weights in terms of the ω. We shall not delve further into this question of equivalences.

8.2 The transfer matrix and the symmetries of the self-dual model

8.2.1 The transfer matrix

The self-dual eight-vertex model is that of a ferroelectric in which the energy of any vertex configuration is invariant under reversal of the orientation of the dipoles formed by the hydrogen bonds; under this hypothesis of zero electric field, horizontal and vertical (see Chapter 7, Section 7.1), we thus have the equalities

$$\omega_1 = \omega_2 = a, \qquad \omega_3 = \omega_4 = b. \tag{8.8}$$

In contrast, the partition function Z is exclusively a function of the products $c^2 = \omega_5 \omega_6$, $d^2 = \omega_7 \omega_8$. It is sufficient to show that the 7 and 8 vertices appear in equal numbers in any configuration Γ. Considering Γ as an electrical network, we associate a current $+1$ with each oriented link; vertices of type 1 to 6 conserve the current, those of type 8 are sources of flow $+4$ and vertices of type 7 are sinks of flow -4. The equality of incoming and outgoing flows leads to that of the number of sources and wells, or said otherwise of the ferroelectric sites with 4 or 0 protons. We can make the same remark concerning vertices 5 and 6: changing the orientation of all the horizontal arrows exchanges vertices 5 and 8, 6 and 7.

Definition (8.1) of Z allows us to write

$$Z = \operatorname{Tr} T^M \tag{8.9}$$

in which T is a square matrix of dimension 2^N whose elements are indexed by two 'vertical' arrow configurations along two successive 'horizontal' rows (Figure 8.4). If we refer to Section 7.2, we write the elements of matrix T as

$$\langle \alpha_1 \alpha_2 \ldots \alpha_N | T | \beta_1 \beta_2 \ldots \beta_N \rangle = \sum_{\{\lambda\}} (\alpha_1 \lambda_1 | R | \beta_1 \lambda_2)(\alpha_2 \lambda_2 | R | \beta_2 \lambda_3) \ldots (\alpha_N \lambda_N | R | \beta_N \lambda_1)$$
$$\tag{8.10}$$

in which the state of the arrows is described by the variables

$$\alpha_i, \beta_i, \lambda_i = \pm 1, \qquad i = 1, 2, \ldots, N.$$

The quantities $(\alpha \lambda | R | \beta \mu)$ are the 16 weights of the general 16-vertex model, which can be expressed on the basis of matrices $\sigma^i \otimes \sigma^j$, $i, j = 0, 1, 2, 3$ in which σ^i is the set of two-dimensional Pauli matrices, σ^0 being the unit matrix.

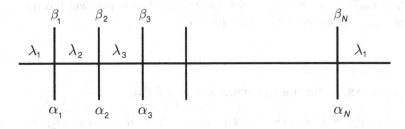

Figure 8.4 Vertical arrow configurations along successive horizontal rows.

Introducing a set of N commuting spin operators $\sigma_1, \sigma_2, \ldots, \sigma_N$, the matrix $(\alpha_1 \lambda | R | \beta_1 \mu)$ will be expanded on the basis of matrices $\sigma^i \otimes \sigma^j$ which we shall denote $\sigma_1^i \tau^j$ to avoid writing tensor products. We shall thus write

$$T = \operatorname*{Tr}_\tau R_1 R_2 \ldots R_N \tag{8.11}$$

in which R_n is a function of σ_n and τ.

For the self-dual model, we obtain by simple observation

$$R = \frac{a+b}{2} + \frac{a-b}{2}\sigma^z \tau^z + \frac{c+d}{2}\sigma^x \tau^x + \frac{c-d}{2}\sigma^y \tau^y, \tag{8.12}$$

which we will write in condensed form as

$$R = \sum_{\alpha=0}^{3} w_\alpha \sigma^\alpha \tau^\alpha \equiv (w\sigma\tau), \tag{8.13}$$

with

$$w_0 = \frac{1}{2}(a+b), \qquad w_3 = \frac{1}{2}(a-b),$$

$$w_1 = \frac{1}{2}(c+d), \qquad w_2 = \frac{1}{2}(c-d). \tag{8.14}$$

We thus have

$$T(w) = \operatorname*{Tr}_\tau (w\sigma_1\tau)(w\sigma_2\tau)\ldots(w\sigma_N\tau), \tag{8.15}$$

which is a very convenient form for what follows. T presents itself as an operator of spin $\sigma_1, \ldots, \sigma_N$ whose greatest eigenvalue t gives the free energy per site according to the formula

$$-\beta f = \lim_{N\to\infty} N^{-1} \ln t.$$

8.2.2 Invariance of the partition function

Proposition: $|Z(w)|$ is invariant under the $2^4 4!$ operations which permute the w and change their sign (hypercubic group).

Proof:

(a) Invariance $w_2 \to -w_2$.

By construction, the transfer matrix is real and thus

$${}^t T(w_2) = T^+(w_2) = \operatorname*{Tr}_\tau R_1(-w_2) \ldots R_N(-w_2) = T(-w_2), \tag{8.16}$$

since the σ_n are Hermitian and commute, and the complex conjugation of the τ changes the sign of τ^y which is equivalent to changing that of w_2. We thus have

$$Z(w_2) = \text{Tr}(^t T(w_2))^M = Z(-w_2). \tag{8.17}$$

(b) Invariance $w_i \to w_i$, $w_j \to -w_j$, $w_k \to w_k$, $(i, j, k) = (1, 2, 3)$.

The canonical transformation of the σ and τ matrices induced by a π rotation around the k axis leaves the total trace invariant ('total' on τ and σ for Z).

(c) Invariance by circular permutation (123) and (12).

Same reasoning as in (b) with a $\dfrac{2\pi}{3}$ rotation around the ternary axis or around an axes bisector.

(d) Invariance $w_0 \leftrightarrow w_1$, $w_2 \leftrightarrow w_3$.

On the given square lattice, let us consider the set of zigzag links shown in Figure 8.5. Given a configuration Γ on the lattice, we associate with it a configuration Γ' obtained by inverting the arrow orientations on the zigzag. In the Γ–Γ' correspondence we have the vertex exchanges

$$1 \leftrightarrow 5, \quad 2 \leftrightarrow 6, \quad 3 \leftrightarrow 8, \quad 4 \leftrightarrow 7,$$

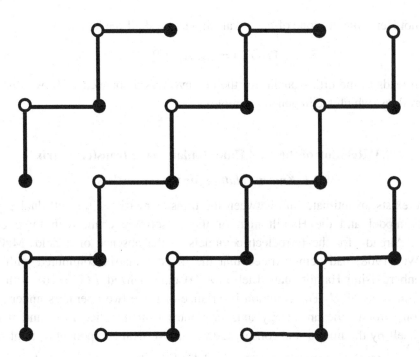

Figure 8.5 Zigzag links.

which induce the transformation $a \leftrightarrow c, b \leftrightarrow d$ or $w_0 \leftrightarrow w_1, w_2 \leftrightarrow w_3$. From paragraphs (c) and (d) we conclude that Z is invariant under the permutations (123), (12) and (01)(23), and thus under the group of the 4! permutations which they generate. From paragraphs (a) and (b) we conclude that the absolute value $|Z|$ is invariant under changes of sign of the w. $\qquad\square$

The involutions of this group can be considered as duality operations which exchange regions of high and low temperature. Their fixed points shall be defined by equating their 'relative invariants' introduced in the next section:

$$J_\alpha = \pm J_\beta \quad (\alpha, \beta = 1, 2, 3),$$

which will indeed determine the transition lines (Wannier, 1950; Sutherland, 1970).

We can further make the dependence of the transfer matrix with respect to w more concrete, by expanding the trace in (8.15) using Wick's theorem for the τ. We obtain

$$T(w) = S(w_\alpha^2) + w_0 w_1 w_2 w_3 A(w_\alpha^2), \tag{8.18}$$

in which S and A are real matrices, functions of the w^2 with symmetry properties

$$^t S = S, \quad ^t A = -A. \tag{8.19}$$

T is not Hermitian as none of the w vanish, but we shall see later that

$$[T, T^+] = [S, A] = 0, \tag{8.20}$$

which leads to the orthogonality of the eigenvectors associated with two distinct eigenvalues (which are in general complex).

8.3 Relation of the *XYZ* Hamiltonian to the transfer matrix

8.3.1 Known relation for the six vertices

There exists an intimate link between the transfer matrix of the self-dual eight-vertex model and the Hamiltonian of the anisotropic chain with three constants. Already for the ferroelectric models in the absence of a field, McCoy and Wu (1968) had demonstrated that the transfer matrix commuted with the Heisenberg–Ising Hamiltonian. Lieb (1967a) diagonalized $T(d = 0)$ with the eigenfunctions of H. The common invariance of these two operators under spin rotations around the anisotropy axis originates from the 'ice rule' and manifests itself by the line-to-line conservation of the vertical component of total spin $S^z = \dfrac{N}{2} - M$. One of the eight-vertex model's difficulties is certainly linked to the absence of this conservation law.

Let us point out here that the transfer matrix of the general six-vertex model (cf. Chapter 7) also commutes with a Hamiltonian of spin–spin interactions between neighbours depending on two parameters (McCoy and Wu, 1968):

$$H_{\Delta,h} = \sum_n \cosh 2\beta h \left(S_n^x S_{n+1}^x + S_n^y S_{n+1}^y \right) + \Delta S_n^z S_{n+1}^z$$

$$+ i \sinh 2\beta h \left(S_n^x S_{n+1}^y - S_n^y S_{n+1}^x \right), \tag{8.21}$$

$$\left[T(a, b, c, h, v), H_{\Delta,h} \right] = 0, \tag{8.22}$$

with $\Delta = (a^2 + b^2 - c^2)/2ab$, the parameters of the six vertices being

$$\omega_1 = ae^{\beta(h+v)}, \qquad \omega_2 = ae^{-\beta(h+v)}, \qquad \omega_5 = \omega_6 = c,$$
$$\omega_3 = be^{\beta(v-h)}, \qquad \omega_4 = be^{\beta(h-v)}.$$

8.3.2 Commutation of H and T

Proposition (Sutherland, 1970): The XYZ Hamiltonian (1.4)

$$H = \frac{1}{2} \sum_n \sum_{\alpha=1}^{3} J_\alpha \sigma_n^\alpha \sigma_{n+1}^\alpha \equiv \sum_n H_{n,n+1}$$

commutes with a one-parameter (J) family of transfer matrices $T(w(J))$ such that

$$w_\alpha = (J + J_\alpha)^{1/2}, \qquad \alpha = 0, 1, 2, 3 \tag{8.23}$$

with

$$J_0 = \sum_{\alpha=1}^{3} J_\alpha.$$

Let us mention the useful relations

$$J_1 = \frac{1}{2}(w_0^2 + w_1^2 - w_2^2 - w_3^2) = \frac{1}{2}(ab + cd),$$

$$J_2 = \frac{1}{2}(w_0^2 + w_2^2 - w_3^2 - w_1^2) = \frac{1}{2}(ab - cd),$$

$$J_3 = \frac{1}{2}(w_0^2 + w_3^2 - w_1^2 - w_2^2) = \frac{1}{2}(a^2 + b^2 - c^2 - d^2),$$

$$J = \frac{1}{2}(-w_0^2 + w_1^2 + w_2^2 + w_3^2) = \frac{1}{2}(c^2 + d^2 - 2ab). \tag{8.24}$$

Proof: This is based on the following identity:

$$\left[H_{n,n+1}, R_n R_{n+1} \right] = \frac{i}{2}\sqrt{P(J)} \left(R_n \frac{dR_{n+1}}{dJ} - \frac{dR_n}{dJ} R_{n+1} \right), \tag{8.25}$$

with

$$P(J) \equiv \prod_{\alpha=0}^{3}(J + J_\alpha).$$

It suffices to calculate the commutator in the form

$$[H, T] = \frac{1}{2}\sum_{n=1}^{N} \operatorname*{Tr}_{\tau} R_1 \dots R_{n-1}\left[H_{n,n+1}, R_n R_{n+1}\right] R_{n+2} \dots R_N,$$

in which we identify R_{N+1} and R_1. According to (8.25), we have

$$[H, T] = \frac{i}{4}\sqrt{P(J)}\left\{-\sum_{n=1}^{N} \operatorname*{Tr}_{\tau} R_1 \dots R_{n-1}\frac{dR_n}{dJ}R_{n+1} \dots R_N\right.$$
$$\left. + \sum_{n=1}^{N} \operatorname*{Tr}_{\tau} R_1 \dots R_{n-1} R_n\frac{dR_{n+1}}{dJ}R_{n+2} \dots R_N\right\} = 0. \quad (8.26)$$

The proof of identity (8.25) is a simple calculational exercise starting from the definitions provided in (8.23):

$$R_1 = \sum_{\alpha=0}^{3} w_\alpha \sigma_1^\alpha \tau^\alpha \Rightarrow 2\frac{dR_1}{dJ} = \sum_{\alpha=0}^{3} w_\alpha^{-1}\sigma_1^\alpha \tau^\alpha,$$

$$H_{1,2} = \frac{1}{2}\sum_{\alpha=1}^{3} J_\alpha \sigma_1^\alpha \sigma_2^\alpha.$$

We shall see in Subsection 8.5.4 how this result of Sutherland originates from Baxter's theorem on the commutation of $T(J)$ matrices among themselves, which can be inferred inversely from (8.23) if H is non-degenerate.

8.3.3 H as a derivative of T

Proposition: Setting

$$T_0(J) = 2^{-N}J^{-(N/2)}T(J)$$

we have

$$H = -\frac{N}{2}J_0 + 2T_0^{-1}\frac{dT_0}{dJ^{-1}}|_{J^{-1}=0},$$

$$T_0^{-1}|_{J^{-1}=0} = C. \quad (8.27)$$

Proof: Defining the operators

$$P_{0n} = \frac{1}{2}\left(1 + \vec{\sigma}_n \cdot \vec{\tau}\right),$$

$$H_{0n} = \frac{1}{2}\left(J_0 + \sum_{\alpha=1}^{3} J_\alpha \sigma_n^\alpha \tau^\alpha\right) \tag{8.28}$$

we verify that

$$H_{0n} P_{0n} = H_{0n}. \tag{8.29}$$

Expanding R_n in the neighbourhood of $J = \infty$,

$$R_n/(2\sqrt{J}) = P_{0n} + \frac{1}{2J} H_{0n} + O\left(\frac{1}{J^2}\right), \tag{8.30}$$

and substituting this in T taking (8.29) into account,

$$T_0(J) = \operatorname*{Tr}_\tau\, P_{01}\left(1 + \frac{H_{01}}{2J}\right) P_{02}\left(1 + \frac{H_{02}}{2J}\right)\ldots + O(J^{-2}), \tag{8.31}$$

we use the properties of the permutations P_{0n} to obtain

$$T_0(J) = \operatorname*{Tr}_\tau\, P_{01}\ldots P_{0N}\left(1 + \frac{H_{21}}{2J}\right)\left(1 + \frac{H_{32}}{2J}\right)\ldots\left(1 + \frac{H_{0N}}{2J}\right) + \ldots \tag{8.32}$$

But

$$\operatorname*{Tr}_\tau\, P_{01}\ldots P_{0N} = C^{-1}, \tag{8.33}$$

in which C is the circular permutation $Cn = (n+1)$, $\forall n$. We thus have

$$T_0(J) = C^{-1}\left(1 + \frac{1}{2J}\sum_{n=1}^{N} H_{n,n+1} + O(J^{-2})\right),$$

in other words result (8.27).

8.4 One-parameter family of commuting transfer matrices

8.4.1 Proposition (Baxter, 1971a)

$$\left[T(J), T(J')\right] = 0, \qquad \forall J, J'. \tag{8.34}$$

Proof: Considering $T(w)$ as a function of the w, we look for sufficient conditions guaranteeing $[T(w), T(w')] = 0$. Let us introduce a system of matrices τ' commuting with the τ. We have

$$T = T(w) = \operatorname*{Tr}_{\tau} \, R_1 R_2 \ldots R_N,$$

$$T' = T(w') = \operatorname*{Tr}_{\tau'} \, R'_1 R'_2 \ldots R'_N, \tag{8.35}$$

with

$$R_n = (w\sigma_n\tau), \qquad R'_n = (w'\sigma_n\tau'). \tag{8.36}$$

The product TT' is thus a trace over 4×4 matrices of type $\tau\tau'$:

$$TT' = \operatorname*{Tr}_{\tau\tau'} \, R_1 R'_1 R_2 R'_2 \ldots R_N R'_N. \tag{8.37}$$

From the invariance of the trace under similarity transformations, the product TT' will be identical to $T'T$ if there exists an invertible matrix X which can be expanded on the basis of the 16 matrices $\tau^\alpha \tau'^\beta$ such that

$$X^{-1} R_n R'_n X = R'_n R_n, \qquad \forall n. \tag{8.38}$$

We look for X in a form analogous to that of R:

$$X = (x\tau\tau') \equiv \sum_{\alpha=0}^{3} x_\alpha \tau^\alpha \tau'^\alpha, \tag{8.39}$$

in such a way that we have the ternary relation

$$(w\sigma\tau)(w'\sigma\tau')(x\tau\tau') = (x\tau\tau')(w'\sigma\tau')(w\sigma\tau), \tag{8.40}$$

or explicitly

$$\sum_{\alpha,\beta,\gamma} w_\alpha w'_\beta x_\gamma \left(\sigma^\alpha \sigma^\beta \cdot \tau^\alpha \tau^\gamma \cdot \tau'^\beta \tau'^\gamma - \sigma^\beta \sigma^\alpha \cdot \tau^\gamma \tau^\alpha \cdot \tau'^\gamma \tau'^\beta \right) = 0. \tag{8.41}$$

By virtue of the obvious relations $\tau^\alpha \tau^\beta = \pm \tau^\beta \tau^\alpha$, $\forall \alpha, \beta$, only terms $\alpha \neq \beta \neq \gamma$ contribute on the left-hand side of (8.41) to give the six terms of type $\sigma^3 \tau^2 \tau'^1$. By equating their coefficients to zero, one obtains six homogeneous relations between the w, w' and x:

$$w_1 w'_2 x_3 - w_0 w'_3 x_2 + w_3 w'_0 x_1 - w_2 w'_1 x_0 = 0 \tag{8.42}$$

and five analogous relations. Their expression becomes very symmetrical at the cost of a slight change of notation

$$w_4 = -w_0, \qquad w'_4 = w'_0, \qquad x_4 = -x_0, \tag{8.43}$$

which allows us to write the whole set of relations in condensed form

$$\{I + (12)(34) + (13)(24) + (14)(23)\}\, w_i w'_j x_k = 0 \qquad (8.44)$$

in which i, j, k are three distinct indices chosen from 1 to 4. The four permutations which operate on the given indices, i, j, k, are those of the invariant subgroup $\mathbb{Z}_2 \oplus \mathbb{Z}_2$ of S_4 (symmetric group of order 4!). We thus have $24/4 = 6$ distinct relations (8.44).

Considered as a linear homogeneous system of six equations for four unknowns x_j, system (8.44) admits a nontrivial solution if three determinants of order 4 vanish, which gives after calculation the conditions for w and w':

$$\frac{w_i^2 - w_j^2}{w_k^2 - w_l^2} = \frac{w_i'^2 - w_j'^2}{w_k'^2 - w_l'^2} = \text{same in } x, \qquad \forall \text{ distinct } i, j, k, l \qquad (8.45)$$

which reduces to two independent relations.

The general solution of equations (8.45) leads us precisely to the parametrization introduced in (8.23):

$$w_\alpha^2 = J_\alpha + J, \qquad w_\alpha'^2 = J_\alpha + J'. \qquad (8.46)$$

The w can be chosen arbitrarily. The J and J_0 are determined uniquely ($J_0 = J_1 + J_2 + J_3$); the w' thus depend on a parameter J'. We shall show in the next subsection how to build the solution x by uniformizing the algebraic relations (8.44) or (8.40) in a completely symmetrical fashion.

Let us finally point out that the origin of the S_3 symmetry of system (8.44) does not appear in the derivation we have presented. The problem of the ternary relations of type (8.40) can be presented at different levels of generalization. We present one of these in Appendix H, see also Section 13.4. The interest of the ternary relations, also known as star–triangle or Yang–Baxter relations, is underlined in Chapter 10 with a view towards the construction of integrable Hamiltonian systems.

8.4.2 Elliptic uniformization

The explicit solution of the algebraic system (8.42) led Baxter naturally to the introduction of elliptic functions. These intervene via the integral

$$V = \int dJ [P(J)]^{-1/2},$$

whose differential element we have already seen appear in relation (8.25). This is the natural parameter allowing us to express the Fan matrix as the derivative of R, namely $\dfrac{dR}{dV}$. Once we have recognized the existence of elliptic curves on our algebraic surfaces, it is straightforward to also recognize a perfect similarity between relations (8.42) and the quartic Jacobi relations between four θ functions (Whittaker and Watson, 1927). Appendix I will be devoted to a brief review of the definitions, notation and properties of elliptic functions.

The quartic Jacobi relations allow the complete uniformization of system (8.44). Indeed, for four numbers of vanishing sum

$$z_1 = -V, \quad z_2 = V', \quad z_3 = V - V' - \zeta, \quad z_4 = \zeta, \tag{8.47}$$

in which V, V', ζ are arbitrary, relation (I.12) of Appendix I is written

$$S \cdot \Theta_i(-V)\Theta_j(V')\Theta_k(V - V' - \zeta)\Theta_l(\zeta) = 0, \tag{8.48}$$

in which the operator

$$S = I + (12)(34) + (13)(24) + (14)(23) \tag{8.49}$$

acts on the four distinct indices i, j, k, l constituting a permutation of $1, 2, 3, 4$. After dividing (8.48) by $\prod_i \Theta_i(\zeta)$, we obtain a structure identical to (8.44). Going back to indices $0, 1, 2, 3$ (cf. (8.43)) and taking into account the fact that only Θ_1 is odd, we obtain the solution in the form of ratios:

$$w_\alpha = \frac{\Theta_{\alpha+1}(V)}{\Theta_{\alpha+1}(\zeta)}, \quad w'_\alpha = \frac{\Theta_{\alpha+1}(V')}{\Theta_{\alpha+1}(\zeta)}, \quad x_\alpha = \frac{\Theta_{\alpha+1}(\zeta + V' - V)}{\Theta_{\alpha+1}(\zeta)}. \tag{8.50}$$

We thus have three essential parameters at our disposal: ζ, V and the period ratio which we will denote $\tau_l = K'_l/K_l$ in terms of the modulus l, to keep Baxter's notation. The variable parameter of the commuting family is V, corresponding to J. How to determine these three parameters as a function of the w? According to (8.50), we have

$$\frac{w_0}{w_3} = \frac{\operatorname{sn}(V; l)}{\operatorname{sn}(\zeta; l)},$$

$$\frac{w_1}{w_3} = \frac{\operatorname{cn}(V; l)}{\operatorname{cn}(\zeta; l)},$$

$$\frac{w_2}{w_3} = \frac{\operatorname{dn}(V; l)}{\operatorname{dn}(\zeta; l)}, \tag{8.51}$$

from which we deduce successively the modulus l, then V by inverting sn V, and finally ζ:

$$l^2 = \frac{w_2^2 - w_3^2}{w_2^2 - w_0^2} : \frac{w_1^2 - w_3^2}{w_1^2 - w_0^2}, \tag{8.52}$$

$$\operatorname{sn}\zeta = \left(\frac{w_1^2 - w_3^2}{w_1^2 - w_0^2}\right)^{1/2}, \quad \operatorname{sn}\zeta = \frac{w_3}{w_0}\operatorname{sn}V. \tag{8.53}$$

Given the symmetry in w of the partition function (8.22), we can restrict the domain of study to $w_0^2 < w_3^2 < w_2^2 < w_1^2$, which leads to $0 < l < 1$, ζ and V real.

Let us finally give the differential relation between dJ and dV:

$$\frac{1}{2}(J_1 + J_2)\frac{dJ}{\sqrt{P(J)}} = \frac{\text{cn}\,\zeta\,\text{dn}\,\zeta}{\text{sn}\,\zeta}dV. \tag{8.54}$$

8.4.3 Parameter transformation

It will be more useful in what follows to work with the four parameters a, b, c, d than with the w, and to obtain a factorized expression for them in terms of θ factors, as was done for the w. The following change of modulus and of variables achieves this goal:

$$\tau_l = K_l'/K_l \rightarrow \tau = K'/K = 2/\tau_l, \tag{8.55}$$

in which K, K', τ are associated with the modulus $k = \dfrac{1-l}{1+l}$. We go from (V, ζ, l) to new parameters (v, η, k) defined as

$$\frac{\eta}{K} = \frac{i\zeta}{K_l'}, \quad \frac{v}{K} = \frac{iV}{K_l'}, \quad \frac{K'}{K} = 2\frac{K_l}{K_l'}. \tag{8.56}$$

The transformation is made in two simple steps.

- First, the *Jacobi* transformation $\tau_l \rightarrow \dfrac{1}{\tau_1} = \dfrac{\tau}{2}$. According to (I.17), we go from (8.50)

$$w_\alpha = \theta_{\alpha+1}\left(\frac{\pi V}{2K_l}\bigg|\frac{K_l'}{K_l}\right)\bigg/\theta_{\alpha+1}\left(\frac{\pi\zeta}{2K_l}\bigg|\frac{K_l'}{K_l}\right) \tag{8.57}$$

to the following proportion:

$$w_0 : w_1 : w_2 : w_3 = \frac{\theta_1\left(\frac{\pi v}{2K}\big|\frac{K'}{2K}\right)}{\theta_1\left(\frac{\pi\eta}{2K}\big|\frac{K'}{2K}\right)} : \frac{\theta_4}{\theta_4} : \frac{\theta_3}{\theta_3} : \frac{\theta_2}{\theta_2}. \tag{8.58}$$

- Second, the *Landen* transformation to go from $\tau/2$ to τ (I.16):

$$w_0 : w_1 : w_2 : w_3 = \frac{\Theta_1\left(v|\frac{\tau}{2}\right)}{\Theta_1\left(\eta|\frac{\tau}{2}\right)} : \ldots = \frac{\Theta_1\Theta_4(v|\tau)}{\Theta_1\Theta_4(\eta|\tau)} : \frac{\Theta_4^2 + \Theta_1^2}{\cdots} : \frac{\Theta_4^2 - \Theta_1^2}{\cdots} : \frac{\Theta_2\Theta_3}{\cdots}. \tag{8.59}$$

From now on we shall write in an abridged manner

$$\Theta_\alpha(v) \equiv \Theta_\alpha(v|\tau) = \theta_\alpha\left(\frac{\pi v}{2K}\bigg|\frac{K'}{K}\right). \tag{8.60}$$

We finally obtain, according to definition (8.14),

$$
\begin{aligned}
a : b : c : d &= w_0 + w_3 : w_0 - w_3 : w_1 + w_2 : w_1 - w_2 \\
&= \frac{\Theta_1\Theta_4(v) \cdot \Theta_2\Theta_3(\eta) + \Theta_1\Theta_4(\eta) \cdot \Theta_2\Theta_3(v)}{\Theta_1\Theta_2\Theta_3\Theta_4(\eta)} : \ldots : \ldots : \ldots \\
&= \frac{\Theta_1(v+\eta)\Theta_4(v-\eta)}{\Theta_1(2\eta)\Theta_4(0)} : \frac{\Theta_1(v-\eta)\Theta_4(v+\eta)}{\Theta_1(2\eta)\Theta_4(0)} \\
&\quad : \frac{\Theta_4(v+\eta)\Theta_4(v-\eta)}{\Theta_4(2\eta)\Theta_4(0)} : \frac{\Theta_1(v+\eta)\Theta_1(v-\eta)}{\Theta_4(2\eta)\Theta_4(0)},
\end{aligned}
$$

$$(8.61)$$

which is Baxter's parametrization. We should duly note that these are proportions. The normalization, if needed, can be determined with the help of relations (8.24), for example $J_1 + J_2 = ab$. We shall come back to this point in the next section.

8.5 A representation of the symmetric group π_N

8.5.1 The X_{ij} operators

The ternary relation (8.40) on which the commutation of transfer matrices is based can be presented in a very symmetric form in the three intervening matrices. It suffices to perform the change of notation

$$
\begin{aligned}
\tau &\to \sigma_1, & V &\to \zeta + V_1 - V_3, \\
\tau' &\to \sigma_2, & V' &\to \zeta + V_2 - V_3, \\
\sigma &\to \sigma_3, & w_{12} &= w(\zeta + V_1 - V_2), \text{ etc.}
\end{aligned}
$$

$$(8.62)$$

to obtain

$$
(w_{13}\sigma_1\sigma_3)(w_{23}\sigma_2\sigma_3)(w_{21}\sigma_1\sigma_2) = (w_{21}\sigma_1\sigma_2)(w_{23}\sigma_2\sigma_3)(w_{13}\sigma_1\sigma_3). \quad (8.63)
$$

With the help of the quartic relations, we verify that

$$
(w(V)\sigma_1\sigma_2) \cdot (w(-V)\sigma_1\sigma_2) = \frac{4}{\Theta_1^2(2\zeta)}\Theta_1(2\zeta + V)\Theta_1(2\zeta - V)
$$

(function Θ relative to modulus l, or to τ_l). We are led to normalizing our operators by defining

$$
X_{12} = \frac{1}{2}\frac{\Theta_1(2\zeta)}{\Theta_1(2\zeta + V_1 - V_2)}(w(V_1 - V_2)\sigma_1\sigma_2). \quad (8.64)
$$

If we give ourselves N commuting spin operators $\sigma_1, \sigma_2, \ldots, \sigma_N$ and N distinct arbitrary quantities $V_1, V_2, \ldots, V_N$, the X_{ij} verify the following relations:

$$X_{ij}X_{ji} = 1,$$
$$X_{ji}X_{jk}X_{ik} = X_{ik}X_{jk}X_{ji},$$
$$X_{ij}X_{kl} = X_{kl}X_{ij}, \quad i \neq j \neq k \neq l \tag{8.65}$$

which were formulated by Yang (1967) in the context of another model handled in Chapter 12.

8.5.2 'Inhomogeneous' transfer matrix

The vertex model on the square lattice can also be solved in certain inhomogeneous cases in which the weights are site-dependent. The so-called 'inhomogeneous' transfer matrix is defined as the partial trace

$$T(V_0) = \operatorname*{Tr}_{0} \; X_{10}X_{20}\ldots X_{N0}, \tag{8.66}$$

in which we have introduced a matrix $\vec{\sigma}_0$ and a supplementary parameter V_0 to complete the set of X.

Proposition: $T(V)$ is a commuting family,

$$[T(V_a), T(V_b)] = 0. \tag{8.67}$$

The proof is simple. With two matrices σ_a and σ_b independent of $\sigma_1, \ldots, \sigma_N$, we define

$$T(V_a) = \operatorname*{Tr}_{a} \; X_{1a}\ldots X_{Na}, \quad T(V_b) = \operatorname*{Tr}_{b} \; X_{1b}\ldots X_{Nb},$$

and use the ternary relation

$$X_{ab}X_{an}X_{bn} = X_{bn}X_{an}X_{ab}$$

as in Subsection 8.4.1.

8.5.3 Regular representation of π_N

We build using the X_{ij} a basis of generators of π_N, $Y_{12}, Y_{23}, \ldots, Y_{N-1N}$ isomorphic to the transpositions $P_{12}, P_{23}, \ldots, P_{N-1N}$ considered as operators in the regular representation of π_N. Note that we already have the representation

$$P_{12} = X_{12}(0) = \frac{1}{2}(1 + \vec{\sigma}_1 \cdot \vec{\sigma}_2) = (12).$$

In order to introduce without too much arbitrariness a basis for the space of representations, let us consider the matrix

$$\overset{\circ}{T}_P = \bar{P} X_{P1} X_{P2} \dots X_{PN} P \quad (X_j \equiv X_{j0}), \tag{8.68}$$

which is a function of a running perturbation P. We thus have up to a normalization

$$\overset{\circ}{T}_P \sim (w_{P1}\sigma_1\sigma_0) \dots (w_{PN}\sigma_N\sigma_0). \tag{8.69}$$

We denote by

$$|P\rangle \equiv |P_1 P_2 \dots P_N\rangle$$

an eigenvector of $\overset{\circ}{T}_P$ having eigenvalue $\overset{\circ}{t}$. Let us calculate $\overset{\circ}{T}_{P(34)}$ in which (34) is a neighbour transposition ((34) or $(n, n+1)$):

$$\overset{\circ}{T}_{P(34)} = P_{34}\bar{P}X_{P1}\dots X_{P4}X_{P3}\dots X_{PN}PP_{34} = P_{34}\bar{P}X_{P3,P4}P\,\overset{\circ}{T}_P\,\bar{P}X_{P4,P3}PP_{34}, \tag{8.70}$$

in which we used the ternary relation (8.65). Let us define the operators

$$Y_{34}^{ij} = X_{34}(V_i - V_j) \cdot P_{34} \tag{8.71}$$

in which the notation $X_{34}(V)$ is obvious. We thus have $\bar{P}X_{P3,P4}P = X_{34}(V_{P3} - V_{P4})$ and therefore the similarity

$$\overset{\circ}{T}_{P(34)} = Y_{34}^{P3P4}\,\overset{\circ}{T}_P\,Y_{34}^{P4P3}, \quad \forall P \tag{8.72}$$

with

$$Y_{34}^{ij} Y_{34}^{ji} = 1. \tag{8.73}$$

We deduce from (8.72) and the definition of $|P\rangle$ that an eigenvector $|P(34)\rangle$ belongs to operator $T_{P(34)}$ with eigenvalue t which is common to the two similar operators, in such a way that

$$|P(34)\rangle = Y_{34}^{P3P4}|P\rangle. \tag{8.74}$$

We see better the action of the transposition between neighbours of operator Y_{34} on basis $|P\rangle$ by writing

$$Y_{34}^{kl}|ijkl\dots\rangle = |ijlk\dots\rangle. \tag{8.75}$$

We easily verify, with the help of the ternary relations in X and from definition (8.71), the rule

$$Y_{12}^{jk} Y_{23}^{ik} Y_{12}^{ij} = Y_{23}^{ij} Y_{12}^{ik} Y_{23}^{jk}, \tag{8.76}$$

which is nothing but the relation

$$(Y_{12} \cdot Y_{23})^3 = 1, \tag{8.77}$$

in the sense defined by the linear transformation (8.75).

Relations (8.73) show that the $(N - 1)$ linear transformations $Y_{12}, Y_{23}, \ldots$ are *involutions* or *reflections*; that the product of two such reflections is a rotation of second or third order (8.77). According to Coxeter and Moser (1972) the group generated by these operators is isomorphic to the permutation group π_N.

8.5.4 Elliptic expression of the relative invariants J

In Section 8.3 we showed how the *XYZ* Hamiltonian can be obtained by derivation of the transfer matrix. Let us show here how the key formula of Fan and Sutherland (8.25) itself originates naturally from the ternary relation (8.40), and, with the help of the elliptic parametrization of operators X and Y, let us calculate the 'relative invariants' constituted by the J_α ($\alpha = 1, 2, 3$) coefficients.

Taking the product $X_{12}P_{12}$ in (8.71), we obtain with the help of the quartic relations

$$Y_{12}^{12} = \frac{\Theta_1(V_1 - V_2)}{2\Theta_1(2\zeta + V_1 - V_2)} \sum_{\alpha=0}^{3} \frac{\Theta_{\alpha+1}\left(2\zeta + \frac{V_1 - V_2}{2}\right)}{\Theta_{\alpha+1}\left(\frac{V_1 - V_2}{2}\right)} \sigma_1^\alpha \sigma_2^\alpha. \qquad (8.78)$$

The ternary relation can be written in the form

$$(w(V_1)\sigma_1\tau)(w(V_2)\sigma_2\tau)Y_{12}^{12}(V_1 - V_2) = Y_{12}^{12}(V_1 - V_2)(w(V_2)\sigma_1\tau)(w(V_1)\sigma_2\tau). \qquad (8.79)$$

By derivation with respect to V_1 evaluated in $V_1 = V_2$, we obtain

$$\left[R_1 R_2, \frac{dY_{12}^{12}}{dV}\right]_{V=0} = \left(R_1 \frac{dR_2}{dV} - \frac{dR_1}{dV} R_2\right)_{V=0}, \qquad (8.80)$$

in the notation of Subsection 8.4.1. The calculation of the derivative of Y_{12}^{12} gives us

$$\frac{dY_{12}^{12}}{dV}|_{V=0} = -\frac{1}{2}\frac{\Theta_1'(2\zeta)}{\Theta_1(2\zeta)} + \frac{1}{2}\frac{\Theta_1'(0)}{\Theta_1(2\zeta)} \sum_{\alpha=1}^{3} \frac{\Theta_{\alpha+1}(2\zeta)}{\Theta_{\alpha+1}(0)} \sigma_1^\alpha \sigma_2^\alpha. \qquad (8.81)$$

From this, (8.25) and (8.81), we deduce the following proposition, which defines the 'relative invariants' with respect to parameter V, in other words the coefficients J_α ($\alpha = 1, 2, 3$):

$$J_\alpha \doteq \frac{\Theta_{\alpha+1}(2\zeta | \tau_l)}{\Theta_{\alpha+1}(0 | \tau_l)}. \qquad (8.82)$$

After the Jacobi transformation (I.17) and switch to parameters η, $\tau/2$, (8.56),

$$J_1 : J_2 : J_3 = \frac{\theta_4\left(-\frac{i\pi\zeta}{K_l'}\big|\frac{K_l}{K_l'}\right)}{\theta_4\left(0\big|\frac{K_l}{K_l'}\right)} : \frac{\theta_3}{\theta_3} : \frac{\theta_2}{\theta_2}$$

$$= \frac{\Theta_4\left(2\eta|\frac{\tau}{2}\right)}{\Theta_4\left(0|\frac{\tau}{2}\right)} : \frac{\Theta_3\left(2\eta|\frac{\tau}{2}\right)}{\Theta_3\left(0|\frac{\tau}{2}\right)} : \frac{\Theta_2\left(2\eta|\frac{\tau}{2}\right)}{\Theta_2\left(0|\frac{\tau}{2}\right)}, \qquad (8.83)$$

and after the Landen transformation (I.16), $\dfrac{\tau}{2} \to \tau$,

$$J_1 : J_2 : J_3 = (\Theta_4^2 + \Theta_1^2)(2\eta) : (\Theta_4^2 - \Theta_1^2)(2\eta) : \frac{\Theta_4^2(0)}{\Theta_2\Theta_3(0)}(\Theta_2\Theta_3)(2\eta)$$

$$= 1 + k\,\mathrm{sn}^2(2\eta; k) : 1 - k\,\mathrm{sn}^2(2\eta; k) : \mathrm{cn}\,(2\eta; k)\,\mathrm{dn}\,(2\eta; k). \quad (8.84)$$

Let us finally give the expression of the *XYZ* Hamiltonian as a derivative of the transfer matrix at the point $V = \zeta$. With the help of (8.27) and (8.54), we obtain

$$H = -\frac{N}{2}J_0 + 2\frac{\mathrm{sn}\,\zeta}{\mathrm{cn}\,\zeta\,\mathrm{dn}\,\zeta}\frac{d}{dV}\ln T|_{V=\zeta} \qquad (8.85)$$

in which T is built with the $w_\alpha = \Theta_{\alpha+1}(V|\tau_l)/\Theta_{\alpha+1}(\zeta|\tau_l)$.

8.6 Diagonalization of the transfer matrix

8.6.1 A remarkable identity

We have noticed, during the course of our study of models solved by Bethe's method, that the coupled equations allowing the calculation of the momenta $k(\varphi_j)$, for example in the Heisenberg chain, have the following structure:

$$g(\varphi_j + i\Phi)Q(\varphi_j - 2i\Phi) + g(\varphi_j - i\Phi)Q(\varphi_j + 2i\Phi) = 0$$

with

$$Q(\varphi_j) = 0, \quad j = 1, 2, \dots, M \qquad (8.86)$$

in which g and Q are trigonometric polynomials (cf. (1.40)). The first is given, and the second is unknown with M zeroes per period. The origin of system (8.86) is completely clear, if we remember that in the six-vertex model, the following relation exists between the eigenvalue of the suitably normalized transfer matrix $T(\varphi)$ and the Q polynomial (formula (7.85)):

$$T(\varphi)Q(\varphi) = g(\varphi - i\Phi)Q(\varphi + 2i\Phi) + g(\varphi + i\Phi)Q(\varphi - 2i\Phi). \qquad (8.87)$$

Such a relation persists in the inhomogeneous ferroelectric models in which Lieb's solution was generalized by Baxter (1971a). From this comes the idea of reproducing this type of relation for the eight-vertex model by building a new family of matrices $Q(\varphi)$ commuting with the first and verifying a relation analogous to (8.87).

To establish identity (8.87), we shall essentially follow Baxter's method, albeit with some minor formal modifications. In particular, we shall try to postpone as for as possible the unavoidable elliptic parametrization, in order to better appreciate the algebraic structure.

Considering the transfer matrix $T(w)$ as a homogeneous polynomial of $w(a, b, c, d)$, we look for a matrix $Q(w)$, a numerical function $\Phi(w)$ and an invertible translation operation, denoted t, $t : w \rightarrow tw = w'$, whose inverse is $\bar{t}$, $\bar{t} : w = \bar{t}w'$, such that we have

$$T(w)Q(w) = \Phi^N(w)Q(tw) + \Phi^N(tw)Q(\bar{t}w), \qquad (8.88)$$

$$[Q(w), Q(tw)] = 0. \qquad (8.89)$$

This last condition is destined to ensure the simultaneous diagonalization of Q and T.

8.6.2 Mechanism of the fundamental identity

As a first step, we construct a matrix $S^R(w)$, defined modulo right-multiplication by a constant matrix, in the form of a trace of a product, such that the structure of Q^R is analogous to that of T:

$$\langle \alpha_1 \ldots | Q^R | \beta_1 \ldots \rangle = \operatorname*{Tr}_0 \, (\alpha_1 | S | \beta_1)(\alpha_2 | S | \beta_2) \ldots (\alpha_N | S | \beta_N) \qquad (8.90)$$

in which each $(\alpha | S | \beta)$ is a matrix with elements

$$(\alpha | S_{nm} | \beta) \neq 0 \quad \text{iff} \quad |n - m| = 1. \qquad (8.91)$$

Baxter chooses matrices of finite dimension L. We shall make a different choice here by taking infinite-dimensional matrices, but modifying the definition of the trace such that our trace (8.90) means $\operatorname{Tr} \rho_0 \ldots$ in which ρ_0 is a projector onto a given diagonal element n_0, which is anyway arbitrary. In other words, the partial trace is an 'expectation value in state n_0'. Owing to the property postulated in (8.91), the trace operation is a finite sum and despite the fact that we are dealing with infinite matrices, no convergence problem occurs.

The product $T Q^R$ can be represented as a total trace of a product of U ($2L \times 2L$ in Baxter's work) matrices

$$\langle \alpha | T Q^R | \beta \rangle = \operatorname{Tr} \, \rho_0 (\alpha_1 | U | \beta_1) \ldots (\alpha_N | U | \beta_N) \qquad (8.92)$$

in which each $(\alpha|U|\beta)$ is a matrix $n \to m$, $\lambda \to \nu$:

$$(\alpha|U_{nm}|\beta)_{\lambda\nu} = \sum_{\gamma}(\alpha\lambda|R|\gamma\nu)(\gamma|S_{nm}|\beta) \tag{8.93}$$

or

$$(\alpha|U|\beta) = \sum_{\gamma}(\alpha|R|\gamma)(\gamma|S|\beta). \tag{8.94}$$

Each element $(\alpha|R|\gamma)$ is a 2×2 matrix, which according to the definition of Subsection 8.2.1 is written

$$(+|R|+) = \begin{pmatrix} a & 0 \\ 0 & b \end{pmatrix}, \qquad (-|R|+) = \begin{pmatrix} 0 & c \\ d & 0 \end{pmatrix},$$

$$(+|R|-) = \begin{pmatrix} 0 & d \\ c & 0 \end{pmatrix}, \qquad (-|R|-) = \begin{pmatrix} b & 0 \\ 0 & a \end{pmatrix}. \tag{8.95}$$

Baxter's idea consists of decomposing the $2L$-dimensional trace (8.92) into a sum of two L-dimensional traces in order to obtain two objects having the same structure as Q^R, in the analogy (8.87). We thus look for a transformation of U which reduces it to a triangular form $\begin{pmatrix} L & 0 \\ L & L \end{pmatrix}$ such that the product of various U entering into TQ^R is also triangular. Considering $(\alpha|U|\beta)$ as an $L \times L$ (here, infinite) matrix with quaternionic elements, we look for the transformation in the form

$$M_n^{-1}U_{nm}M_m \equiv \begin{pmatrix} A_{nm} & 0 \\ C_{nm} & B_{nm} \end{pmatrix}, \tag{8.96}$$

in which M_α is an invertible quaternionic matrix independent of the spin indices α_j, β_j appearing in U. Owing to the similarity transformation invariance of the trace and since M commutes with ρ_0, we shall have

$$(\alpha|TQ^R|\beta) = \mathrm{Tr}\ \rho_0(\alpha_1|A|\beta_1)\ldots(\alpha_N|A|\beta_N) + \mathrm{Tr}\ \rho_0(\alpha_1|B|\beta_1)\ldots(\alpha_N|B|\beta_N). \tag{8.97}$$

Let us calculate $(\alpha|U_{nm}|)$ according to (8.94):

$$(+|U_{nm}|) = \begin{pmatrix} a(+|S_{nm}) & d(-|S_{nm}|) \\ c(-|S_{nm}) & b(+|S_{nm}|) \end{pmatrix},$$

$$(-|U_{nm}|) = \begin{pmatrix} b(-|S_{nm}) & c(+|S_{nm}|) \\ d(+|S_{nm}|) & a(-|S_{nm}|) \end{pmatrix}. \tag{8.98}$$

To realize (8.96), it is sufficient to take M_n triangular:

$$M_n = \begin{pmatrix} 1 & p_n \\ 0 & 1 \end{pmatrix} \tag{8.99}$$

and perform the product of three matrices. From this, we deduce the two conditions

$$(ap_m - bp_n)(+|S_{nm}|) + (d - cp_n p_m)(-|S_{nm}|) = 0,$$
$$(c - dp_n p_m)(+|S_{nm}|) + (bp_m - ap_n)(-|S_{nm}|) = 0 \tag{8.100}$$

and the expressions for A and B:

$$(+|A_{nm}|) = a(+|S_{nm}|) - cp_n(-|S_{nm}|),$$
$$(+|B_{nm}|) = b(+|S_{nm}|) + cp_m(-|S_{nm}|),$$
$$(-|A_{nm}|) = b(-|S_{nm}|) - dp_n(+|S_{nm}|),$$
$$(-|B_{nm}|) = a(-|S_{nm}|) + dp_m(+|S_{nm}|). \tag{8.101}$$

From the homogeneous system (8.100), we deduce the following 'selection rule' such that the matrix elements $(|S_{nm}|)$ are non-vanishing:

$$(c - dp_n p_m)(d - cp_n p_m) = (bp_m - ap_n)(ap_m - bp_n)$$

or further still

$$(c^2 + d^2 - a^2 - b^2)p_n p_m + ab(p_n^2 + p_m^2) - cd(1 + p_n^2 p_m^2) = 0. \tag{8.102}$$

We notice that this relation depends on the three parameters $J_1 : J_2 : J_3$ of the *XYZ* Hamiltonian (formula (8.24)). We can rewrite it in the form

$$4J_3 = J_2(p_n + p_n^{-1})(p_m + p_m^{-1}) - J_1(p_n - p_n^{-1})(p_m - p_m^{-1}) \tag{8.103}$$

if

$$(|S_{nm}|) \neq 0.$$

We then solve (8.100) by introducing an unknown $S_{nm}(w)$:

$$(+|S_{nm}|) = S_{nm}(w)(d - cp_n p_m),$$
$$(-|S_{nm}|) = S_{nm}(w)(bp_n - ap_m), \tag{8.104}$$

with which the matrices A and B can be expressed as

$$(+|A_{nm}|) = S_{nm}(w)(ad - cbp_n^2),$$
$$(-|A_{nm}|) = S_{nm}(w)((b^2 - d^2)p_n + (cdp_n^2 - ab)p_m),$$

$$(+|B_{nm}|) = S_{nm}(w)(bd - acp_m^2),$$
$$(-|B_{nm}|) = S_{nm}(w)((d^2 - a^2)p_m + (ab - cdp_m^2)p_n). \tag{8.105}$$

8.6.3 Construction of S

Relation (8.102) is quadratic separately in p_n and p_m, and symmetrical in both these arguments. These two properties are at the origin of the selection rule $|n - m| = 1$. Let us, for example, consider the line n_0. To a given p_{n_0} correspond two possible solutions of the quadratic equation in p_m and therefore two possible matrix elements which we shall associate with columns $(n_0 + 1)$ and $(n_0 - 1)$, thus defining $S_{n_0, n_0 \pm 1}$. Let us then go to the line $(n_0 + 1)$ with associated parameter $p_{(n_0+1)}$. With it correspond two solutions, one of which is necessarily p_{n_0} due to the symmetry; the other shall be called $p_{(n_0+2)}$ and so on. The chain thus constructed

$$\cdots \ p_{n_0-1} \ \mathrel{\reflectbox{$\curvearrowright$}}\mkern-16mu\curvearrowleft \ p_{n_0} \ \mathrel{\reflectbox{$\curvearrowright$}}\mkern-16mu\curvearrowleft \ p_{n_0+1} \ \mathrel{\reflectbox{$\curvearrowright$}}\mkern-16mu\curvearrowleft \ p_{n+2} \ \cdots$$

can be infinite or cyclic. It will be cyclic for exceptional values of the parameters, these however forming a dense set. This is the case in which Baxter places himself. We shall choose the infinite case, which can include the cyclic case as periodic element. The assumed selection rule $|n - m| = 1$ is thus proved.

It is now necessary to try to identify the trace over A and the trace over B in (8.97) with each of the two terms on the right-hand side of (8.88). It is sufficient to have

$$(\alpha|A_{nm}|\beta) = \Phi(w)x_n^{-1}x_m(\alpha|S_{nm}|\beta)_{tw},$$
$$(\alpha|B_{nm}|\beta) = \Phi(tw)x_n x_m^{-1}(\alpha|S_{nm}|\beta)_{\bar{t}w} \tag{8.106}$$

in which x_n is undetermined. We shall note $tw = w'$, $\bar{t}w = w''$, $|n - m| = 1$.

Substituting expressions (8.105) in the left-hand side of (8.106), taking (8.104) into account, leads us to the four equations

$$\frac{x_m}{x_n}\Phi(w)\frac{S_{nm}(tw)}{S_{nm}(w)} = \frac{ad - cbp_n^2}{d' - c'p_n p_m} = \frac{(b^2 - d^2)p_n + (cdp_n^2 - ab)p_m}{b'p_n - a'p_m},$$
$$\frac{x_n}{x_m}\Phi(tw)\frac{S_{nm}(\bar{t}w)}{S_{nm}(w)} = \frac{bd - acp_m^2}{d'' - c''p_n p_m} = \frac{(d^2 - a^2)p_m + (ab - cdp_m^2)p_n}{b''p_n - a''p_m}. \tag{8.107}$$

We have made the important hypothesis that the coefficients in relation (8.102) or (8.103) defining p_n are *relative invariants* of the transformation $w \to tw = w'$, such that the set p_n is unaltered by the translations t. We must thus have

$$\frac{a'b'}{ab} = \frac{c'd'}{cd} = \frac{c'^2 + d'^2 - a'^2 - b'^2}{c^2 + d^2 - a^2 - b^2}, \quad \text{etc.} \tag{8.108}$$

This can be verified by determining transformation $w \to w'$ with the help of the second equations (8.107). It is sufficient to identify the equation

$$(ad - cbp_n^2)(b'p_n - a'p_m) - (d' - c'p_n p_m)((b^2 - d^2)p_n + (cdp_n^2 - ab)p_m) = 0 \tag{8.109}$$

with the quartic relation (8.102). We find that

$$(8.109) \equiv cc'dp_n \times (8.102)$$

under the condition of determining t by the ratios

$$t : w \rightarrow w' \quad a' : b' : c' : d' = \frac{b}{b^2 - d^2} : \frac{a}{a^2 - c^2} : \frac{c}{a^2 - c^2} : \frac{d}{b^2 - d^2} \quad (8.110)$$

and similarly, for the inverse

$$\bar{t} : w \rightarrow w'' \quad a'' : b'' : c'' : d'' = \frac{b}{c^2 - b^2} : \frac{a}{d^2 - a^2} : \frac{c}{c^2 - b^2} : \frac{d}{d^2 - a^2}. \quad (8.111)$$

Finally, there remain the first equations (8.107) to determine $\Phi(w)$ and $S_{n,m}(w)$. After the translation $w \rightarrow tw$ in the second line of equations (8.107) and term-by-term multiplication with the first line, we eliminate $S(w)$ to obtain the functional relation

$$\Phi(w)\Phi(t^2 w) = \frac{(ad - cbp_n^2)(d'b' - a'c'p_m^2)}{(d' - c'p_n p_m)(d - cp_n p_m)}. \quad (8.112)$$

Taking (8.110) and (8.102) into account, we notice the fortunate elimination of the p_n to obtain

$$\Phi(w)\Phi(t^2 w) = \frac{c'}{c}(a^2 - c^2). \quad (8.113)$$

We notice that if we define the transformation t no longer by the ratio but rather by the strict equality $c' = \dfrac{c}{a^2 - c^2}$, etc., we can choose

$$\Phi(w) \equiv 1. \quad (8.114)$$

The fundamental relation

$$T(w)Q^R(w) = Q^R(tw) + Q^R(\bar{t}w) \quad (8.115)$$

is thus established in a purely algebraic way. We shall not push this method any further and shall introduce already in the next subsection the elliptic parametrization in order to calculate explicitly the set p_n, the matrix S and Φ.

8.7 The coupled equations for the spectrum

8.7.1 Elliptic uniformization

Since the relative invariants J_1, J_2, J_3 determine the two *elliptic* parameters η and k (formula (8.84)), the translation t (8.110) can only operate on the third parameter

v. If we adopt the convention of normalizing $w(a, b, c, d)$ by replacing ratio (8.61) by the equality

$$a = \frac{\Theta_1(v + \eta)\Theta_4(v - \eta)}{\Theta_1(2\eta)}, \quad \text{etc.} \tag{8.116}$$

using one of the relations (8.110), we immediately find that t is a simple translation of v:

$$t : v \to v + 2\eta.$$

With the help of (8.113), we obtain

$$\Phi(w)\Phi(t^2 w) = \frac{\Theta_1(v - \eta)\Theta_4(v - \eta)\Theta_1(v + 3\eta)\Theta_4(v + 3\eta)}{\Theta_1^2(2\eta)\Theta_4^2(2\eta)},$$

from which we get the obvious solution

$$\Phi(w) = \frac{\Theta_1(v - \eta)\Theta_4(v - \eta)}{\Theta_1(2\eta)\Theta_4(2\eta)} = \frac{\Theta_1\left(v - \eta|\frac{\tau}{2}\right)}{\Theta_1\left(2\eta|\frac{\tau}{2}\right)}. \tag{8.117}$$

• *The set p_n.*

We look for the solution in the form

$$p_n = \frac{\Theta_1(v_n)}{\Theta_4(v_n)} = \sqrt{k} \, \text{sn} \, (v_n; k) \tag{8.118}$$

and calculate with the help of formulas (I.16)

$$\frac{1}{2}(p_n + p_n^{-1}) = \frac{\Theta_4\left(v_n|\frac{\tau}{2}\right)\Theta_3\left(0|\frac{\tau}{2}\right)}{\Theta_1\left(v_n|\frac{\tau}{2}\right)\Theta_2\left(0|\frac{\tau}{2}\right)},$$

$$\frac{1}{2}(p_n - p_n^{-1}) = -\frac{\Theta_3\left(v_n|\frac{\tau}{2}\right)\Theta_4\left(0|\frac{\tau}{2}\right)}{\Theta_1\left(v_n|\frac{\tau}{2}\right)\Theta_2\left(0|\frac{\tau}{2}\right)}. \tag{8.119}$$

Using expression (8.83) for the relative invariants and substituting in (8.103), we obtain

$$\Theta_1(v_n)\Theta_1(v_m)\Theta_2(0)\Theta_2(2\eta) = \Theta_4(v_n)\Theta_4(v_m)\Theta_3(0)\Theta_3(2\eta)$$
$$- \Theta_3(v_n)\Theta_3(v_m)\Theta_4(0)\Theta_4(2\eta), \quad (8.120)$$

in which the functions Θ are relative to the ratio $\frac{\tau}{2}$. Relation (8.120) is but an application of the quartic relation (I.13), provided we have the rule

$$v_n - v_m = \pm 2\eta, \quad \text{if} \quad n - m = \pm 1. \tag{8.121}$$

From this, we deduce that

$$v_n = v_0 + 2n\eta, \tag{8.122}$$

and from this p_n. Here v_0 is arbitrary; we can take it to be zero.

- $(|S_{nm}|)$.

The last relation to satisfy is the first (8.107):

$$\frac{x_m}{x_n}\Phi(w)\frac{S_{nm}(tw)}{S_{nm}(w)} = \frac{ad - bcp_n^2}{d' - c'p_np_m},$$

in other words, taking (8.110)–(8.118) into account:

$$x_n^{-1}x_m\frac{S_{nm}(v + 2\eta)}{S_{nm}(v)} = \frac{\Theta_4(v_m)}{\Theta_4(v_n)}\frac{\Theta_4(2\eta)}{\Theta_4(0)}$$

$$\times\frac{\Theta_1^2(v + \eta)\Theta_4^2(v_n) - \Theta_4^2(v + \eta)\Theta_1^2(v_n)}{\Theta_1(v + 3\eta)\Theta_1(v + \eta)\Theta_4(v_n)\Theta_4(v_m) - \Theta_4(v + 3\eta)\Theta_4(v + \eta)\Theta_1(v_n)\Theta_1(v_m)}. \quad (8.123)$$

We shall take $x_n = \Theta_4(v_n)$.

Let us treat the case $m = n + 1$. With the help of the quartic relation (I.13):

$$(x = v + 2\eta, \, y = \eta, \, z = v_n + \eta, \, t = \eta + iK'),$$

we find

$$\frac{S_{n,n+1}(v + 2\eta)}{S_{n,n+1}(v)} = \frac{\Theta_4(2\eta)}{\Theta_4(0)}\frac{\Theta_4^2(0)\Theta_1(v + v_n + \eta)\Theta_1(v - v_n + \eta)}{\Theta_4(0)\Theta_4(2\eta)\Theta_1(v + v_n + 3\eta)\Theta_1(v - v_n + \eta)}$$

$$\equiv \frac{\Theta_1(v + v_n + \eta)}{\Theta_1(v + v_n + 3\eta)}. \quad (8.124)$$

This leads to the obvious solution

$$S_{n,n+1}(v) \doteq 1/\Theta_1(v + v_n + \eta); \quad (8.125)$$

we would similarly find

$$S_{n+1,n}(v) \doteq 1/\Theta_1(v - v_n - \eta). \quad (8.126)$$

The elements $(\alpha|S_m|)$ are given by (8.104). We finally find the solution

$$(\pm|S_{n,n+1}|) \doteq \Theta_{1,4}(v - v_n - \eta),$$
$$(\pm|S_{n+1,n}|) \doteq \Theta_{1,4}(v + v_n + \eta), \quad (8.127)$$

which thus completes the determination of a matrix $Q^R(v)$ satisfying relation (8.88):

$$T(v)Q^R(v) = \left(\frac{\Theta_1\left(v - \eta|\frac{\tau}{2}\right)}{\Theta_1\left(v|\frac{\tau}{2}\right)}\right)^N Q^R(v + 2\eta) + \left(\frac{\Theta_1\left(v + \eta|\frac{\tau}{2}\right)}{\Theta_1\left(v|\frac{\tau}{2}\right)}\right)^N Q^R(v - 2\eta).$$

$$(8.128)$$

8.7.2 Construction of Q

We have built a matrix Q^R defined modulo right-multiplication. We can equivalently build a matrix Q^L modulo left-multiplication verifying $Q^L T = Q^L + Q^L$. It is sufficient to perform a transposition. Indeed, according to (8.16) and (8.14), transposing $T(v)$ is equivalent to exchanging c and d. But the involution $v \to 2K + iK' - v$ operates the transformation

$$\Theta_{1,4}(2K + iK' - u) = -iq^{1/4}e^{-(i\pi u/2K)}\Theta_{4,1}(u),$$

and thus

$$T(2K + iK' - v) = (-1)^N q^{N/2} e^{-\frac{\pi i v N}{K}}\, {}^tT(v). \tag{8.129}$$

We could define $Q^L(v)$ in the following manner:

$$Q^L(v) = i^N q^{-N/4} e^{\frac{\pi i v N}{2K}}\, {}^tQ^R(2K + iK' - v). \tag{8.130}$$

According to (8.127) we can write matrix S in a unified manner, defining either Q^R or Q^L, by taking $v_0 = 0$:

$$Q^R : (\alpha|S_{nm}^R|) = \left(\frac{\Theta_1}{\Theta_4}\right)(v + (n^2 - m^2)\eta),$$

$$Q^L : (|S_{nm}^L|\alpha) = \left(\frac{\Theta_4}{\Theta_1}\right)(v + (m^2 - n^2)\eta), \tag{8.131}$$

with $n - m = \pm 1$.

Proposition:

$$Q^L(v)Q^R(v') \equiv Q^L(v')Q^R(v). \tag{8.132}$$

Proof: Given the trace structure (Tr $\ldots$) of Q^L and Q^R, it suffices to show the existence of diagonal transformation coefficients $A_{n,n'}$ such that

$$\sum_\alpha A_{n,n'}(|S_{nm}^L(v)|\alpha)(\alpha|S_{n'm'}^R(v')|) = \sum_\alpha (|S_{nm}^L(v')|\alpha)(\alpha|S_{n'm'}^R(v)|)A_{m,m'}. \tag{8.133}$$

The summation over α on the left-hand side gives

$$\Theta_1(v + (n^2 - m^2)\eta)\Theta_4(v' + (m'^2 - n'^2)\eta) + \Theta_4(\ldots)\Theta_1(\ldots)$$

$$= \Theta_2\left(\frac{v + v'}{2} + \frac{m'^2 - n'^2 + n^2 - m^2}{2}\eta\right)\Theta_2\left(\frac{v - v'}{2} + \frac{n^2 + n'^2 - m^2 - m'^2}{2}\eta\right).$$

It is therefore sufficient to realize

$$\frac{A_{n,n'}}{A_{m,m'}} = \frac{\Theta_2\left(\frac{v-v'}{2} + \frac{n^2+n'^2-m^2-m'^2}{2}\eta\right)}{\Theta_2\left(\frac{v'-v}{2} + \frac{n^2+n'^2-m^2-m'^2}{2}\eta\right)}. \tag{8.134}$$

for $|n-m| = |n'-m'| = 1$. Specializing conditions (8.134) into four possible cases, we obtain only two independent recursions on the sum $n + n'$ and the difference $n - n'$:

$$\frac{A_{n+1,n'+1}}{A_{n,n'}} = f_{n+n'}, \qquad \frac{A_{n+1,n'-1}}{A_{n,n'}} = f_{n-n'}, \qquad (8.135)$$

with

$$f_n = \frac{\Theta_2\left(\frac{v-v'}{2} + (n+1)\eta\right)}{\Theta_2\left(\frac{v'-v}{2} + (n+1)\eta\right)}. \qquad (8.136)$$

The $A_{n,n'}$ are thus uniquely determined and relation (8.132) is proven.

Up to now, we have not needed to specify the matrices Q^R and Q^L more precisely. This must however be done to define their inverses so as to obtain a matrix $Q(v)$ commuting with $T(v)$, via the expression

$$Q(v) = Q^R(v)\left\{Q^R(v_0)\right\}^{-1} = \left\{Q^L(v_0)\right\}^{-1} Q^L(v) \qquad (8.137)$$

in which v_0 is a particular value of v such that the matrix is invertible. We deduce from (8.137) and (8.132) the commutation relation

$$\left[Q(v), Q(v')\right] = 0.$$

To specify Q^R, we can start from a matrix $S(v)$ depending on a parameter v_0 and having the properties

$$S^R(v, v_0) = S^L(v_0, v). \qquad (8.138)$$

It is then sufficient to choose, according to (8.131),

$$(\alpha|S^R_{n,m}(v, v_0)|\beta) = \Theta_{1,4}(v + (n^2 - m^2)\eta)\Theta_{4,1}(v_0 + (m^2 - n^2)\eta). \qquad (8.139)$$

For the particular value $v = v_0$, we thus have

$$Q^R(v = v_0) = Q^L(v = v_0). \qquad (8.140)$$

We shall make the hypothesis that there exists at least one value v_0 for which this matrix is invertible. The construction of matrix $Q(v)$ is then completed and the simultaneous diagonalization of the matrices appearing in identity (8.88) is possible.

8.7.3 Equations for the spectrum of the transfer matrix

The commutativity properties demonstrated in the preceding sections lead to the fact that the matrices $T(v)$, $Q(v)$, $Q(v \pm 2\eta)$ can be simultaneously diagonalized in a basis independent of v. The eigenvalues, which we will also write $T(v)$, $Q(v)$, etc., are entire functions of variable v, since so are the matrix elements and the basis of eigenvectors does not depend on v. The 'periodicity' properties of $Q(v)$ will complete the determination of the form of this function. According to (8.131), we obtain

$$Q(v + 2K) = (-1)^{v'} Q(v) \tag{8.141}$$

in which v' is a conserved quantum number, which modulo 2 equals the number of arrows $\downarrow$ or spins -1 in the set $\{\alpha\}$ (or $\{\beta\}$) defining the matrix element $\langle \alpha | Q | \beta \rangle$. In contrast, the translation $v \to v + iK'$ on elements (8.131) gives us the property

$$\langle \alpha | Q(v + iK') | \beta \rangle = q^{N/4} e^{-\frac{i\pi(v+K)N}{2K}} \langle \bar{\alpha} | Q(v) | \beta \rangle \tag{8.142}$$

in which

$$\bar{\alpha}_j = -\alpha_j, \quad \forall j.$$

If we thus introduce the quantum number v'' defining the parity of an eigenvector under the operation $\alpha \to \bar{\alpha}$, we will have between the associated eigenvalues

$$Q(v + iK') = (-1)^{v''} q^{N/4} e^{-\frac{i\pi(v+K)N}{2K}} Q(v). \tag{8.143}$$

We deduce from (8.141) and (8.143) that the holomorphic function $Q(v)$ possesses exactly $\dfrac{N}{2}$ zeroes in the $2K, iK'$ parallelogram of the plane of v, and that there exists, according to the fundamental theorem on elliptic functions (Appendix I), a factorization of the form

$$Q(v) = e^{-(iv'\pi v/2K)} \prod_{j=1}^{N/2} \Theta_1 \Theta_4 (v - v_j), \tag{8.144}$$

in which the set v_j is that of the zeroes in the rectangle $2K, iK'$ obeying the condition originating from (8.143):

$$i \sum_{j=1}^{N/2} v_j = v'' K + iv' \frac{K}{2} \quad (\text{mod } 2K, iK'). \tag{8.145}$$

These zeroes are determined by writing $Q(v_j) = 0$ in relation (8.128), which gives the $N/2$ coupled equations

$$\left(\frac{\Theta_1 \left(v_j - \eta | \frac{\tau}{2} \right)}{\Theta_1 \left(v_j + \eta | \frac{\tau}{2} \right)} \right)^N = e^{-(iv'\pi\eta/K)} \prod_{i \, (\neq j)} \frac{\Theta_1 \left(v_j - v_i - 2\eta | \frac{\tau}{2} \right)}{\Theta_1 \left(v_j - v_i + 2\eta | \frac{\tau}{2} \right)}. \tag{8.146}$$

We have used one of the formulas (I.16):

$$\Theta_1 \left(v | \frac{\tau}{2} \right) \sim \Theta_1(v|\tau) \Theta_4(v|\tau).$$

We will address in the next chapter, Section 9.3, some consequences of system (8.146) for the thermodynamics of the eight-vertex model.

Appendix H

This appendix is concerned with an attempt to extend the ternary relations starting from a Clifford algebra.

If $\gamma_{(\alpha)}$ designates a complete basis of matrix algebra of arbitrary dimension, we define the tensor products

$$\begin{cases} \sigma_{(\alpha)} = 1 \otimes 1 \otimes \gamma_{(\alpha)} \\ \tau_{(\alpha)} = 1 \otimes \gamma_{(\alpha)} \otimes 1 \\ \tau'_{(\alpha)} = \gamma_{(\alpha)} \otimes 1 \otimes 1 \end{cases} \tag{H.1}$$

and the notation

$$(w\sigma\tau) = \sum_{(\alpha,\beta)} w_{(\alpha)(\beta)}\sigma_{(\alpha)}\tau_{(\beta)}. \tag{H.2}$$

The general problem of ternary relations

$$(w\sigma\tau)(w'\sigma\tau')(x\tau\tau') = (x\tau\tau')(w'\sigma\tau')(w\sigma\tau) \tag{H.3}$$

defining an algebraic variety in w, w' and x is probably difficult. Let us pose a restricted problem, by choosing as basis a Clifford algebra Γ_n of order 2^n, and by making the diagonal hypothesis

$$(w\sigma\tau) = \sum_{(\alpha)} w_{(\alpha)}\sigma_{(\alpha)}\tau_{(\alpha)}. \tag{H.4}$$

This is a direct generalization of the microtransfer matrix of Subsection 8.4.2.

We construct the Γ_n algebra by introducing the n generators γ_j, $j = [1, n]$, verifying

$$[\gamma_j, \gamma_l]_+ = 2\delta_{jl}. \tag{H.5}$$

We set

$$\gamma_{(\alpha)} = \gamma_1^{\alpha_1}\gamma_2^{\alpha_2}\ldots\gamma_n^{\alpha_n}, \qquad \alpha_j = 1, 2; \quad \mod 2$$

$$(\alpha) = \{\alpha_1\alpha_2\ldots\alpha_n\}. \tag{H.6}$$

We have the multiplication rule

$$\gamma_{(\alpha)}\gamma_{(\beta)} = \varepsilon(\alpha, \beta)\gamma_{(\alpha+\beta)}, \tag{H.7}$$

with

$$\varepsilon(\alpha, \beta) = (-1)^{\sum_{j>k}\alpha_j\beta_k}. \tag{H.8}$$

We define the scalar product

$$(\alpha, \beta) = \alpha_0 \beta_0 + \sum_j \alpha_j \beta_j, \qquad \alpha_0 + \sum_j \alpha_j = 0, \tag{H.9}$$

to define the sign

$$\varepsilon(\alpha, \beta)\varepsilon(\beta, \alpha) = (-1)^{(\alpha,\beta)}, \tag{H.10}$$

which occurs in the commutator between two elements of Γ_n:

$$\gamma_{(\alpha)}\gamma_{(\beta)} = (-1)^{(\alpha,\beta)}\gamma_{(\beta)}\gamma_{(\alpha)}. \tag{H.11}$$

From this comes the rule

$$(\alpha, \beta) = 1 \ (\text{mod } 2) \Leftrightarrow \gamma_{(\alpha)} \text{ and } \gamma_{(\beta)} \text{ anticommute} \Leftrightarrow \sum_j \alpha_j \beta_j = 1 \ (\text{mod } 2)$$

except if α_0 and β_0 are both odd, in which case we have

$$\sum_j \alpha_j \beta_j = 0 \ (\text{mod } 2). \tag{H.12}$$

The direct application of rules (H.7) and (H.11) to the expansion of the ternary relation gives us the algebraic relations

$$\sum_{\substack{(\alpha)+(\beta)=(c) \\ (\beta)+(\gamma)=(a) \\ (\gamma)+(\alpha)=(b)}} w_{(\alpha)} w'_{(\beta)} x_{(\gamma)} \varepsilon(\alpha, \beta)\varepsilon(\alpha, \gamma)\varepsilon(\beta, \gamma) = 0 \tag{H.13}$$

in which the summation is constrained by two conditions:

(1) (a), (b), (c) are given, $(a + b + c) = 0 \ (\text{mod } 2)$,

(2) $(\alpha, \beta) + (\alpha, \gamma) + (\beta, \gamma) = 1 \ (\text{mod } 2)$. \hfill (H.14)

The sum thus runs over a single set, for example (γ):

$$(\alpha) = (b + \gamma), \qquad (\beta) = (a + \gamma). \tag{H.15}$$

Condition (H.14) is rewritten as

$$(b + \gamma, a + \gamma) + (b + \gamma, \gamma) + (a + \gamma, \gamma) = 1 \ (\text{mod } 2).$$

Further,

$$(\gamma, \gamma) = 0, \qquad \forall \gamma \Rightarrow (b, a) = 1 \ (\text{mod } 2), \tag{H.16}$$

which is a restriction on the pairs a, b.

With the help of definition (H.8), we can show that the sign

$$\varepsilon(\alpha, \beta)\varepsilon(\alpha, \gamma)\varepsilon(\beta, \gamma)$$

changes as $\varepsilon(\gamma + a, \gamma + a)$. If we perform a slight change of definition

$$\tilde{w}_{(\beta)} = \varepsilon(\beta, \beta) w'_{(\beta)}, \tag{H.17}$$

the ternary relations are rewritten as

$$\sum_{(\gamma)} x_{(\gamma)} w_{(b+\gamma)} \tilde{w}_{(a+\gamma)} = 0, \tag{H.18}$$

each time we have

$$(a, b) = 1 \pmod 2.$$

The sign of each term of the sum is positive after redefinition (H.17). Condition (H.1) implies

$$(a) \neq (b) \neq 0.$$

System (H.18) remains overdetermined with respect to the available variables, and we do not know if there exist more extensive solutions than those already encountered. Let us suppose that the sum (H.18) only contains four nonzero terms of indices $(A), (B), (C), (D)$. These are necessarily of the form

$$(\gamma), \quad (\gamma + a), \quad (\gamma + a + b), \quad (\gamma + b),$$

all distinct, with the only relation among them being

$$(A + B + C + D) = 0 \pmod 2.$$

We thus have up to a permutation of the four indices

$$\begin{cases} (A, B) = (B, C) = (A, C) = 1 \\ (A, D) = (B, D) = (C, D) = 0 \pmod 2. \end{cases} \tag{H.19}$$

According to remark (H.11), this leads to the fact that the four matrices of Γ_n, $\gamma_{(A)}, \gamma_{(B)}, \gamma_{(C)}, \gamma_{(D)}$, constitute an algebra isomorphic to that of quaternions $\tau^1, \tau^2, \tau^3, \tau^0$.

The solution thus obtained is that of Subsection 8.4.2 and we have not gained anything by increasing the dimension. However, the symmetry of system (8.44), which is that of $\mathbb{Z}_2 + \mathbb{Z}_2$, is manifest in the form (H.18). The six pairs verifying $(a, b) = 1 \pmod 2$ are, for $n = 2$, (01, 11), (10, 11), (01, 10) *and their transposes*.

Appendix I

Brief reminder on elliptic functions

Doubly periodic meromorphic functions can easily be constructed as ratios of entire functions called theta functions, repeating themselves up to a multiplier under the group of periods. There is essentially a single function $\theta(z|\tau)$ of periods π and $i\pi\tau$ with $\Re\tau > 0$, but the functions obtained under translation by half-periods play a similar role. We thus obtain four functions, depending on a parameter $q = e^{-\pi\tau}$:

$$\theta_3(z) = \sum_{n\in\mathbb{Z}} q^{n^2} e^{2inz},$$

$$\theta_4(z) = \theta_3\left(z + \frac{\pi}{2}\right),$$

$$\theta_2(z) = \sum_{m+\frac{1}{2}\in\mathbb{Z}} q^{m^2} e^{2imz} = q^{\frac{1}{4}} e^{iz} \theta_3\left(z + \frac{i\pi\tau}{2}\right),$$

$$\theta_1(z) = -\theta_2\left(z + \frac{\pi}{2}\right) = -iq^{\frac{1}{4}} e^{iz} \theta_3\left(z + \frac{\pi}{2} + \frac{i\pi\tau}{2}\right). \tag{I.1}$$

$\theta_1(z)$ is odd; $\theta_2, \theta_3, \theta_4$ are even.

Under the group of periods, we have

$$\theta_3(z + \pi) = \theta_3(z),$$
$$\theta_3(z + i\pi\tau) = q^{-1} e^{-2iz} \theta_3(z). \tag{I.2}$$

Factorization of $\theta_1(z)$:

$$\theta_1(z) = 2q^{\frac{1}{4}} \sin z \prod_{n=1}^{\infty} (1 - q^{2n})(1 - q^{2n} e^{2iz})(1 - q^{2n} e^{-2iz}). \tag{I.3}$$

Let us mention the useful formula

$$\frac{\theta_1(z)}{\theta_1(z')} = \prod_{n=-\infty}^{+\infty} \frac{\sin(z + in\pi\tau)}{\sin(z' + in\pi\tau)}, \tag{I.4}$$

in which the product is convergent under association of the n and $-n$ terms.

Elliptic functions having a prescribed set of zeroes and poles in a period parallelogram can be constructed as θ-quotients. In this way we can define the three Jacobi functions sn u, cn u, dn u:

$$\text{sn } (u; k) = \frac{\theta_3(0|\tau)}{\theta_2(0|\tau)} \frac{\theta_1\left(\frac{\pi u}{2K}|\tau\right)}{\theta_4\left(\frac{\pi u}{2K}|\tau\right)} = \frac{\Theta_3(0)}{\Theta_2(0)} \frac{\Theta_1(u)}{\Theta_4(u)}, \tag{I.5}$$

in which we have defined

$$K = \frac{\pi}{2} \theta_3^2(0|\tau), \qquad \Theta_\alpha(u) = \theta_\alpha\left(\frac{\pi u}{2K}|\tau\right). \tag{I.6}$$

Jacobi employs the notation

$$H(u) = \Theta_1(u), \qquad \Theta(u) = \Theta_4(u). \tag{I.7}$$

The number k is called the modulus,

$$k = \theta_2^2(0)/\theta_3^2(0). \tag{I.8}$$

K is the complete elliptic integral, function of k or τ. We set $K' = \tau K$. Finally

$$\operatorname{cn}(u; k) = \frac{\Theta_4(0)}{\Theta_2(0)} \frac{\Theta_2(u)}{\Theta_4(u)}, \qquad \operatorname{dn}(u; k) = \frac{\Theta_4(0)}{\Theta_3(0)} \frac{\Theta_3(u)}{\Theta_4(u)}. \tag{I.9}$$

sn u and cn u have periods $4K$ and $4iK'$; dn u has periods $2K$ and $2iK'$.

The crucial theorem of the theory is the following. Between two elliptic functions belonging to the same period group, there exists an algebraic relation of type 1. As a consequence, there exist homogeneous algebraic relations between the various theta functions. Examples:

$$\operatorname{sn}^2 u + \operatorname{cn}^2 u = 1,$$
$$k^2 \operatorname{sn}^2 u + \operatorname{dn}^2 u = 1,$$
$$\frac{d}{du} \operatorname{sn} u = \operatorname{cn} u \operatorname{dn} u,$$
$$\operatorname{sn}(u + v) = \frac{\operatorname{sn} u \operatorname{cn} v \operatorname{dn} v + \operatorname{sn} v \operatorname{cn} u \operatorname{dn} u}{1 - k^2 \operatorname{sn}^2 u \operatorname{sn}^2 v}. \tag{I.10}$$

There exist many methods to prove these relations. First, Euler's algebraic method (cf. Siegel, 1969). For the quadratic relations, one can also calculate the squares θ^2 directly starting from the series (I.1). Finally, the most general method consists of constructing an elliptic function without pole in a period parallelogram and deducing from Liouville's theorem that this function reduces to a constant.

Quartic Jacobi relations: Consider four complex numbers z_1, z_2, z_3, z_4. We take

$$z_1' = \frac{1}{2}(z_1 + z_2 + z_3 + z_4),$$

$$z_2' = \frac{1}{2}(z_1 + z_2 - z_3 - z_4),$$

$$z_3' = \frac{1}{2}(z_1 - z_2 + z_3 - z_4),$$

$$z_4' = \frac{1}{2}(z_1 - z_2 - z_3 + z_4), \tag{I.11}$$

which defines an involution. We have the remarkable identity

$$\Theta_1(z_1')\Theta_2(z_2')\Theta_3(z_3')\Theta_4(z_4') = \frac{1}{2}S\Theta_1(z_1)\Theta_2(z_2)\Theta_3(z_3)\Theta_4(z_4), \tag{I.12}$$

in which S is the index symmetrizer defined in (8.49):

$$S \equiv 1 + (12)(34) + (13)(24) + (14)(23).$$

Note that the right-hand side vanishes under the sum z_1'.

Another very useful quartic relation is the following:

$$\Theta_1(x+y)\Theta_1(x-y)\Theta_1(z+t)\Theta_1(z-t)$$
$$-\Theta_1(x+t)\Theta_1(x-t)\Theta_1(z+y)\Theta_1(z-y)$$
$$= \Theta_1(y-t)\Theta_1(y+t)\Theta_1(x+z)\Theta_1(x-z). \qquad \text{(I.13)}$$

We can deduce this from Liouville's theorem, or from the following relations deriving from Jacobi's identity:

$$\theta_1(y+z)\theta_1(y-z)\theta_4^2(0) = \theta_3^2(y)\theta_2^2(z) - \theta_2^2(y)\theta_3^2(z),$$
$$\theta_4(y+z)\theta_4(y-z)\theta_4^2(0) = \theta_3^2(y)\theta_3^2(z) - \theta_2^2(y)\theta_2^2(z). \qquad \text{(I.14)}$$

Transformation of theta functions: The Landen transformation, operating the period doubling $\tau \to 2\tau$:

$$\theta_3(z|\tau)\theta_4(z|\tau) = \theta_4(2z|2\tau)\theta_4(0|2\tau), \qquad \text{(I.15)}$$

or further still

$$\theta_1\theta_4(z|\tau) = \frac{1}{2}\theta_1\left(z\Big|\frac{\tau}{2}\right)\theta_2\left(0\Big|\frac{\tau}{2}\right),$$
$$\theta_2\theta_3(z|\tau) = \frac{1}{2}\theta_2\left(z\Big|\frac{\tau}{2}\right)\theta_2\left(0\Big|\frac{\tau}{2}\right),$$
$$\theta_4^2(z|\tau) - \theta_1^2(z|\tau) = \theta_3\left(z\Big|\frac{\tau}{2}\right)\theta_4\left(0\Big|\frac{\tau}{2}\right),$$
$$\theta_4^2(z|\tau) + \theta_1^2(z|\tau) = \theta_4\left(z\Big|\frac{\tau}{2}\right)\theta_3\left(0\Big|\frac{\tau}{2}\right). \qquad \text{(I.16)}$$

The Jacobi transformation, or inversion of the period ratio $\tau \to \tau^{-1}$:

$$\theta_{3,2,4}(z|\tau) = \tau^{-\frac{1}{2}}e^{-\frac{z^2}{\pi\tau}}\theta_{3,4,2}\left(\frac{z}{\tau}\Big|\frac{1}{\tau}\right),$$
$$\theta_1(z|\tau) = -i\tau^{-\frac{1}{2}}e^{-\frac{z^2}{\pi\tau}}\theta_1\left(\frac{z}{\tau}\Big|\frac{1}{\tau}\right). \qquad \text{(I.17)}$$

Let us finally mention the relations between θ-constants, which are identities for q, for example

$$\theta_2^4(0) + \theta_4^4(0) = \theta_3^4(0), \qquad \text{(I.18)}$$

which can be derived from (I.12) and which express the relation $k^2 + k'^2 = 1$ between the complementary moduli k and k' such that

$$K'(k) = \tau K(k) = K(k'). \qquad \text{(I.19)}$$

9

The eight-vertex model: Eigenvectors and thermodynamics

9.1 Reduction to an Ising-type model

9.1.1 Problem setting

Let us again write the transfer matrix T, the purpose of this chapter being to construct its eigenvectors:

$$\langle \alpha | T | \beta \rangle \equiv \langle \alpha_1 \alpha_2 \ldots \alpha_N | T | \beta_1 \beta_2 \ldots \beta_N \rangle$$
$$= \text{Tr } \langle \alpha_1 | R | \beta_1 \rangle \langle \alpha_2 | R | \beta_2 \rangle \ldots \langle \alpha_N | R | \beta_N \rangle, \quad \alpha_i, \beta_i = \pm 1, \quad (9.1)$$

in which for each pair (α_i, β_i), $\langle \alpha_i | R | \beta_i \rangle$ is a 2×2 matrix defined as

$$\langle + | R | + \rangle = \begin{pmatrix} a & 0 \\ 0 & b \end{pmatrix}, \ \langle + | R | - \rangle = \begin{pmatrix} 0 & d \\ c & 0 \end{pmatrix},$$

$$\langle - | R | + \rangle = \begin{pmatrix} 0 & c \\ d & 0 \end{pmatrix}, \ \langle - | R | - \rangle = \begin{pmatrix} b & 0 \\ 0 & a \end{pmatrix}. \quad (9.2)$$

The objective of this section is to transform the eight vertex-type transfer matrix T into a transfer matrix W for an Ising-type model that will be defined at the end of the transformation. Having discovered, thanks to his diagonalization method, a sort of invariance of traces of products of certain matrices under the effect of multiplication by T, Baxter analyses the detail of this operation and deduces from this a basis of vectors invariant under T. The 2^N component vectors labelled $\{\beta_1 \beta_2 \ldots \beta_N\}$ are indexed by a set of integers $\{l_1 l_2 \ldots l_N\}$ and have the structure of the elements intervening in the traces of products defining the Q^R or Q^L matrices; more precisely

$$\langle \beta | \psi | l \rangle = \langle \beta_1 | \psi | l_1 l_2 \rangle \langle \beta_2 | \psi | l_2 l_3 \rangle \ldots \langle \beta_N | \psi | l_N l_1 \rangle. \quad (9.3)$$

We thus wish to construct a matrix W, transform of T by ψ such that

$$T\psi = \psi W, \quad (9.4)$$

179

in other words

$$\sum_{\beta} \langle \alpha | T | \beta \rangle \langle \beta | \psi | l \rangle = \sum_{m} \langle \alpha | \psi | m \rangle \langle m | W | l \rangle \tag{9.5}$$

with

$$\langle m | W | l \rangle = \prod_{j=1}^{N} \langle m_j m_{j+1} | W | l_j l_{j+1} \rangle, \tag{9.6}$$

and the previously encountered restrictions

$$|m_{j+1} - m_j| = 1, \qquad |l_{j+1} - l_j| = 1, \qquad j = 1, \dots, N, \tag{9.7}$$

$$m_{N+1} = m_1. \tag{9.8}$$

9.1.2 A model of Ising type

Supposing for a moment the existence of W, matrix (9.6) is interpreted as a row-to-row transfer matrix for an Ising model in which the 'spin' $l_{i,j}$ is associated with the link of the square lattice with coordinates (i, j) (j labelling the columns).

The values of successive spins on a same row differ by one unit. For reasons of symmetry, we are thus led to postulate the same condition from row to row. We shall show that this supplementary condition, which is written

$$\langle m | W | l \rangle = 0 \quad \text{if} \quad |l_j - m_j| \neq 1, \tag{9.9}$$

is sufficient to determine the basis Ψ and the six nonzero coefficients

$$\omega_1(l) = \langle l, l + 1 | W | l - 1, l \rangle,$$
$$\omega_2(l) = \langle l, l - 1 | W | l + 1, l \rangle,$$
$$\omega_3(l) = \langle l - 1, l | W | l, l + 1 \rangle,$$
$$\omega_4(l) = \langle l + 1, l | W | l, l - 1 \rangle,$$
$$\omega_5(l) = \langle l + 1, l | W | l, l + 1 \rangle,$$
$$\omega_6(l) = \langle l - 1, l | W | l, l - 1 \rangle. \tag{9.10}$$

The six corresponding spin configurations are represented in Figure 9.1.

The 'spin' configurations around a vertex of the square lattice constitute six vertices numbered from 1 to 6.

9.1.3 Baxter's transformation

If we substitute expression (9.1) for matrix T on the left-hand side of (9.5), we obtain the equality to be established in the form

$$\text{Tr} \ \langle \alpha_1 | S | l_1 l_2 \rangle \dots \langle \alpha_N | S | l_N l_1 \rangle = \langle \alpha | \psi W | l \rangle, \tag{9.11}$$

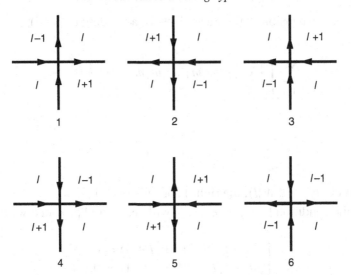

Figure 9.1 Spin configurations around the vertices of the square lattice.

with

$$\langle\alpha|S|ll'\rangle = \sum_{\beta}\langle\alpha|R|\beta\rangle\langle\beta|\psi|ll'\rangle, \tag{9.12}$$

in other words

$$\langle+|S|ll'\rangle = \begin{pmatrix} a\langle+|\psi|ll'\rangle & d\langle-|\psi|ll'\rangle \\ c\langle-|\psi|ll'\rangle & b\langle+|\psi|ll'\rangle \end{pmatrix}, \tag{9.13}$$

$$\langle-|S|ll'\rangle = \begin{pmatrix} b\langle-|\psi|ll'\rangle & c\langle+|\psi|ll'\rangle \\ d\langle+|\psi|ll'\rangle & a\langle-|\psi|ll'\rangle \end{pmatrix}. \tag{9.14}$$

Equality (9.11) shall be proven if it is possible to determine 2×2 matrices M_l such that we have, for each $\alpha = \pm 1$,

$$M_l^{-1}\langle\alpha|S|ll'\rangle M_{l'} \equiv \langle\alpha|\tilde{S}|ll'\rangle$$
$$\equiv \begin{pmatrix} \langle\alpha|\psi|l+1, l'+1\rangle \times & \langle\alpha|\psi|l+1, l'-1\rangle \times \\ \times\langle l+1, l'+1|W|ll'\rangle & \times\langle l+1, l'-1|W|ll'\rangle \\ \langle\alpha|\psi|l-1, l'+1\rangle \times & \langle\alpha|\psi|l-1, l'-1\rangle \times \\ \times\langle l-1, l'+1|W|ll'\rangle & \times\langle l-1, l'-1|W|ll'\rangle \end{pmatrix}. \tag{9.15}$$

Indeed, the left-hand side of (9.11) is equal to

$$\text{Tr} \prod_{j=1}^{N} M_{l_j}\langle\alpha_j|S|l_j l_{j+1}\rangle M_{l_{j+1}}$$

and the explicit expansion of the trace gives us, according to (9.15),

$$\sum_{\{m\}} \prod_{j=1}^{N} \langle \alpha_j | \psi | m_j m_{j+1} \rangle \langle m_j m_{j+1} | W | l_j l_{j+1} \rangle$$

with

$$m_j = l_j \pm 1,$$

which is precisely $\langle \alpha | \psi W | l \rangle$, the right-hand side of (9.11).

Finally, the condition $|m_{j+1} - m_j| = 1$ will be ensured provided we can choose

$$\begin{cases} \langle l-1, l+2 | W | l, l+1 \rangle = 0, \\ \langle l+1, l-2 | W | l, l-1 \rangle = 0. \end{cases} \tag{9.16}$$

To this end, we write the unimodular 2×2 matrix $M_{l'}$,

$$M_{l'} = \begin{pmatrix} s_{l'} q_{l'} & r_{l'} p_{l'} \\ s_{l'} & r_{l'} \end{pmatrix}, \tag{9.17}$$

with

$$r_{l'} s_{l'} = (q_{l'} - p_{l'})^{-1},$$

and obtain according to (9.15)

$$\langle +|\tilde{S}|ll'\rangle = \begin{pmatrix} r_l s_{l'} \{(aq_{l'} - bp_l)\langle +|\psi|ll'\rangle + & r_l r_{l'} \{(ap_{l'} - bp_l)\langle +|\psi|ll'\rangle + \\ + (d - cp_l q_{l'})\langle -|\psi|ll'\rangle\} & + (d - cp_l p_{l'})\langle -|\psi|ll'\rangle\} \\ s_l s_{l'} \{(bq_l - aq_{l'})\langle +|\psi|ll'\rangle + & s_l r_{l'} \{(bq_l - ap_{l'})\langle +|\psi|ll'\rangle + \\ + (cq_l q_{l'} - d)\langle -|\psi|ll'\rangle\} & + (cq_l p_{l'} - d)\langle -|\psi|ll'\rangle\} \end{pmatrix} \tag{9.18}$$

and an analogous formula for $\langle -|\tilde{S}|ll'\rangle$ obtained by the exchange $+ \leftrightarrow -$, $a \leftrightarrow b$, $c \leftrightarrow d$.

The existence conditions of the W coefficients satisfying (9.15) are obtained by elimination, by performing the quotients of the corresponding matrix elements in (9.18) for $(+|\tilde{S}|)$ and its analogon for $(-|\tilde{S}|)$.

Let us define

$$R(l, l') = \frac{\langle +|\psi|ll'\rangle}{\langle -|\psi|ll'\rangle}. \tag{9.19}$$

We obtain four relations

$$R(l+1, l'+1) = \frac{(aq_{l'} - bp_l)R(l, l') + (d - cp_l q_{l'})}{(bq_{l'} - qp_l) + (c - dp_l q_{l'})R(l, l')}, \tag{9.20}$$

$$R(l-1, l-'1) = \frac{(bq_l - ap_{l'})R(l, l') + (cq_l p_{l'} - d)}{(aq_l - bp_{l'}) + (dq_l p_{l'} - c)R(l, l')}, \tag{9.21}$$

$$R(l+1, l = l'-1) = \frac{(ap_{l'} - bp_l)R(l, l') + (d - cp_l p_{l'})}{(bp_{l'} - ap_l) + (c - dp_l p_{l'})R(l, l')}, \tag{9.22}$$

$$R(l-1, l = l'+1) = \frac{(bq_l - aq_{l'})R(l, l') + (cq_l q_{l'} - d)}{(aq_l - bq_{l'}) + (dq_l q_{l'} - c)R(l, l')}. \tag{9.23}$$

In contrast, conditions (9.16) give us the four supplementary conditions

$$R(l, l-1) = \frac{d - cp_l p_{l-1}}{bp_l - ap_{l-1}} = \frac{ap_l - bp_{l-1}}{c - dp_l p_{l-1}}, \tag{9.24}$$

$$R(l, l+1) = \frac{d - cq_l q_{l+1}}{bq_l - aq_{l+1}} = \frac{aq_l - bq_{l+1}}{c - dq_l q_{l+1}}. \tag{9.25}$$

We conclude from the second equations (9.24) and (9.25) that the quantities p_{l-1} and p_l on the one hand, and q_l and q_{l+1} on the other hand, are linked by the same fundamental relation already obtained in (8.102) or (8.103). We can thus write as per (8.118) in the notation of Chapter 8

$$p_l = \frac{\Theta_1(s + 2l\eta)}{\Theta_4(s + 2l\eta)}, \qquad q_l = \frac{\Theta_1(t + 2l\eta)}{\Theta_4(t + 2l\eta)}, \tag{9.26}$$

in which w and t are two undetermined but distinct parameters, so that M_l is invertible.

The ratios $R(l, l')$ are determined by (9.24), (9.25), and it remains to be verified that the four relations (9.20) to (9.23) are satisfied. We see easily that (9.22) and (9.23) are trivial identities. There remain (9.20) and (9.21), which are both amenable to the following type:

$$R(l, l+1) = \frac{(b^2 - c^2)q_l + (cdq_l^2 - ab)q_{l-1}}{bdq_l^2 - ac} = \frac{d - cq_l q_{l+1}}{bq_l - aq_{l+1}}. \tag{9.27}$$

Equation (9.27) leads to the cancelling of a certain polynomial in $a, b, c, d, q_{l-1}, q_{l+1}$. If we are looking to have only the invariants appear, it is sufficient to show the two identities

$$\frac{q_{l-1}q_{l+1} - q_l^2}{q_{l-1}q_l^2 q_{l+1} - 1} \equiv \frac{cd}{ab} \tag{9.28}$$

and

$$(q_{l+1} + q_{l-1})(cdq_l - abq_l^{-1}) \equiv c^2 + d^2 - a^2 - b^2. \tag{9.29}$$

Their final verification necessitates the elliptic parametrization (8.61) and (9.27) and is performed without difficulty.

9.1.4 Explicit determination of ψ and W

Taking (8.61) and (9.26) into account, we obtain directly for the ratios $R(l, l')$, in the notation of Chapter 8,

$$\begin{cases} R(l, l+1) = \sqrt{k}\, \text{sn}\, (t - v + (2l+1)\eta), \\ R(l+1, l) = \sqrt{k}\, \text{sn}\, (s + v + (2l+1)\eta), \end{cases} \tag{9.30}$$

which leads us to choosing $\langle \alpha | \psi | ll' \rangle$ in the following way:

$$\langle + | \psi | l, l+1 \rangle = \Theta_1(t - v + (2l+1)\eta),$$
$$\langle - | \psi | l, l+1 \rangle = \Theta_4(t - v + (2l+1)\eta),$$
$$\langle + | \psi | l+1, l \rangle = \Theta_1(s + v + (2l+1)\eta),$$
$$\langle - | \psi | l+1, l \rangle = \Theta_4(s + v + (2l+1)\eta). \tag{9.31}$$

Finally, we deduce from (9.17) and (9.26) the form of matrix M:

$$M_l = \Theta_1\left(\frac{t-s}{2} \middle| \frac{\tau}{2}\right)^{-\frac{1}{2}} \begin{pmatrix} \rho_l \dfrac{\Theta_1(t + 2l\eta)}{\Theta_1\left(w_l | \frac{\tau}{2}\right)} & \dfrac{1}{\rho_l}\Theta_1(s + 2l\eta) \\[3mm] \rho_l \dfrac{\Theta_4(t + 2l\eta)}{\Theta_1\left(w_l | \frac{\tau}{2}\right)} & \dfrac{1}{\rho_l}\Theta_4(s + 2l\eta) \end{pmatrix}, \tag{9.32}$$

in which ρ_l is undetermined and quantity W is

$$w_l = \frac{1}{2}(s + t) + 2l\eta - K, \qquad \tau = \frac{K'}{K}. \tag{9.33}$$

The fact that M_l is unimodular comes from the relation

$$\Theta_1(t|\tau)\Theta_4(s|\tau) - \Theta_1(s|\tau)\Theta_4(t|\tau) = \Theta_1\left(\frac{t-s}{2} \middle| \frac{\tau}{2}\right)\Theta_2\left(\frac{t+s}{2} \middle| \frac{\tau}{2}\right), \tag{9.34}$$

$$\Theta_2\left(t | \frac{\tau}{2}\right) \equiv \Theta_1\left(t - K | \frac{\tau}{2}\right).$$

The explicit determination of W is performed by comparing matrices (9.18) and (9.15), for a given fixed value of α. If we use the elliptic parametrization of coefficients a, b, c, d given by (8.61), we have, for a convenient choice of the global inhomogeneity factor of the transfer matrix,

$$a = \frac{\Theta_1(v + \eta)\Theta_4(v - \eta)}{\Theta_1(2\eta)\Theta_4(0)}, \qquad \text{etc.}$$

We shall set

$$h(v) = \Theta_1\left(v\Big|\frac{\tau}{2}\right),\tag{9.35}$$

$$\Phi(v) = \frac{h(v)}{h(2\eta)}.\tag{9.36}$$

We then obtain the following results:

$$\omega_1(l) = \langle l, l+1|W|l-1, l\rangle = \Phi(v+\eta)\frac{\rho_l}{\rho_{l-1}},$$

$$\omega_2(l) = \langle l, l-1|W|l+1, l\rangle = \Phi(v+\eta)\frac{h(w_{l+1})}{h(w_l)}\frac{\rho_{l+1}}{\rho_l},$$

$$\omega_3(l) = \langle l-1, l|W|l, l+1\rangle = \Phi(v-\eta)\frac{\rho_l}{\rho_{l+1}},$$

$$\omega_4(l) = \langle l+1, l|W|l, l-1\rangle = \Phi(v-\eta)\frac{h(w_{l-1})}{h(w_l)}\frac{\rho_{l-1}}{\rho_l},$$

$$\omega_5(l) = \langle l+1, l|W|l, l+1\rangle = \frac{h(w_l+\eta-v)}{h(w_l)h(w_{l+1})}\frac{1}{\rho_l\rho_{l+1}},$$

$$\omega_6(l) = \langle l-1, l|W|l, l-1\rangle = h(w_l-\eta+v)\rho_l\rho_{l-1}.\tag{9.37}$$

For a reason that will later become clear (see equation (9.69)), we must choose $\rho_l = 1$, which we will do from now on to somewhat lighten the equations.

9.2 Equivalence to a six-vertex model

9.2.1 The inhomogeneous six-vertex model

The transformation (9.4) brings the diagonalization of the transfer matrix T back to that of a transfer matrix W for an equivalent Ising model. But this can furthermore be interpreted as an inhomogeneous ferroelectric-type model.

If we decompose an eigenvector of T on the basis of the functions ψ defined in (9.3):

$$\langle\alpha| = \sum_{\{m\}}\langle\alpha|\psi|m\rangle\langle m|,\tag{9.38}$$

the eigenvalue Λ and amplitudes $\langle m|$ equation is written according to (9.5) and (9.6):

$$\sum_{\{l\}}\langle m|W|l\rangle\langle l| = \Lambda\langle m|.\tag{9.39}$$

We shall now describe a convenient representation of the element $\langle m|W|l\rangle$, as suggested by the graphical representation (Figure 9.1) of the weights $\langle m_1m_2|W|l_1l_2\rangle$. This representation associates one of the six vertices of the ice model with each of the six functions $\omega_i(l)$. The rule is the following: since the 'spins' l_j, l_{j+1} of the equivalent Ising model associated with two neighbouring links j and $j+1$ on a same row differ by only one unit (8.7), we split these two links by a down arrow ↓ if $l_{j+1} = l_j - 1$, and by an up arrow ↑ if $l_{j+1} = l_j + 1$. We repeat this process along a column. The result is that, along a given line of links, the set of integers l_j is determined by the input of l_1 and a configuration of arrows ↓ on this line, of which the increasing abscissas will be $x_1, x_2, \ldots, x_M$.

An amplitude $\langle l|\rangle$ will thus be expressible as

$$\langle l|\rangle = \langle l_1 l_2 .. l_N|\rangle = f(l_1|x_1, x_2, \ldots, x_M), \tag{9.40}$$

with

$$l_{j+1} = \begin{cases} l_j - 1, & j \in \{X\} = \{x_1 x_2 \ldots\}, \\ l_j + 1, & j \notin \{X\}. \end{cases} \tag{9.41}$$

A matrix element $\langle m|W|l\rangle$ will thus be defined by the input of two sets $\{m_1|x_1 x_2 \ldots\}$ and $\{l_1|y_1 y_2 \ldots\}$, which will completely define a horizontal configuration of compatible vertices. These configurations have already been described in Chapter 7 and are illustrated in Figure 9.2. The matrix elements are divided into two categories:

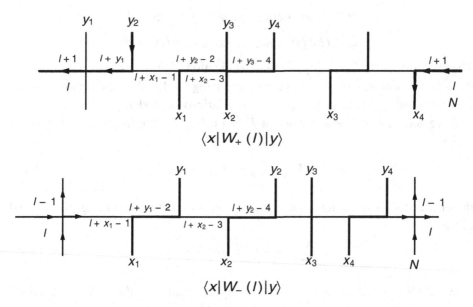

$$\langle x|W_+(l)|y\rangle$$

$$\langle x|W_-(l)|y\rangle$$

Figure 9.2 The two types of W matrix elements; thick lines indicate the ↓ or ← arrows.

for $l_1 = m_1 + 1 = l + 1$ $\quad \langle m|W|l\rangle = \langle x|W_+(l)|y\rangle,$

for $l_1 = m_1 - 1 = l - 1$ $\quad \langle m|W|l\rangle = \langle x|W_-(l)|y\rangle.$ (9.42)

The cyclic invariance of W will be ensured by the condition

$$l_{N+1} - l_1 = N - 2M.$$ (9.43)

9.2.2 *Explicit form of the elements of the inhomogeneous transfer matrix*

According to the method already applied in Chapter 7, formulas (7.28), (7.29) and by inspection of Figure 9.2, the matrix elements of W are a product of statistical weights ω_i defined by each vertex, which can be decomposed in the form

$$\langle x|W_+(l)|y\rangle = D_l(0, y_1)\bar{D}_{l-1}(y_1, x_1)D_{l-2}(x_1, y_2)\ldots\bar{D}_{l-2M+1}(y_M, x_M)$$
$$\times D_{l-2M}(x_M, N+1),$$

with (Figure 9.3)

$$0 < y_1 \leq x_1 \leq y_2 \leq \ldots \leq x_M < N + 1,$$ (9.44)

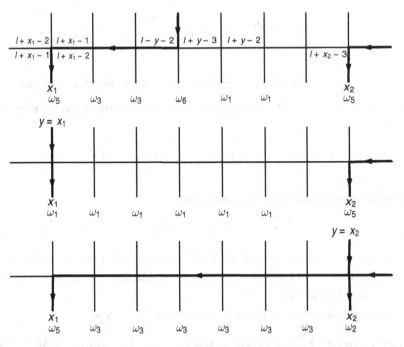

Figure 9.3 Statistical weights associated with different configurations defined by $x_1 \leq y \leq x_2$.

the y (and x) remaining distinct:

$$\langle x|W_-(l)|y\rangle = \bar{D}_{l-1}(0, x_1)D_{l-2}(x_1, y_1)\bar{D}_{l-3}(y_1, x_2)\ldots\bar{D}_{l-2M-1}(y_M, N+1),$$
$$0 < x_1 \leq y_1 \leq \ldots \leq y_M < N+1. \tag{9.45}$$

With the definitions of the D functions:

$$D_l(x, y) = \omega_3(l+x+1)\omega_3(l+x+2)\ldots\omega_3(l+y-1)\omega_6(l+y),$$
$$D_l(x, x) = \omega_4(l+x)/\omega_5(l+x), \tag{9.46}$$
$$\bar{D}_l(y, z) = \omega_1(l+y+1)\omega_1(l+y+2)\ldots\omega_1(l+z-1)\omega_5(l+z-1),$$
$$\bar{D}_l(y, y) = \omega_2(l+y)/\omega_6(l+y+1). \tag{9.47}$$

When the second variable takes the value $N+1$, we eliminate the factor ω_5 or ω_6.

Taking expressions (9.37) for the ω_i into account, we obtain for functions D and $\bar{D}$:

$$D_l(x, y) = \Phi^{y-x-1}(v-\eta)h(w_{l+y} - \eta + v),$$
$$D_l(x, x) = \Phi(v-\eta)\frac{h(w_{l+x-1})h(w_{l+x+1})}{h(w_{l+x} + \eta - v)},$$
$$\bar{D}_l(y, z) = \Phi^{z-y-1}(v+\eta)\frac{h(w_{l+z-1} + \eta - v)}{h(w_{l+z-1})h(w_{l+z})},$$
$$\bar{D}_l(z, z) = \Phi(v+\eta)\frac{h(w_{l+z+1})}{h(w_{l+z})h(w_{l+z} + \eta + v)}. \tag{9.48}$$

9.2.3 Expansion of the amplitude using the Bethe–Lieb method

In the notation (9.40), (9.42), the eigenvalue equation of the transfer matrix (9.39) is written

$$\Lambda f(l|x) = \sum_{\{y\}}\langle x|W_+(l)|y\rangle f(l+1|y) + \langle x|W_-(l)|y\rangle f(l-1|y), \tag{9.49}$$

in which the summation runs over the domain

$$1 \leq y_1 \leq y_2 \leq \ldots \leq y_M \leq N. \tag{9.50}$$

Baxter achieves the solution of the linear system (9.49) by a generalization of the method of Bethe and Lieb analogous to the one he applied to the solution of an inhomogeneous ferroelectric model.

For the amplitude, we look for an expansion of Bethe type

$$f(l|x_1 x_2 \ldots x_M) = \sum_P A(P)g_{P1}(l, x_1)g_{P2}(l-2, x_2)\ldots g_{PM}(l-2M+2, x_M), \tag{9.51}$$

in which the sum runs over the permutations P of $1, 2, \ldots, M$.

The cyclic invariance of the transfer matrix leads to the imposition of the periodic boundary condition

$$f(l|x_1x_2\ldots x_M) = f(l-2|x_2x_3\ldots x_M, x_1+N), \qquad (9.52)$$

which allows us to extend the definition of $f(l|x)$ outside the $[1, N]$ interval, and will prove useful in what follows. Indeed, as we have already seen in Chapter 7, this condition allows us to group the two terms on the right-hand side of (9.49) into a single one, but summed over an extended interval

$$x_M \le y_M \le x_{M+1}, \qquad x_{M+1} \equiv x_1 + N. \qquad (9.53)$$

Let us consider the second term first. With the help of (9.47), we verify that the relation $N - 2M = 0$ leads to

$$\bar{D}_{l-1}(0, x_1)\bar{D}_{l-2M-1}(y, N+1) = \bar{D}_{l-2M-1}(y, x_{M+1}) \qquad (9.54)$$

for $x_M \le y < N + 1$.

We deduce from this that

$$\sum_{\{y\}}\langle x|W_-(l)|y\rangle f(l-1|y) = \sum_{\{y\}}\left\{\prod_{j=1}^{M} D_{l-2j}(x_j, y_j)\bar{D}_{l-2j-1}(y_j, x_{j+1})\right\} f(l-1|y),$$
$$x_M \le y_M \le N. \qquad (9.55)$$

Let us then consider the first term in (9.49). We easily verify the identity

$$D_l(0, y_1)\bar{D}_{l-1}(y_1, x_1)D_{l-2M}(x_M, N+1) \equiv D_{l-2M}(x_M, y_1 + N)$$
$$\times \bar{D}_{l-2M-1}(y_1 + N, x_1 + N) \qquad (9.56)$$

such that, when setting $y_{m+1} = y_1 + N$, we have

$$\sum_{\{y\}}\langle x|W_+(l)|y\rangle f(l+1|y)$$

$$= \sum_{\{y\}}\left\{\prod_{j=1}^{M} D_{l-2j}(x_y, y_{j+1})\bar{D}_{l-2j-1}(y_{j+1}, x_{j+1})\right\} f(l+1|y). \quad (9.57)$$

The periodicity condition (9.52) allows us to replace $f(l + 1|y_1\ldots y_M)$ by $f(l - 1|y_2\ldots y_{M+1})$ on the right-hand side of (9.57). Afterwards, the change of summation variables $y_j \to y_{j-1}$ makes the integrand of (9.57) identical to (9.55). Only the domain differs: after the translation $y_{M+1} \to y_M$, it becomes

$$N < y_M \le x_{M+1}.$$

The union of the two domains is thus

$$x_M \leq y_M \leq x_{M+1}. \tag{9.58}$$

Taking Ansatz (9.51) into account, equation (9.49) is rewritten as

$$\Lambda f(l|x) = \sum_{\{y\}} \sum_{P} A(P) \prod_{j=1}^{M} g_{Pj}(l - 2j + 1, y_j) D_{l-y}(x_j, y_j) \bar{D}_{l-y-1}(y_j, x_{j+1}). \tag{9.59}$$

The translation of the periodicity condition (9.52) remains, which gives us

$$A(P)g_1(l+1, y) = A(PC)g_{P1}(l - 2M + 1, y + N) \tag{9.60}$$

in which C is the cyclic permutation $(1 \rightarrow 2 \rightarrow \ldots M)$.

9.2.4 Summation

The next step would obviously consist of determining the functions g. We shall content ourselves with the verification of Baxter's result. After summation over the y_j, the right-hand side of equation (9.59) has the structure already defined in Chapter 7 (Subsection 7.3.2):

$$\sum_P A(P)(X_{P1} + Z_{P1})(X_{P2} + Z_{P2})(X_{P3} + Z_{P3}) \quad \cdots \quad (X_{PM} + Z_{PM}) \tag{9.61}$$

and in order that the two subsisting 'diagonal' terms reproduce the Bethe sum appearing on the left-hand side, it is sufficient that the X_i (and the Z_i) are functions $g(\ldots, y_i)$. The linear recursions thus obtained lead to a solution of the type

$$g_j(l+1, y) = \left(\frac{h(v_j - \eta)}{h(v_j + \eta)}\right)^y \frac{h(w_{l+y-1} + \eta - v_j)}{h(w_{l+y-1})h(w_{l+y})}, \tag{9.62}$$

in which the $v_j = 1, \ldots, M$ are still undetermined.

Lieb's method now leads us to relax the condition $y_i < y_{i+1}$ and sum independently over the y_i in the domains

$$x_i \leq y_i \leq x_{i+1}; \quad i = 1, \ldots, M, \tag{9.63}$$

leaving us with the task of eliminating all the products of dangerous non-diagonal terms, symbolized by $Z_{P3}X_{P4}$, thanks to a convenient choice for $A(P)$.

We are thus led to perform sums of the type

$$X_a + Z_a \equiv \sum_{y=x}^{z} D_l(x, y) \bar{D}_{l-1}(y, z) g_\alpha(l + 1, y), \tag{9.64}$$

which by virtue of (9.48) and (9.62) are rewritten as

$$X_a + Z_a = \Phi(v - \eta)\Phi^{z-x-1}(v + \eta)\left(\frac{h(v_\alpha + \eta)}{h(v_\alpha - \eta)}\right)^x$$

$$\times \frac{h(w_{l+x+1})h(w_{l+x+1} + \eta - v_\alpha)h(w_{l+z-1} - \eta - v)}{h(w_{l+x})h(w_{l+x} + \eta - v)h(w_{l+z-1})h(w_{l+z-2})}$$

$$+\Phi(v + \eta)\Phi^{z-x-1}(v - \eta)\left(\frac{h(v_\alpha + \eta)}{h(v_\alpha - \eta)}\right)^z \frac{h(w_{l+z-1} + \eta - v_\alpha)}{h^2(w_{l+z-1})}$$

$$+\frac{\Phi^{z-1}(v + \eta)}{\Phi^{x+1}(v - \eta)} \frac{h(w_{l+z-1} - \eta - v)}{h(w_{l+z-1})h(w_{l+z-2})}$$

$$\times \sum_{y=x+1}^{z-1} \left(\frac{h(v - \eta)}{h(v + \eta)}\right)^y h(w_{l+y} - \eta + v)g_\alpha(l + 1, y). \tag{9.65}$$

The two 'dangerous terms' originating from the confluence $y = x$ or $y = z$ have been isolated. The last sum in (9.65) is easily performed by introducing the function

$$\Gamma(x) = \left(\frac{h(v - \eta)h(v_\alpha + \eta)}{h(v + \eta)h(v_\alpha - \eta)}\right)^x \frac{h(w_{l+x-1} + v - v_\alpha)}{h(w_{l+x-1})}. \tag{9.66}$$

The difference $\Gamma(x + 1) - \Gamma(x)$ is performed with the help of the quartic relation (9.31), and from definition (9.62) we obtain

$$\Gamma(x+1)-\Gamma(x)=\left(\frac{h(v - \eta)}{h(v + \eta)}\right)^{x+1} \frac{1}{\Phi(v - \eta)}h(w_{l+x-1}-\eta+v)\frac{h(v - v_\alpha)}{h(v_\alpha - \eta)}g_\alpha(l+1, x). \tag{9.67}$$

The sum from $x + 1$ to $z - 1$ in (9.65) thus contributes by the two terms

$$X'_\alpha + Z'_\alpha = \frac{h(v_\alpha - \eta)}{h(v - v_\alpha)}$$

$$\times \left\{\Phi^{z-x-1}(v - \eta)\left(\frac{h(v_\alpha + \eta)}{h(v_\alpha - \eta)}\right)^z \frac{h(w_{l+z-1} - \eta - v)h(w_{l+z-1} + v - v_\alpha)}{h(w_{l+z-2})h^2(w_{l+z-1})}\right.$$

$$\left.-\Phi^{z-x-1}(v + \eta)\left(\frac{h(v_\alpha + \eta)}{h(v_\alpha - \eta)}\right)^{x+1} \frac{h(w_{l+z-1} - \eta - v)h(w_{l+x} + v - v_\alpha)}{h(w_{l+z-1})h(w_{l+z-2})h(w_{l+x})}\right\}. \tag{9.68}$$

Collecting all terms in x of summation (9.65), we have

$$X_\alpha = \Phi^{z-x-1}(v + \eta)\left(\frac{h(v_\alpha + \eta)}{h(v_\alpha - \eta)}\right)^x \frac{h(w_{l+z-1} - \eta - v)}{h(w_{l+x})h(w_{l+z-1})h(w_{l+z-2})}$$

$$\times \{h(w_{l+x} + 2\eta)h(v - \eta)h(w_{l+x} - \eta - v_\alpha)h(v - v_\alpha)$$

$$-h(w_{l+x} + v - v_\alpha)h(v_\alpha + \eta)h(w_{l+x} + \eta - v)h(2\eta)\}$$

$$\times \frac{1}{h(2\eta)h(w_{l+x} + \eta - v)h(v - v_\alpha)}. \tag{9.69}$$

According to the quartic relation (9.31), the quantity in parentheses in the numerator of (9.69) equals

$$h(v - v_\alpha + 2\eta)h(w_{l+x} + \eta - v_\alpha)h(w_{l+x})h(v + \eta).$$

It should be noted here that this combination was rendered possible only by the choice $\rho_l = 1$, (9.32)–(9.37).

We thus obtain

$$X_\alpha(l) = g_\alpha(l + 2, x)\Phi^{z-x}(v + \eta)\frac{h(v - v_\alpha - 2\eta)h(w_{l+x})h(w_{l+x+1})h(w_{l+z-2} + \eta - v)}{h(v - v_\alpha)h(w_{l+z-1})h(w_{l+z-2})h(w_{l+z} + \eta - v)}. \tag{9.70}$$

In the same way, we find

$$Z_\alpha(l) = \frac{h(v - v_\alpha + 2\eta)}{h(v - v_\alpha)}g_\alpha(l, z)\Phi^{z-x}(v - \eta). \tag{9.71}$$

9.2.5 Calculation of $A(P)$

The $A(P)$ coefficients are determined from the condition

$$\sum_P A(P)\dots \underbrace{Z_{Pj}(l - 2j)X_{P(j+1)}(l - 2j - 2)}\dots = 0 \tag{9.72}$$

in which we have represented one of the factors intervening in the expansion of expression (9.61) that we are currently evaluating, with the exception of two diagonal terms $X_{P1}X_{P2}\dots X_{PM}$ and $Z_{P1}Z_{P2}\dots Z_{PM}$. The ⌐⌐ symbol under the ZX products means, as in Chapter 7, that we have removed from this product some 'dangerous terms', since such a product should not occur on the sum over y on the right-hand side of equality (9.59).

According to (9.70), (9.71) and (9.65), after having set

$$l - 2j = n, \qquad Pj = \alpha, \qquad P(j + 1) = \beta,$$

$$x_j = x, \qquad x_{j+1} = y, \qquad x_{j+2} = z,$$

we have

$$Z_\alpha(n)X_\beta(n - 2) = \Phi^{z-y}(v + \eta)\Phi^{y-x}(v + \eta)\left(\frac{h(v_\alpha + \eta)h(v_\beta + \eta)}{h(v_\alpha - \eta)h(v_\beta - \eta)}\right)^y$$

$$\times \{h(w_{n+y-2} + \eta - v_\alpha)h(w_{n+y-2} + \eta - v_\beta)h(v - v_\alpha + 2\eta)h(v - v_\beta - 2\eta)$$

$$-h(w_{n+y-1} + \eta - v_\alpha)h(w_{n+y-3} + \eta - v_\beta)h(v - v_\alpha)h(v - v_\beta)\}$$

$$\times\frac{h(w_{n+z-4} + \eta - v)}{h(w_{n+y-2} + \eta - v)h(w_{n+y-2})h(w_{n+y-1})h(w_{n+z-3})h(w_{n+z-4})}. \tag{9.73}$$

By virtue of the quartic relation (9.31), the quantity in parentheses in the numerator of (9.73) is equal to

$$h(v_\alpha - v_\beta - 2\eta)h(2\eta)h(w_{n+y-2} + \eta - v)h(w_{n+y-2} + \eta + v - v_\alpha - v_\beta).$$

In this way we can write

$$Z_\alpha(n)X_\beta(n-2) = h(v_\alpha - v_\beta - 2\eta)S(\alpha, \beta), \tag{9.74}$$

with

$$S(\alpha, \beta) = S(\beta, \alpha). \tag{9.75}$$

Condition (9.72) will be fulfilled, provided we have

$$A(P)\, Z_\alpha(n)X_\beta(n-2) + A(P(j, j+1))\, Z_\beta(n)X_\alpha(n-2) = 0 \tag{9.76}$$

or in other words, according to (9.74) and (9.75),

$$\frac{A(P(j, j+1))}{A(P)} = -\frac{h(v_{Pj} - v_{P(j+1)} - 2\eta)}{h(v_{P(j+1)} - v_{Pj} - 2\eta)}. \tag{9.77}$$

The unique solution of equations (9.77) is

$$A(P) = \chi(P) \prod_{1 \leq i < j \leq M} h(2\eta + v_{Pj} - v_{Pi}), \tag{9.78}$$

in which $\chi(P)$ is the sign of the parity of P.

9.2.6 The coupled equations

We must now complete the calculation of the right-hand side of (9.59). A similar argument to that used in (7.60) and (7.61) allows us to show that the choice (9.78) for $A(P)$ and the periodicity conditions (9.60) make the following quantity vanish:

$$\sum_P A(P) \ldots X_{P1} Z_{PM} \ldots$$

We are thus left with the two diagonal terms and we obtain, with the help of (9.70) and (9.71), equality (9.59) in the form

$$\Lambda f(l|x) = \sum_P A(P) \left\{ \Lambda_+ \prod_{j=1}^M g_{Pj}(l - 2j + 2, x_j) + \Lambda_- \prod_{j=1}^M g_{Pj}(l - 2j, x_{j+1}) \right\}, \tag{9.79}$$

with

$$\Lambda_\pm = \Phi^N(v \pm \eta) \prod_{j=1}^M \frac{h(v - v_j \mp 2\eta)}{h(v - v_j)}. \tag{9.80}$$

By virtue of the periodic boundary conditions (9.52), equality (9.79) is an identity in x, provided we have

$$\Lambda = \Lambda_+ + \Lambda_-. \qquad (9.81)$$

Eigenvectors and eigenvalues of the transfer matrix are thus determined functions of the M parameters v_j. It remains to calculate these by using the periodicity condition (9.60) and the solution (9.78) for $A(P)$:

$$\frac{A(PC)}{A(P)} = \prod_{j \neq P1} \frac{h(v_{P1} - v_j + 2\eta)}{h(v_{P1} - v_j - 2\eta)}, \qquad (9.82)$$

and according to (9.62),

$$\frac{g_{P1}(l - 2M + 1, y + N)}{g_{P1}(l + 1, y)} = \left(\frac{h(v_{P1} - \eta)}{h(v_{P1} + \eta)}\right)^N. \qquad (9.83)$$

From this, we deduce the coupled equations for the v_j:

$$\prod_{j \neq i} \frac{h(v_i - v_j + 2\eta)}{h(v_i - v_j - 2\eta)} = \left(\frac{h(v_i + \eta)}{h(v_i - \eta)}\right)^N, \qquad (9.84)$$

which completes our exposition of the tour-de-force accomplished by Baxter.

9.3 The thermodynamic limit

9.3.1 General method

We end this chapter with a rapid exposition of some results obtained starting from equations (8.146), concerning the thermodynamic limit of the eight-vertex model or of the linear chain.

Let us first note that this system of equations is completely analogous to that already obtained for the six-vertex model ($d = 0$) or the Heisenberg–Ising chain ($J_1 = J_2 = 1$, $J_3 = \cosh \Phi$). In this limit, we have among others according to (8.84),

$$d = 0, \quad k = 0, \quad q = 0, \quad K = \frac{\pi}{2}, \quad K' = \infty \qquad (9.85)$$

and according to (I.4),

$$\lim_{q \to 0} \frac{\Theta_1 \left(v - \eta | \frac{\tau}{2}\right)}{\Theta_1 \left(v + \eta | \frac{\tau}{2}\right)} = \frac{\sin \frac{1}{2}(\varphi - i\Phi)}{\sin \frac{1}{2}(\varphi + i\Phi)}, \qquad (9.86)$$

in which the correspondence between Baxter's and our notation is the following:

$$\varphi = i\alpha = \frac{\pi v}{K} = \frac{i\pi V}{K'_l}, \quad i\Phi = i\lambda = \frac{\pi \eta}{K} = \frac{i\pi \zeta}{K'_l},$$

$$z = e^{i\varphi}, \quad x = e^{-\Phi}, \quad q = e^{-2\tau'} = e^{-\pi\tau}, \quad \tau = \frac{K'}{K}. \qquad (9.87)$$

We thus fall back on system (1.40) of Chapter 1 for $v' = 0$.

It is interesting for the purposes of calculation to notice that we go from the limiting case $\tau = \infty$ ($q = 0$) to the general case of τ finite by replacing all ratios of type

$$\frac{\sin \frac{1}{2}(\varphi - i\Phi)}{\sin \frac{1}{2}(\varphi + i\Phi)}$$

by the infinite product

$$\prod_{n=-\infty}^{+\infty} \frac{\sin \frac{1}{2}(\varphi_n - i\Phi)}{\sin \frac{1}{2}(\varphi_n + i\Phi)}, \qquad \varphi_n = \varphi + in\pi\tau,$$

with remark (I.4).

The standard method for calculating a thermodynamic limit ($N \to \infty$) is then as follows.

(a) Take the logarithm on both sides of the coupled equations, which introduces a set of $N/2$ quantum numbers $\{\lambda\}$:

$$\ln \frac{h(v_i - \eta)}{h(v_i + \eta)} = 2\pi \frac{\lambda_i}{N} + \frac{1}{N} \sum_j{}' \ln \frac{h(v_i - v_j - 2\eta)}{h(v_i - v_j + 2\eta)},$$

$$h(v) = \theta_1 \left(\frac{\pi v}{2K} \Big| \frac{\tau}{2} \right). \tag{9.88}$$

Write, in a form very similar to (1.32) or (1.82),

$$\sum_n Nk(\varphi_j + in\pi\tau) = 2\pi\lambda_j + \sum_j{}' \sum_n \psi(\varphi_j - \varphi_i + in\pi\tau). \tag{9.89}$$

(b) Use the principle of continuity to associate with each root system $\{v_j\}$ a unique set of quantum numbers: continuity in τ starting from $\tau = \infty$ thus gives the same sets corresponding to the 'largest eigenvalue' of T or to the 'ground state' of H, or to the 'excited states', etc.

(c) Assume that the roots become dense on a segment of the real axis (for the largest eigenvalue) in the limit $N = \infty$. Define the density ρ such that

$$N\rho(\varphi)d\varphi = \text{number of roots } \varphi_j \text{ in interval } [\varphi, \varphi + d\varphi].$$

(d) The limit form of the coupled equations for φ_j is a linear integral equation for $\rho(\varphi)$. By virtue of remark (9.89), a calculation performed in the restricted case of the Heisenberg–Ising chain or of the six-vertex model can be transposed safely to the eight-vertex case. This is, for example, the way in which one obtains the following remarkable result. The density $\rho(\varphi)$ (for the largest eigenvalue of T or the ground state of the XYZ Hamiltonian) is independent of q (or of τ) and remains given by expression (2.20) from Chapter 2:

$$\rho(\varphi) = \mathrm{Dn}\ (\varphi) = \frac{K_x}{\pi}\ \mathrm{dn}\ \left(\frac{K_x\varphi}{\pi}\Big|\Phi\right), \qquad x = e^{-\Phi}. \tag{9.90}$$

Indeed the integral equation for ρ is obtained, according to remark (9.89), starting from equation (2.18) (with $\tilde{\varphi} = \pi$) by replacing φ by $\varphi_n = \varphi + in\pi\tau$ in the inhomogeneous term and the kernel in which we perform the summation over n.

To show (9.90), it is then sufficient to verify that

$$\frac{\sinh\Phi}{\cosh\Phi - \cos\varphi_n} = \frac{1}{2\pi}\int_{-\pi}^{+\pi}\frac{\sinh 2\Phi}{\cosh 2\Phi - \cos(\varphi_n - \varphi')}\rho(\varphi')d\varphi' \tag{9.91}$$

irrespective of $n \neq 0$. We shall put ourselves in region $\pi\tau \geq 2\Phi$. In the half-plane $\Im\varphi_n > 2\Phi$, we expand in a convergent Fourier series and (9.91) is reduced to the identities

$$\sinh p\Phi = \sinh 2p\Phi\ \rho(p), \tag{9.92}$$

in which $\rho(p)$ is precisely the Fourier coefficient of $\rho(\varphi)$ calculated in Chapter 2.

9.3.2 The free energy

Given the symmetry properties of $T(w)$, we can put ourselves in the following regime:

$$\omega_0^2 \leq \omega_3^2 \leq \omega_2^2 \leq \omega_1^2,$$

which corresponds to the antiferroelectric regime in which C dominates, $C \geq a + b + d$. We thus have, in the notation of Subsection 8.4.2,

$$0 \leq l \leq 1.$$

V and ζ are real; $V \leq \zeta \leq K_l$, by virtue of (8.51) or

$$\sqrt{q} \leq x \leq z, \qquad 0 < \alpha \leq \lambda \leq \tau', \tag{9.93}$$

which justifies the hypothesis that $\pi\tau \geq 2\Phi$ made in Subsection 9.3.1.

We get for the free energy

$$- \beta f = \ln t = \lim_{N \to \infty} N^{-1} \ln |T_{max}|$$

$$= \underset{\pm}{\mathrm{Max}} \lim_{N \to \infty} \left\{ \ln \frac{h(v \pm \eta)}{h(2\eta)} + N^{-1} \sum_j \ln \frac{h(v - v_j \mp 2\eta)}{h(v - v_j)} \right\}$$

$$= \ln c + \ln \frac{\theta_1(v + \eta)\theta_4(0)}{\theta_4(v - \eta)\theta_1(2\eta)}$$

$$+ \frac{1}{2\pi} \int_{-\pi}^{+\pi} \ln \frac{\theta_1\theta_4(v - v' - 2\eta)}{\theta_1\theta_4(v - v')} \rho(\varphi')d\varphi', \quad \text{where } \varphi' = \pi v'/2K. \quad (9.94)$$

We see that $t = c$ for $v = \eta$ ($a = b = d = 0$). The calculation is performed using Fourier series and one obtains the result

$$\ln \frac{t}{c} = 2 \sum_{p=1}^{\infty} \frac{\sinh^2 p \left(\frac{\pi \tau}{2} - \Phi \right) (\cosh p\Phi - \cosh p\alpha)}{p \sinh p\pi\tau \cosh p\Phi} \quad (9.95)$$

in region $0 \le \alpha \le \Phi \le \frac{1}{2}\pi\tau$.

We can easily check that formula (9.95) reduces to expression (7.116) for $d = 0$ ($q = 0$, $\tau = +\infty$), which is the case for the F model (six vertex in zero field),

$$\ln t|_{d=0} = \ln a + z_1$$

in the notation of Chapter 7.

Baxter obtained expression (9.95) for the free energy by a direct method, without integration, starting from equations (8.146). We can present his result in the form

$$t(z) = T_{max}^{1/N} = t_0 \frac{L(x^{-1}z)L(x^{-1}z^{-1})}{L(xz)L(xz^{-1})}, \quad (9.96)$$

with the definitions

$$t_0 = (1 - x^2) \prod_{n=1}^{\infty} (1 - x^2 q^n)(1 - x^{-2}q^n),$$

$$L(z) = \prod_{n=1}^{\infty} \prod_{m=1}^{\infty} (1 - q^n z)(1 - x^{4m}z)(1 - q^n x^{4m}z)^2. \quad (9.97)$$

The function $t(z)$, now completely factorized, is meromorphic in the complex plane except at the origin and at infinity.

9.3.3 Singularity of the free energy

In the real domain defined by the inequalities $w_0^2 \le w_3^2 \le w_2^2 \le w_1^2$, series (9.95) converges normally and defines an analytic function of Φ, α and τ. The eventual singularities are thus to be looked for on the real line. We easily see that the three sections $\Phi = \alpha$, $\Phi = \frac{1}{2}\pi\tau$ or $\tau = \infty$ are not singular. It remains to examine the neighbourhood of $\tau = 0$, in which Φ and α are also close to zero according to the inequalities. By virtue of (8.55), we have simultaneously

$$\begin{cases} \tau \to 0, \quad k \to 1, \quad K \to \infty, \quad K' \to \dfrac{\pi}{2}, \\ \tau_l \to \infty, \quad l \to 0, \quad K_l \to \dfrac{\pi}{2}, \quad K'_l \to \infty. \end{cases} \tag{9.98}$$

According to (8.51),

$$\frac{\omega_2}{\omega_3} = \frac{\mathrm{dn}\,(V, l)}{\mathrm{dn}\,(\zeta, l)} \to 1.$$

The parameters V and ζ tend to limiting values V_c and ζ_c such that

$$\sin^2 \zeta_c = \frac{\omega_1^2 - \omega_2^2}{\omega_1^2 - \omega_0^2}, \quad \sin V_c = \frac{\omega_0}{\omega_2} \sin \zeta_c \tag{9.99}$$

with

$$\lim \frac{\Phi}{\tau} = \zeta_c, \quad \lim \frac{\alpha}{\tau} = V_c. \tag{9.100}$$

The point $w_2 = w_3$, frontier of the low-temperature region in the chosen regime $c > a+b+d$, defines a temperature T_c such that in the neighbourhood of $T_c = \beta_c^{-1}$ we have

$$\omega_2^2 - \omega_3 \sim l^2 = e^{-2K'_l} = e^{-\frac{2\pi}{\tau}}. \tag{9.101}$$

Series (9.95) obviously tends to the integral

$$- \beta_c f_c = \ln c + \int_{-\infty}^{\infty} \frac{\sinh^2 \left[\left(\frac{\pi}{2} - \zeta_c \right) x \right] (\cosh \zeta_c x - \cosh V_c x)}{x \sinh \pi x \cosh \zeta_c x} dx, \tag{9.102}$$

which does not have any singularities in ζ_c and V_c in the domain $0 \le V_c \le \zeta_c$. The eventual singularity of βf thus originates from the difference between the discrete sum and the integral. The sum can be transformed into an integral over a holomorphic function of p in the vicinity of the real axis, multiplied by $(e^{2\pi i p} - 1)^{-1}$. The dominant singularity is obtained by calculating the contribution of the pole lying closest to the real axis. However, one can verify that the pole at $p = 1/i\tau$ and its congruents do not give a singular contribution. There remains the pole at $p = i\pi/2\Phi$, with asymptotic behaviour of the residue given by

$$4e^{-\pi^2/\Phi} \cos \frac{\pi\alpha}{\Phi} \cot \frac{\pi^2\tau}{4\Phi},$$

in other words, taking (9.101) and (9.100) into account, the singular part of the free energy

$$-\beta f \sim 4\cos\frac{\pi V_c}{2\zeta_c}\cot\frac{\pi^2}{4\zeta_c}|T - T_c|^{\pi/2\zeta_c}. \qquad (9.103)$$

If the exponent $\pi/2\zeta_c$ is an even integer, the singularity becomes logarithmic, as will be the case for the Ising model in which one of the invariants vanishes (equivalence to the XY model). The critical parameters are indeed expressed as functions of the invariants J_α. The critical condition $w_2^2 = w_3^2$ gives

$$J_2 = J_3 \qquad (9.104)$$

and on the contrary, according to (9.99) and (8.24),

$$\cos 2\zeta_c = \frac{2\omega_2^2 - \omega_1^2 - \omega_0^2}{\omega_1^2 - \omega_0^2} = \frac{J_1}{J_2}. \qquad (9.105)$$

For $J_1 = 0$, $\zeta_c = \dfrac{\pi}{4}$,

$$-\beta f \propto (T - T_c)^2 \ln|T - T_c|.$$

The critical exponent for the specific heat

$$C = -\beta^2\frac{\partial^2}{\partial\beta^2}\beta f \sim |T - T_c|^{-\alpha} \qquad (9.106)$$

thus takes the value

$$\alpha = 2 - \frac{\pi}{2\zeta_c}, \qquad 0 \le \zeta_c \le \frac{\pi}{2}. \qquad (9.107)$$

It can take a continuous set of values between 0 and ∞ depending on the energies of each of the vertices. The compatibility of this result with the general hypothesis of universality of critical phenomena was shown by Kadanoff and Wegner (1971).

9.4 Various results on the critical exponents

Let us mention without proof some simple results often obtained after very elaborate calculations in the framework of the eight-vertex model.

9.4.1 Correlation length

If T_{sec} designates the second largest eigenvalue of the transfer matrix after T_{Max}, the arrow-to-arrow correlation on the same column of the lattice admits the asymptotic behaviour $G(R) \propto (T_{sec}/T_{Max})^R$ and the correlation length is thus $-\ln(T_{sec}/T_{Max})$. The evaluation of this quantity was performed by Johnson *et al.* (1973) using a method analogous to that for calculating the first excited state of the linear magnetic chain. One thus obtains the critical exponent

$$v = \frac{\pi}{4\zeta_c}. \tag{9.108}$$

9.4.2 Surface tension

An argument due to Fisher (1969) allows us to relate the boundary energy σ of the two phases present under the critical temperature to the ratio (T_{sec}/T_{Max}):

$$\beta\sigma = \lim_{N\to\infty} N^{-1} \ln |\ln(T_{sec}/T_{max})|. \tag{9.109}$$

Baxter (1973a) obtains, after a difficult analysis,

$$\beta\sigma = \frac{1}{2} \ln k_2, \tag{9.110}$$

in which k_2 is the modulus associated with the ratio of periods τ_2 such that

$$x^2 = e^{-\pi\tau}, \qquad \tau_2 = 2\Phi/\pi.$$

The critical exponent μ, such that $\sigma = (T_c - T)^\mu$, takes the value $\mu = \pi/4\zeta_c$.

9.4.3 The order parameter in the Ising limit

According to formulas (8.6) linking the weights to the interaction constants of the Ising model ($K = 0$), we have

$$b^{-1} = a = e^{K_3+K_4}, \qquad d^{-1} = c = e^{K_3-K_4}. \tag{9.111}$$

The formula of Onsager (1944) and Yang (1952) for the spontaneous magnetization M_0 is

$$M_0^8 = 1 - (\sinh 2K_3 \sinh 2K_4)^{-2} = 1 - \frac{J_1^2}{J_3^2} \quad \text{with } J_2 = 0, \tag{9.112}$$

in which we have used (8.24), (8.52) and (9.111). For convenience we will interchange the roles of J_1 and J_2 by changing the sign of b, which does not modify the associated vertex model.

According to (8.52) and (9.112), we then have

$$M_0^8 = 1 - \frac{J_2^2}{J_3^2} = l^2, \qquad J_1 = 0. \tag{9.113}$$

According to (8.82), this is equal to $\zeta = \dfrac{1}{2}K_l$. We thus have

$$M_0 = l^{1/4} = \left(\frac{\theta_2(0|\tau_l)}{\theta_3(0|\tau_l)}\right)^{1/2} = \left(\frac{\theta_4(0|\tau_l')}{\theta_3(0|\tau_l')}\right)^{1/2} = \prod_{m=1}^{\infty} \frac{1 - q'^{2m-1}}{1 + q'^{2m-1}}, \qquad (9.114)$$

with

$$\tau_l' = \frac{1}{\tau_l} = \frac{K_l}{K_l'}, \qquad q' = e^{-\pi \tau_l'}. \qquad (9.115)$$

But the parameter $x = e^{-\Phi}$ takes in this case the value

$$x = e^{i\pi \eta/K} = e^{-\pi \zeta/K_l'} = e^{-\pi K_l/2K_l'} = \sqrt{q'}. \qquad (9.116)$$

From this comes the formula for the magnetization as a function of x,

$$M_0(x) = \prod_{m=1}^{\infty} \frac{1 - x^{4m-2}}{1 + x^{4m-2}}. \qquad (9.117)$$

9.4.4 Order parameter of the eight-vertex model

Barber and Baxter (1973) have used perturbation theory to calculate the first terms in the power series in c^{-1} for the spontaneous magnetization of the Ising model (K_3, K_4, K) equivalent to the eight-vertex model. They find

$$M_0 = 1 - 2x^2 + 2x^4 + O(x^6), \qquad (9.118)$$

which to this order is independent of z and q.

The independence with respect to z comes from the fact that M_0 is a matrix element of σ^z between two eigenstates of the transfer matrix. The independence with respect to q is, however, remarkable. Since the perturbative expansion (9.118) coincides with that of $M_0(x)$ in (9.117), Barber and Baxter make the conjecture that this expression is valid for all x. From this they easily deduce the critical exponent β, such that $M_0 = |T - T_c|^{\beta}$, and which takes the value $1/8$ for the ordinary Ising model $(K = 0)$.

Indeed the comparison of (9.117) with identity (9.114) allows us to write

$$M_0 = \lambda^{1/4}, \qquad x^2 = q_{\lambda}' = e^{-\pi K_{\lambda}/K_{\lambda}'}, \qquad (9.119)$$

but according to the definition of $x = e^{-\pi \zeta/K_l'}$ we have, in the vicinity of $T = T_c$,

$$l, \lambda \to 0, \qquad K_l, K_{\lambda} \to \infty, \qquad \lim \frac{K_{\lambda}'}{K_l'} = \frac{\pi}{4\zeta_c},$$

which leads to

$$M_0 = \lambda^{1/4} \propto e^{-K_{\lambda}'/4} \propto e^{-\pi K_l'/8\zeta_c},$$

and according to (9.101)

$$M_0 \propto |T_c - T|^{\pi/32\zeta_c}. \tag{9.120}$$

From this one gets the value of the critical exponent $\beta = \pi/32\zeta_c$.

9.4.5 Scaling laws and other exponents

- $\alpha = 2 - d\nu$ is verified: $d = 2$, $\alpha = 2 - (\pi/2\zeta_c)$ according to (9.107), $\nu = (\pi/4\zeta_c)$ according to (9.108).
- $\beta = \pi/32\zeta_c$ (magnetization in zero field).
- $\gamma = d\nu - 2\beta = \dfrac{7\pi}{16\zeta_c}$ (susceptibility).
- $\eta = 2 - \dfrac{\gamma}{\nu} = \dfrac{1}{4}$ (correlation length at the critical temperature).
- $\delta = 1 + \dfrac{\gamma}{\beta} = 15$ ($M = H^{1/\delta}$).

10

Identical particles with δ-interactions

10.1 The Bethe hypothesis

10.1.1 Elementary solutions

Let us consider N identical particles, that is particles having identical mass and internal symmetry, interacting on an indefinite axis via a two-body delta function potential. The Hamiltonian is thus the same as that defined in Section 4.1:

$$H = -\sum_{j=1}^{N} \frac{\partial^2}{\partial x_j^2} + 2c \sum_{i<j} \delta(x_j - x_i), \qquad (10.1)$$

in which x_j represents the coordinate of particle number j and where $2c$ is the interaction intensity, which can be attractive or repulsive. We sometimes consider $\{x\}$ as a vector in $\mathbb{R}_N$.

We shall call an *elementary solution* of the Schrödinger equation $H\Psi = E\Psi$ a function Ψ which is continuous in $\mathbb{R}_N$, irrespective of its behaviour at infinity. The Hamiltonian being symmetric, the solutions can be classified according to their symmetry type, defined by the Young tableau $\mathscr{T}$. The completely symmetric case of the bosonic system has been treated in Chapter 4.

The analogy between the Schrödinger equation and the wave equation for a particle in $\mathbb{R}_N$ in the presence of $\frac{1}{2}N(N-1)$ infinitely refracting thin plates was exploited (McGuire, 1964; Gaudin, 1972; Derrida, 1976). The essential result of this analogy is the following: the total wave Ψ is a finite superposition of incoming waves, transmitted or reflected geometrically. The absence of diffracted waves, due to the specially symmetric configuration of the plates, is the physical content of what is improperly called the Bethe 'hypothesis' (Ansatz) for the structure of the wavefunction (see also Section 5.2).

10.1.2 Gluing conditions

The Schrödinger equation is reduced to the free wave equation

$$-\sum_{j=1}^{N} \frac{\partial^2 \Psi}{\partial x_j^2} - E\Psi = 0, \tag{10.2}$$

when $\{x\}$ belongs to any of the open sets, or sectors of $\mathbb{R}_N$, denoted D_Q in which Q is a permutation of order N:

$$D_Q : x_{Q1} < x_{Q2} < \ldots < x_{QN}. \tag{10.3}$$

The initial equation is equivalent to (10.2) supplemented by gluing conditions at the boundaries of sectors D_Q which contain:

(a) *the continuity conditions*

$$\Psi|_{x_{Q4}-x_{Q3}=+0} \equiv \Psi|_{x_{Q4}-x_{Q3}=-0} \quad 3,4 \Leftrightarrow j, j+1, \quad j = [1, N-1]; \tag{10.4}$$

(b) *the conditions on the normal derivatives* deduced from Green's theorem (see (5.34))

$$\left(\frac{\partial \Psi}{\partial x_{Q4}} - \frac{\partial \Psi}{\partial x_{Q3}}\right)\bigg|_{x_{Q4}-x_{Q3}=+0} - \left(\frac{\partial \Psi}{\partial x_{Q4}} - \frac{\partial \Psi}{\partial x_{Q3}}\right)\bigg|_{x_{Q4}-x_{Q3}=-0} \equiv 2c\Psi|_{x_{Q4}-x_{Q3}=\pm 0}. \tag{10.5}$$

Bethe's hypothesis consists of supposing the existence of an elementary solution, associated with a given vector $\{k\}$ of $\mathbb{C}_N$, which in each sector D_Q of $\mathbb{R}_N$ has the form of a superposition of plane waves

$$\Psi|_{\{x\}\in D_Q} = \sum_{P\in\pi_N} \langle Q||P\rangle e^{i(\bar{P}k, \bar{Q}x)}, \tag{10.6}$$

with the notation $(\bar{P}k)_j = k_{Pj}$, $\bar{P} = P^{-1}$. The energy associated with $\Psi_{\{k\}}$ is $E_{\{k\}}$:

$$E = \sum_{j=1}^{N} k_j^2. \tag{10.7}$$

The Schrödinger equation is thus satisfied in the open sets D_Q. There remain the gluing conditions:

(a) *continuity*

$$\langle Q||P\rangle + \langle Q||P(34)\rangle = \langle Q(34)||P\rangle + \langle Q(34)||P(34)\rangle; \tag{10.8}$$

(b) *jump of the normal derivative*

$$i(k_{P4} - k_{P3})(\langle Q||P\rangle - \langle Q||P(34)\rangle + \langle Q(34)||P\rangle - \langle Q(34)||P(34)\rangle)$$
$$= 2c(\langle Q||P\rangle + \langle Q||P(34)\rangle). \tag{10.9}$$

Taking

$$x_{jl} = \frac{ic}{k_j - k_l}, \tag{10.10}$$

we deduce from (10.8) and (10.9) the following relation for passing from sector Q to the adjacent one $Q(34)$:

$$\langle Q(34)||P\rangle = x_{P3,P4}\langle Q||P\rangle + (1 + x_{P3,P4})\langle Q||P(34)\rangle. \tag{10.11}$$

Relations (10.11) lead to (10.8) and (10.9). (34) is written for $(j, j + 1)$; there thus exist $N - 1$ connection relations (10.11). The task is now to prove the compatibility of the $(N - 1)(N!)^2$ linear equations (10.11) for the $(N!)^2$ unknowns $\langle Q||P\rangle$.

10.1.3 Proof of Bethe's hypothesis for the class of elementary solutions

(McGuire, 1964; Brézin and Zinn-Justin, 1966)
In the $N!$-dimensional vector space constituted by the ring of the permutation group π_N, we can consider relations (10.11) as defining $(N - 1)$ linear mappings $T_{12}, T_{23}, \ldots, T_{N-1,N}$ such that

$$\langle Q(34)||P\rangle = \sum_{P'}\langle Q||P'\rangle\langle P'|T_{34}|P\rangle \tag{10.12}$$

with the only non-vanishing matrix elements

$$\langle P|T_{34}|P\rangle = x_{P3,P4}, \quad \langle P(34)|T_{34}|P\rangle = 1 + x_{P3,P4}. \tag{10.13}$$

Let us show that equations (10.12) admit a nontrivial solution, depending linearly on $N!$ arbitrary parameters which are, if one wants, the coefficients $\langle I||P\rangle$ ($I =$ identity in π_N) of the wavefunction in the 'initial' sector.

It is clear that the coefficients $\langle Q||P\rangle$ ($Q \neq I$) are then determined, since a permutation Q can be constructed as a product of transpositions between neighbours $(j, j + 1)$ which form a basis of generators of π_N, and the repeated use of (10.12) determines $\langle Q||P\rangle$. These quantities are however overdetermined, since the factorization of Q in generators is not unique. The result $\langle Q||P\rangle$ will however be unique if the $(N - 1)$ operations $T_{j,j+1}$ generate under multiplication a free group which is isomorphic to the group generated by the $(j, j + 1)$, in other words isomorphic to π_N. This is what has to be checked.

Let us first note that this is a reflection group:

$$T_{34} \cdot T_{34} = I \tag{10.14}$$

which results from the identity

$$x_{P3,P4}^2 + (1 + x_{P3,P4})(1 + x_{P4,P3}) = 1. \tag{10.15}$$

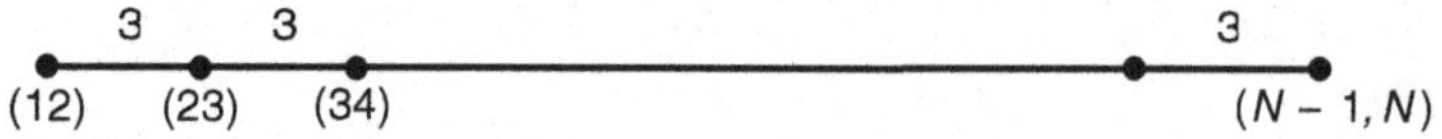

Figure 10.1 Permutations.

We obviously have

$$T_{23} \cdot T_{56} = T_{56} \cdot T_{23}. \tag{10.16}$$

Finally, we shall later verify the relations

$$(T_{12} \cdot T_{23})^3 = I. \tag{10.17}$$

The product of two arbitrary generating reflections is thus a rotation of order 2 or 3. According to Coxeter and Moser (1972), relations (10.14) to (10.17) uniquely determine the diagram of the reflection group, and the group itself (see Section 5.2), which is here π_N. See Figure 10.1.

The representation of π_N thus obtained is isomorphic to the regular representation to which it will be continuously reduced for $c = 0$.

The verification of (10.17) is also that of

$$T_{12} \cdot T_{23} \cdot T_{12} = T_{23} \cdot T_{12} \cdot T_{23} \tag{10.18}$$

(whose common value defines the transposition T_{13}).

We shall define for convenience the ring element

$$|\bar{P}\rangle = \sum_{Q \in \pi_N} \bar{Q} \langle Q || P \rangle, \tag{10.19}$$

which allows us to rewrite (10.12) in the form

$$(34)|\bar{P}\rangle = \sum_{P'} |\bar{P}'\rangle \langle P'|T_{34}|P\rangle = x_{P3,P4}|\bar{P}\rangle + (1 + x_{P3,P4})|\overline{P(34)}\rangle. \tag{10.20}$$

It is thus sufficient to calculate

$$\begin{aligned}
(12)(23)(12)|\bar{P}\rangle = {} & x_{P1,P3}(1 + x_{P1,P2}x_{P2,P3})|\bar{P}\rangle \\
& + x_{P1,P3}x_{P2,P3}(1 + x_{P1,P2})|\overline{P(12)}\rangle + x_{P1,P2}x_{P1,P3}(1 + x_{P2,P3})|\overline{P(23)}\rangle \\
& + (1 + x_{P1,P2})(1 + x_{P1,P3})(1 + x_{P2,P3})|\overline{P(13)}\rangle \\
& + x_{P1,P2}(1 + x_{P1,P3})(1 + x_{P2,P3})|\overline{P(123)}\rangle \\
& + x_{P2,P3}(1 + x_{P1,P3})(1 + x_{P1,P2})|\overline{P(132)}\rangle.
\end{aligned} \tag{10.21}$$

Manifestly, the calculation of $(23)(12)(23)\bar{P}$ gives the same result as (10.21): formally exchanging 1 and 3, followed by the re-establishment of the original order for writing the indices 1, 2, 3.

The structure of wavefunction (10.6) is thus entirely justified. The number of linearly independent parameters is $N!$, which leads to the idea that there exists exactly one elementary wavefunction corresponding to each numbered Young tableau.

10.2 Yang's representation

10.2.1 Yang's representation for π_N

The connection formulas (10.11) or (10.20) give rise to an orthogonal representation of the generators of transpositions, which is distinct from the representation $T_{j,j+1}$ of the preceding section. Yang considers the operators $Y_{j,j+1}$ such that

$$Y_{34}|\bar{P}\rangle = |\overline{P(34)}\rangle = \frac{(34) - x_{P3,P4}}{1 + x_{P3,P4}}|\bar{P}\rangle. \qquad (10.22)$$

These play the same role with respect to the permutations P as the T play with respect to Q. If we represent the vector $|\bar{P}\rangle$ in the following way:

$$|\bar{P}\rangle \equiv |P_1, P_2, \ldots, P_N\rangle \equiv |ijk\ldots\rangle, \qquad (10.23)$$

we shall write according to (10.22)

$$Y_{12}^{ij}|ijk\ldots\rangle = |jik\ldots\rangle \qquad (10.24)$$

with

$$Y_{12}^{ij} = \frac{(12) - x_{ij}}{1 + x_{ij}}. \qquad (10.25)$$

Notation (10.22) is justified by the coherence conditions in π_N:

$$Y_{12} \cdot Y_{12} = I, \qquad Y_{12}Y_{23}Y_{12}|\bar{P}\rangle = Y_{23}Y_{12}Y_{23}|\bar{P}\rangle.$$

A direct calculation allows us to verify

$$Y_{12}^{ji}Y_{12}^{ij} = 1, \qquad (10.26)$$

$$Y_{12}^{jk}Y_{23}^{ik}Y_{12}^{ij} = Y_{23}^{ij}Y_{12}^{ik}Y_{23}^{jk}, \qquad (10.27)$$

which falls back on the π_N relations and the identity

$$x_{ik}x_{ij} + x_{ik}x_{jk} = x_{ij}x_{jk}.$$

10.2.2 The scattering matrix for distinguishable particles

The construction of the scattering matrix, or S matrix, is a first application of the Y operators. We shall suppose that the particles are distinguishable to avoid doing a reduction according to the type of symmetry. In the repulsive case, the wavefunction of a scattering state is necessarily associated with a real set $\{k\}$, which we will order as

$$k_1 > k_2 > \ldots > k_N. \tag{10.28}$$

In each sector D_Q, there exists a single term of the elementary wavefunction (10.6) representing a purely incoming state

$$\langle Q||I\rangle \exp\{i(k, \bar{Q}x)\}, \tag{10.29}$$

since by virtue of (10.28) the associated wavepacket represents N particles x_{Qj} of momentum k_j which have never scattered in the past. The particles being distinguishable by assumption, we shall associate with the incident particle x_j the momentum k_j such that the purely incoming component exists only in the initial sector I:

$$\langle Q||I\rangle = \begin{cases} 1 & \text{if} \quad Q = I, \\ 0 & \text{if} \quad Q \neq I. \end{cases} \tag{10.30}$$

We thus have $|I\rangle = I$ according to (10.19) and the amplitude $\langle Q||P\rangle$ is determined for all P thanks to the connection formulas (10.22).

An outgoing wave in the arbitrary sector Q is proportional to $\exp i(k_N x_{Q1} + k_{N-1} x_{Q2} + \ldots + k_1 x_{QN})$ and its amplitude is $\langle Q||T\rangle$, in which T is the permutation $Tj = (N - j + 1), \forall j$:

$$\begin{aligned} T = & (1N)(2\,N-1)(3\,N-2)\ldots \equiv (12) \cdot (23)(12) \cdot (34)(23)(12)\ldots \\ & (N-1\,N)\ldots(12). \end{aligned} \tag{10.31}$$

From this, and according to (10.22), we deduce that

$$|\bar{T}\rangle = \sum_Q \bar{Q}\langle Q||T\rangle = Y_{12} \cdot Y_{23}Y_{12} \cdot Y_{34}Y_{23}Y_{12} \ldots Y_{N-1N}\ldots Y_{12} \cdot I \tag{10.32}$$

or further still, in the notation (10.24), taking $N = 5$ for convenience of writing:

$$|54321\rangle = Y_{12}^{45} \cdot Y_{23}^{35}Y_{12}^{34} \cdot Y_{34}^{25}Y_{23}^{24}Y_{12}^{23} \cdot Y_{45}^{15}Y_{34}^{14}Y_{23}^{13}Y_{12}^{12} \cdot |12345\rangle. \tag{10.33}$$

If we define the following elements in the ring of π_N, which will play an important role in the following,

$$X_{ij} = (ij)Y_{ij}^{ij} = \frac{1 - (ij)x_{ij}}{1 + x_{ij}}, \tag{10.34}$$

we can write

$$|54321\rangle = T \cdot X_{45} \cdot X_{35}X_{34} \cdot X_{25}X_{24}X_{23} \cdot X_{15}X_{14}X_{13}X_{12}|12345\rangle. \tag{10.35}$$

Let us label the final states such that the diagonal element of the S matrix sends the incoming state to the outgoing state in which the particle x_j has momentum k_j, in other words in the sector $Q = T$. By definition of S we have

$$|\bar{T}\rangle = T S |I\rangle, \tag{10.36}$$

or in other words

$$S = X_{45} \cdot X_{35} X_{34} \cdot \ldots \cdot X_{15} X_{14} X_{13} X_{12}. \tag{10.37}$$

The coefficient of $T Q^{-1}$ in S is $\langle Q || I \rangle$, or said otherwise the element of the S matrix for the transition from initial sector I to final sector Q, for N distinguishable particles.

If the particles are identical, it is sufficient to decompose S (in other words the elements of π_N) according to the various types of symmetry. The antisymmetric case of tableau $[1^N]$ gives $X_{ij} = 1$ and $S = 1$, the symmetric case $[N]$ gives $X_{ij} = \dfrac{1 - x_{ij}}{1 + x_{ij}}$, from which the bosonic result (4.14) for bosons emerges. For spin-1/2 fermions, we have the spin representation

$$(12) = (12)_x (12)_\sigma = -\frac{1}{2} (1 + \vec{\sigma}_1 \cdot \vec{\sigma}_2)$$

from which we get the representation of S on the basis of spin functions.

In the attractive case, the S matrix still has the form (10.35) between the unbound states. In the holomorphic region of the repulsive case

$$\Im(k_j - k_{j+1}) > 0,$$

there appear poles for $c < 0$, defined by the relations $k_{j+1} - k_j = ic$ for certain j. A bound state of p particles is associated with a set of k such that

$$\frac{2}{c} \Im k_v = -(p - 1), -(p - 3), \ldots, (p - 1). \tag{10.38}$$

The scattering matrix between bound states was obtained by Yang (1968), after McGuire (1964), to whose works one must turn for more complete developments.

We shall return in Subsection 10.3.2 to this factorization of the S matrix into two-body S matrices, which is obviously founded on the ternary relations (Zamolodchikov, 1977; Iagolnitzer, 1978).

10.2.3 Periodicity conditions

After having treated the diffusion problem on the infinite line in the preceding subsection, we tackle here that of particles confined to a given 'volume' L, with an eye towards the thermodynamic limit. In order to avoid boundary effects, we impose periodic boundary conditions in L on the wavefunction, which is equivalent to placing the particles on a circle of length L. We shall see that these periodicity conditions are compatible with the structure of the elementary solutions (10.6). Let us write down these conditions on the amplitudes. For a periodic system, it is convenient to restrict the definitions of sectors as follows:

$$D_Q : \begin{cases} x_{Q1} < x_{Q2} < \ldots < x_{QN} \\ x_{QN} - x_{Q1} < L. \end{cases} \tag{10.39}$$

Periodicity in L for x_{Q1} implies the identity

$$\Psi(x_{Q1}, \ldots, x_{QN}) \equiv \Psi(x_{Q2}, \ldots, x_{QN}, x_{Q1} + L) \tag{10.40}$$

in which the right-hand side is defined in sector QC. Substituting (10.6) into (10.40), we deduce

$$\langle Q||P \rangle = \langle QC||PC \rangle e^{ik_{P1}L} \tag{10.41}$$

or, in the notation of (10.19),

$$C|\overline{PC} \rangle = |\bar{P} \rangle e^{-ik_{P1}L}. \tag{10.42}$$

The representation of the circular permutation C in terms of the generators Y gives us

$$|\overline{PC} \rangle = Y_{N-1N} \ldots Y_{23} Y_{12} |\bar{P} \rangle \tag{10.43}$$

and more explicitly ($N = 5$)

$$\begin{aligned} C|\overline{PC} \rangle &= C Y_{45}^{P_1 P_5} Y_{34}^{P_1 P_4} Y_{23}^{P_1 P_3} Y_{12}^{P_1 P_2} |\bar{P} \rangle \\ &= (15) \cdot Y_{15}^{P_1 P_5} \cdot (14) \cdot Y_{14}^{P_1 P_4} \cdot (13) \cdot Y_{13}^{P_1 P_3} \cdot (12) \cdot Y_{12}^{P_1 P_2} |\bar{P} \rangle. \end{aligned} \tag{10.44}$$

Let us multiply on the left by P and use definition (10.34) of the X, and (10.42):

$$P|\bar{P} \rangle e^{-ik_{P1}L} = X_{P1,P5} X_{P1,P4} \ldots X_{P1,P2} \cdot P|\bar{P} \rangle. \tag{10.45}$$

From this, we formulate the following

Proposition: The periodic boundary conditions for the system of identical particles with δ-function interaction can be expressed in the following eigenvalue problem:

$$Z_P \cdot P|\bar{P} \rangle = e^{ik_{P1}L} P|\bar{P} \rangle \tag{10.46}$$

with

$$Z_P \equiv X_{P2P1} X_{P3P1} \ldots X_{PNP1}. \tag{10.47}$$

10.2.4 Compatibility of the periodicity conditions

The compatibility of relations (10.46) originates from the remarkable properties of the Z_p matrices and of the X_{ij} operators. From definition (10.34) and relations (10.26) between Y generators, one can easily deduce the multiplicative relations between the X, identical to those obtained in Chapter 8 (8.65) for analogous operators:

$$X_{ij}X_{ji} = 1, \quad X_{ij}X_{kl} = X_{kl}X_{ij}, \quad X_{jk}X_{ik}X_{ij} = X_{ij}X_{ik}X_{jk} \quad \forall i \neq j \neq k \neq l. \tag{10.48}$$

The matrices Z_P acting in the ring of π_N are divided into N similar families, distinguished by the index $P1 = j$. This comes from the remark

$$Z_{P(34)} = X_{P3P4}Z_P X_{P3P4}^{-1}, \tag{10.49}$$

and the similarity transformations are precisely those of the Y representation acting on the $|P\rangle$. There thus exist only N distinct eigenvalue equations which determine the $\{k\}$. One can choose

$$P = C^n, \quad n = 0, 1, 2, \ldots, N - 1$$

$$(Z_j - e^{ik_j L})C^j|\bar{C}^j\rangle = 0 \tag{10.50}$$

with

$$Z_j = X_{j+1j}X_{j+2j} \ldots X_{Nj}X_{1j} \ldots X_{j-1j}. \tag{10.51}$$

The eigenvectors can in fact be chosen independently of j by virtue of the following

Lemma: The operators Z_j commute among themselves.

The proof is based on (10.48) and will be given in the following section. It is identical to that of Chapter 8 (8.66).

10.3 Ternary relations algebra and integrability

10.3.1 The transfer matrix Z(k)

Before treating the diagonalization problem (10.50) which will lead us to the general solution of the problem for identical particles, it is interesting to study one more time in more detail the products of X operators intervening in the calculations. This is the purpose of the present section, which constitutes a useful aside. The Z_j operators (10.51) are very analogous, trace operation aside, to the transfer matrix of the eight-vertex model (8.66). In fact, the analogy with a vertex system is complete by virtue of the following

Lemma:

$$Z_j = - \lim_{k \to k_j} Z(k), \tag{10.52}$$

with the definitions

$$Z(k) = \mathop{\mathrm{Tr}}_0 X_{10} X_{20} \ldots X_{N0}, \tag{10.53}$$

$$X_{j0} = \frac{1 - x_{j0}(j0)}{1 + x_{j0}}, \qquad x_{j0} = \frac{ic}{k_j - k}, \tag{10.54}$$

in which we have introduced a supplementary index denoted 0, associated with a supplementary variable k. The partial trace operation $\mathop{\mathrm{Tr}}_0$ in π_{N+1} is defined either in abstract manner as

$$\mathop{\mathrm{Tr}}_0 (j0) = 1, \qquad \forall j \tag{10.55}$$

or by adopting the classical representation

$$(j0) = P_{j0} = \sigma_j \cdot \tau \tag{10.56}$$

with

$$\sigma \cdot \tau = \sum_{\alpha, \beta} \sigma^{\alpha\beta} \tau^{\beta\alpha} \equiv P_{\sigma\tau} \tag{10.57}$$

in which $\sigma^{\alpha\beta}$ is the order-n matrix of which the only nonzero element takes value 1 for the $\beta \to \alpha$ transition. One can check that

$$\mathop{\mathrm{Tr}}_\tau P_{\sigma\tau} \equiv \mathop{\mathrm{Tr}}_0 P_{j0} = \sum_{\alpha=1}^n \sigma^{\alpha\alpha} = 1.$$

This representation generalizes the well-known spin representation, for $n = 2$, to symmetry types having Young tableaux with n rows.

The proof is elementary:

$$\lim_{k \to k_j} X_{i0} = (0j) X_{ij}(0j) \qquad (i \neq j)$$

$$\lim_{k \to k_j} X_{j0} = -(0j) \tag{10.58}$$

$$Z(k_j) = - \mathop{\mathrm{Tr}}_0 X_{1j} \ldots X_{j-1\,j}(0j) X_{j+1\,j} \ldots X_{Nj} = -Z_j. \tag{10.59}$$

The commutation of the Z_j among themselves originates from the *proposition*

$$[Z(k), Z(k')] = 0, \qquad \forall k, k'. \tag{10.60}$$

The proof, which rests on the ternary relations between the X, is identical to that for (8.67).

The $Z(k)$ matrices can be considered as a one-parameter family of transfer matrices for a vertex model. If $n = 2$, we have an inhomogeneous six-vertex model, which is the key to finding the solution of the two-state fermion problem following Yang's method. For arbitrary n, we also have a vertex model in which each row possesses n possible states. This problem could be solved completely using Sutherland's method, which is exposed in Chapter 11.

10.3.2 Ternary conditions as associativity conditions for a field algebra

The commutativity of the transfer matrices $T(v)$ and $Z(k)$, and the factorization of the scattering matrix S into two-body scattering matrices, are properties which follow directly from the ternary relations already encountered in various forms: (8.40), (8.55), (H.3), (10.48).

Without going too far away from the subject of the current chapter, it seems interesting to show here the intimate link between these relations and, on the one hand, the associativity conditions of a field operator algebra considered by Zamolodchikov (1979) for the construction of factorized S matrices and, on the other hand, the algebra of *connection matrices* for completely integrable differential systems considered by Faddeev (1980). It is, however, not our intention to deepen the inverse problem methods for integrable models despite the large number of connections with Bethe's method, and we refer the reader to developments in the recent literature concerning these questions.

Whether we talk about the XYZ model or that for δ-function interacting identical particles, the ternary relation can be written

$$(w\sigma\tau)(w'\sigma\tau') = X(w'\sigma\tau')(w\sigma\tau)X^{-1} \tag{10.61}$$

with

$$X = (x\tau\tau').$$

The products $\sigma_i\tau_j\tau'_k$ form a basis of matrices operating on $\mathbb{R}_N \otimes \mathbb{R}_N \otimes \mathbb{R}_N$, and the quantities $w(w', x)$ depend on an additive parameter v or k. Let us introduce operators $A(k)$ such that

$$(w\sigma\tau) \equiv \tau \cdot A. \tag{10.62}$$

We write (10.61) as

$$\tau \cdot A(k)\tau' \cdot A(k') = X\tau' \cdot A(k')\tau \cdot A(k)X^{-1},$$

which we can rewrite explicitly in the form

$$A_i(k)A_j(k') = \sum S_{ij,\alpha\beta}(k - k')A_\beta(k')A_\alpha(k) \tag{10.63}$$

with the expression for the S matrix appearing as the coefficient of $\tau_i \tau_j'$ in the expansion of $X\tau_\beta'\tau_\alpha X^{-1}$ or, if one prefers,

$$S_{ij,\alpha\beta} = \text{Tr }\tau_i\tau_j' X\tau_\beta'\tau_\alpha X^{-1}, \tag{10.64}$$

provided we choose τ matrices verifying $\text{Tr }\tau_\alpha\tau_\beta = \delta_{\alpha\beta}$.

If we now consider the $A_i(k)$ operators as asymptotic fields with n components in the momentum (or rapidity) representation, formula (10.63) defines the commutation of two fields as well as the two-body scattering matrix. The N-body S matrix will be defined uniquely by the rule ($N = 3$)

$$A_i(k_1)A_j(k_2)A_l(k_3) = \sum S_{ijl,\alpha\beta\gamma}(k_1k_2k_3)A_\gamma(k_3)A_\beta(k_2)A_\alpha(k_1) \tag{10.65}$$

if the product of permutations (12)(13)(12) performed with the help of (10.63) on the left-hand side of (10.65) is equivalent to the product (23)(12)(23), in other words if it produces an identical right-hand side. From this we get the associativity condition

$$S_{ij,\lambda\mu}(k_1 - k_2)S_{\lambda l,\alpha\nu}(k_1 - k_3)S_{\mu\nu,\beta\gamma}(k_2 - k_3)$$
$$= S_{jl,\lambda\mu}(k_2 - k_3)S_{i\mu,\nu\gamma}(k_1 - k_3)S_{\nu\lambda,\alpha\beta}(k_1 - k_2) \equiv S_{ijl,\alpha\beta\gamma}(k_1k_2k_3), \tag{10.66}$$

very faithfully illustrated in Figure 10.2.

The general ternary relation is thus an associativity condition for the field algebra of a totally elastic system admitting an S matrix which factorizes into two-body S matrices:

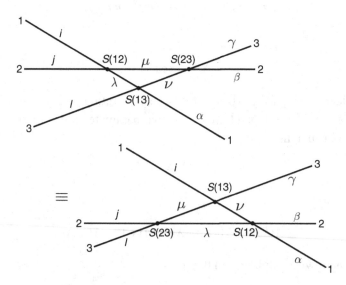

Figure 10.2 'Totally elastic' collision of three particles.

$$S(123) = S(12)S(13)S(23) = S(23)S(13)S(12). \tag{10.67}$$

The $S(12)$ matrices are identical to the X_{12} operators and to Baxter's transfer (micro)matrices R. In terms of scatterings or of the lattice, the illustration in Figure 10.2 translates the invariance (10.67) by expressing the fact that the 'line k_1 can be moved through the crossing point (23)'.

In Chapter 13, we shall come back to the problem posed by the system of algebraic equations (10.66) for the unknowns S, a system for which we already know some particular solutions associated with certain symmetries.

10.3.3 Ternary relation and complete integrability

The ternary relations lead to more elaborate properties than the straightforward commutativity of the transfer matrices for the products of X operators which form their structure. Let us call a *transition matrix* (from site 1 to site j) the product

$$T_j(k) = X_{10}X_{20}X_{30}\ldots X_{j0}, \tag{10.68}$$

in which j is the index of a given site.

Considering $T_j(k)$ as a function of parameter k, it is straightforward to show that the algebra of transition matrices is isomorphic to that of the field operators $A(k)$ considered in the preceding subsection (10.63). One thus obtains a set of realizations of the ternary relations. We shall give a few explicit expressions for these in the next subsection, but here we want to motivate their consideration by showing that the matrices (10.68) naturally intervene in the context of the 'inverse scattering method' for integrable systems in $1 + 1$ dimension.

Indeed, as a function of the site j, the matrix T_j verifies the first-order difference equation

$$T_j = X_{0j}T_{j-1}, \tag{10.69}$$

whose coefficients are operators σ_j depending on the site j. This is reminiscent of the auxiliary differential system of the inverse method, whose coefficients are local operators. If, after imposing an appropriate normalization, X is in the neighbourhood of the unit operator, taking the 'continuum limit' of matrix (10.69) will be possible and inversely, (10.69) could be obtained by discretizing a differential system. The T_j matrix can be interpreted as a spatial transmission operator for a wave in a potential and, in the continuum limit, appropriately normalized, will tend to a scattering operator function of the incoming energy (in other words of the spectral parameter k or v) when the propagation interval L itself becomes infinite. The limit form of the propagation equation will be

$$i\frac{\partial T}{\partial x} = L(x)T, \tag{10.70}$$

in which L is a local operator.

To illustrate this point, let us take the example of Yang's operators (10.54) in the matrix representation of π_N:

$$(0j) = \sigma_j^{\alpha\beta}\sigma^{\beta\alpha} = \sigma_j \cdot \tau. \tag{10.71}$$

It is straightforward to verify, using a completely analogous calculation to that in Subsection 8.3.3, that if all parameters k_j are equal, the transfer matrices $Z(k)$ commute with the generalized exchange Hamiltonian

$$H = \sum_{j=1}^{N}(j, j+1) = \sum_j \sigma_j \cdot \sigma_{j+1}. \tag{10.72}$$

The complete diagonalization of this Hamiltonian will incidentally be performed in Chapter 12. In the continuum limit we would find, after an appropriate normalization,

$$L(x) = \sigma(x) \cdot \tau, \tag{10.73}$$

in which the ultralocal fields $\sigma^{\alpha\beta}(x)$ obey the commutation relations of the Lie algebra associated with the unimodular group. Hamiltonian (10.72) would take the limit form

$$\mathscr{H} = -\frac{1}{2}\int_0^L \frac{d\sigma^{\alpha\beta}}{dx} \cdot \frac{d\sigma^{\beta\alpha}}{dx}dx. \tag{10.74}$$

We know that, for $n = 2$, the system is integrable since it is the isotropic form of the classical Landau and Lifshitz Hamiltonian, or Heisenberg in the quantum case.

Equation (10.70) is the auxiliary equation of the inverse method which, together with a time evolution equation of the type

$$i\frac{\partial T}{\partial t} = M \cdot T, \tag{10.75}$$

will define an integrable system, the 'Lax pair' (L, M) verifying the compatibility condition

$$i\left(\frac{\partial L}{\partial t} - \frac{\partial M}{\partial x}\right) + [L, M] = 0. \tag{10.76}$$

It is not our intent in this exposition to study the continuum limit, which leads to problems in defining the products of local field operators and in renormalization, and we refer the reader to recent developments of the inverse method for integrable systems (Faddeev, 1980 and references cited therein). Going back to the discrete case (equations (10.69) and (10.72)), it is possible to define a Lax pair (Sklyanin, Takhtadzhyan and Faddeev, 1979). If we complement (10.69) with the relation

$$i\frac{dT_j}{dt} = M_j T_j, \tag{10.77}$$

the compatibility condition between (10.69) and (10.77) is written

$$i\frac{dX_{0j}}{dt} = M_j X_{0j} - X_{0j} M_{j-1}. \tag{10.78}$$

Let us show how to determine M_j, the time evolution being governed by the Hamiltonian H (10.72):

$$i\frac{dX_{0j}}{dt} = \left[(j-1, j) + (j, j+1), X_{0j}\right]. \tag{10.79}$$

It is thus sufficient for there to be an $\tilde{X}_j$ such that

$$M_j X_{0j} = \left[(j, j+1), X_{0j}\right] + \tilde{X}_j,$$
$$X_{0j} M_{j-1} = -\left[(j-1, j), X_{0j}\right] + \tilde{X}_j. \tag{10.80}$$

Eliminating M_j between the two relations (10.80), we obtain for the unknown operators $\tilde{X}$:

$$\left[(j, j+1), X_{0j+1} X_{0j}\right] = \tilde{X}_{j+1} X_{0j} - X_{0j+1} \tilde{X}_j, \tag{10.81}$$

which reminds us precisely of Sutherland's relation (8.25), derived from the ternary relation (for the XYZ model) in (8.80). In the specific case with which we are concerned, it is sufficient to differentiate both sides of the relation $X_{12} X_{02} X_{01} = X_{01} X_{02} X_{12}$ with respect to k_1 and set $k_1 = k_2 = 0$; taking into account that $X'_{12}|_{k_1=k_2} = -i(1 + (12))$, we have

$$X'_{12} X_{02} X_{01} - X_{01} X_{02} X'_{12} = (12)(X_{02} X'_{01} - X'_{02} X_{01}) \tag{10.82}$$

or further still

$$[(12), X_{02} X_{01}] = i(X_{02} X'_{01} - X'_{02} X_{01}). \tag{10.83}$$

Comparing relations (10.81) and (10.83) gives us the choice

$$\tilde{X}_j = -i X'_{0j} = -i\frac{1 + (0j)}{(k+i)^2}. \tag{10.84}$$

The operator M_j is then defined by (10.80):

$$M_j = (j, j+1)(1 - X_{0j+1} X_{j0}) - i X'_{0j} X_{j0}$$

or

$$M_j = \frac{1}{k^2+1}\{1 + (j, j+1) + (0j) - (0, j+1) + ik((0, j+1)(0j)$$
$$- (0j)(0, j+1))\} \tag{10.85}$$

or still

$$M_j = \frac{1}{k^2+1}\{\tau \cdot (\sigma_j - \sigma_{j+1})(1 - ik\tau \cdot (\sigma_j + \sigma_{j+1})) + \sigma_j \cdot \sigma_{j+1} + 1\}. \tag{10.86}$$

The existence of an M_j verifying (10.78), which is obtained from the ternary relations by differentiation (see also Chapter 8, Subsection 8.3.2), thus leads to the existence of constants of motion generated by $Z(k)$ and integrability. This last notion is defined for a quantum system by correspondence with the complete integrability of a classical Hamiltonian system, in other words, for N degrees of freedom, by the existence of $N-1$ independent integrals of motion whose Poisson brackets or commutators with H vanish.

10.3.4 Algebra of the transition matrices

In this subsection, we give an explicitly worked-out example of the algebra of ternary relations in the form (10.63).

Yang operators

We put ourselves, for convenience, in the conjugate representation of π_N:

$$X_{j0} = \frac{1 + x_j(j0)}{1 + x_j} \tag{10.87}$$

with

$$x_j = \frac{i}{k - k_j} \quad (c = 1). \tag{10.88}$$

We have the transition matrix of elements

$$T_{\alpha\beta}(k) = \text{Tr } \tau_{\alpha\beta} X_{10} X_{20} \dots X_{N0}. \tag{10.89}$$

We form the product

$$T_{\alpha\beta}(k) T_{\gamma\delta}(k') = \underset{\tau,\tau'}{\text{Tr }} \tau_{\alpha\beta} \tau'_{\gamma\delta} X_{10} X_{10'} X_{20} X_{20'} \dots$$

and apply the ternary relations

$$T_{\alpha\beta}(k) T_{\gamma\delta}(k') = \underset{\tau\tau'}{\text{Tr }} X_{00'} \tau_{\alpha\beta} \tau'_{\gamma\delta} X_{0'0} X_{10'} X_{10} \dots \tag{10.90}$$

with

$$X_{00'} = \frac{1 + x_{00'} \tau \cdot \tau'}{1 + x_{00}}, \quad x_{00'} = \frac{i}{k - k'}. \tag{10.91}$$

It is thus sufficient to calculate

$$(1 + x_{00'} \tau \cdot \tau') \tau_{\alpha\beta} \tau'_{\gamma\delta} (1 - x_{00'} \tau \cdot \tau') \equiv \tau_{\alpha\beta} \tau'_{\gamma\delta} + x_{00'} (\tau_{\gamma\beta} \tau'_{\alpha\delta} - \tau_{\alpha\delta} \tau'_{\gamma\beta}) - x^2_{00'} \tau_{\gamma\delta} \tau'_{\alpha\beta}$$

and substitute this in (10.90) to obtain the result, which we can represent in the form

$$\left[T_{\alpha\beta}(k), T_{\gamma\delta}(k')\right] = \frac{1}{(k-k')^2+1}(T_{\alpha\beta}(k')T_{\gamma\delta}(k) - T_{\gamma\delta}(k')T_{\alpha\beta}(k))$$
$$+\frac{i(k-k')}{(k-k')^2+1}(T_{\alpha\delta}(k')T_{\gamma\beta}(k) - T_{\gamma\beta}(k')T_{\alpha\delta}(k)). \quad (10.92)$$

From this, we can deduce some corollaries:

(a)

$$\left[T_{\alpha\beta}(k), T_{\alpha\beta}(k')\right] = 0. \quad (10.93)$$

(b)

$$\alpha = \beta = \gamma: \quad T_{\gamma\gamma}(k)T_{\gamma\delta}(k') = \frac{1}{1+i(k-k')}T_{\gamma\gamma}(k')T_{\gamma\delta}(k)$$
$$+\frac{i(k-k')}{1+i(k-k')}T_{\gamma\delta}(k')T_{\gamma\gamma}(k). \quad (10.94)$$

(c) Exchanging the role of k and k' in (10.94) we deduce by elimination

$$T_{\gamma\gamma}(k)T_{\gamma\delta}(k') = \left(1+\frac{i}{k-k'}\right)T_{\gamma\delta}(k')T_{\gamma\gamma}(k) - \frac{i}{k-k'}T_{\gamma\delta}(k)T_{\gamma\gamma}(k'). \quad (10.95)$$

One will have noticed the isomorphism of this relation with the transition equation for the amplitudes $\langle Q||P\rangle$ (10.11) (in the conjugate representation).

(d) $n = 2$. Let us set

$$T_{12} = T_+, \quad T_{21} = T_-, \quad T_{11} - T_{22} = T_z, \quad (10.96)$$
$$\left[T_+(k), T_+(k')\right] = \left[T_z(k), T_z(k')\right] = 0 \quad (10.97)$$

and according to (10.95),

$$T_{11}(k)T_+(k') = (1+x_{00'})T_+(k')T_{11}(k) - x_{00'}T_+(k)T_{11}(k'),$$
$$T_{22}(k)T_+(k') = (1-x_{00'})T_+(k')T_{22}(k) + x_{00'}T_+(k)T_{22}(k'). \quad (10.98)$$

These relations were made explicit by Faddeev (1980) in the framework of quantum field theory. We have here a representation of the transpositions between neighbours isomorphic to (10.20), which we had already used to solve the fermion problem (see Subsection 11.3.1 and Gaudin, 1967a,b). The analogous relations for the anisotropic Heisenberg chain are given explicitly in Appendix L and lead to the most elegant solution.

10.4 On the models of Hubbard and Lai

10.4.1 Definition

Before tackling in the next chapter the general solution to the problem defined in the present one, we must mention the Hubbard model, which can be considered as a discrete variant of the system with δ-interactions. It thus pertains to the same methods.

This model consists of a periodic atom chain, but these can bind a variable number of $S = 1/2$ valence electrons: 0, 1 or 2. If two electrons sit on the same atom, necessarily in the singlet state, and only in this case, the interaction energy is U_0 by hypothesis. The kinetic part of the energy is associated with the hoppings of electrons between neighbouring sites without a change in their spin. We thus have a single electronic band whose filling, the total number of electrons M, is given, with $0 \leq M \leq 2N$.

In second quantization the Hamiltonian of the model is thus written

$$H = -\frac{1}{2} \sum_{n,\sigma} b_{n\sigma}^{\dagger} b_{n+1\sigma} + b_{n+1\sigma}^{\dagger} b_{n\sigma} + \frac{1}{2} U_0 \sum_{n,\sigma} b_{n\sigma}^{\dagger} b_{n\sigma} b_{n\bar{\sigma}}^{\dagger} b_{n\bar{\sigma}}. \qquad (10.99)$$

This is therefore a system of spin-1/2 electrons whose interaction is explicitly independent of spin. Two electrons interact only if they are on the same site, necessarily of opposite spin by virtue of Pauli's principle. The amplitude $\Psi(n_1\sigma_1, n_2\sigma_2, \ldots, n_M\sigma_M)$ thus factorizes into orbital and spin functions of complementary symmetry. The orbital amplitude $a(n_1 n_2 \ldots n_M)$, which depends only on the electronic coordinates along the chain $1 \leq n_j \leq N$, obeys the Schrödinger equation issued from Hamiltonian (10.99):

$$\sum_{j=1}^{M} a(\ldots n_j + 1 \ldots) + a(\ldots n_j - 1 \ldots) - 2U_0 \sum_{j<l} \delta(n_j - n_l) a(\ldots)$$
$$= -2Ea(n_1 \ldots n_M), \qquad (10.100)$$

of which it will suffice to look for solutions of a given symmetry type. For spin-1/2 fermions, the symmetry is that of the Young tableau with two columns $[1^S, 2^{M/2-S}]$ in which S is the total spin.

In (10.100), we have the discrete analogue of the equation associated with Hamiltonian (10.1) for identical particles with δ-interactions. Here the 'gluing conditions' and the definition of 'sectors' is somewhat more delicate than in the continuum case. One, however, ends up with similar relations for the coefficients appearing in Bethe's expansion for the amplitude.

In any case, one will have noticed the differences more than the analogies, between equation (10.100) and equation (1.14) for the amplitudes of the magnetic chain: local interaction and not between neighbours; absence of constraint on the coordinates; arbitrary symmetry type.

10.4.2 Method

Following the now familiar method, the amplitude is expanded in each sector D_Q into a sum of plane waves with equal energy

$$E = -\sum_{j=1}^{M} \cos k_j$$

defined by a set of momenta $\{k\}$:

$$a(n_1 n_2 \dots n_M) = \sum_{P \in \pi_M} \langle Q||P \rangle e^{i(\bar{P}k, \bar{Q}n)}. \tag{10.101}$$

The slightly delicate point here is the definition of the sectors, since the n can be equal. One first defines D_Q for distinct n:

$$D_Q : n_{Q1} < n_{Q2} < \dots < n_{QM}.$$

One then extends their definition to the case of equalities: the intersection is then non-empty. Two adjacent sectors D_Q and $D_{Q(23)}$ have a common part, $\dots n_{Q2} = n_{Q3}$. One imposes that the wavefunction (10.101) has the same value in the intersection of D_Q and $D_{Q(23)}$, which gives

$$\langle Q||P \rangle + \langle Q||P(23) \rangle = \langle Q(23)||P \rangle + \langle Q(23)||P(23) \rangle. \tag{10.102}$$

The form (10.101) leads to the fact that the difference equation is verified outside the intersection of sectors. It is sufficient to satisfy it on these intersections. In $D_Q \cap D_{Q(23)}$, $n_{Q2} = n_{Q3} = n$, we isolate the interesting terms in (10.100):

$$\dots + a(\dots n+1, n, \dots) + a(\dots n-1, n, \dots) + a(\dots n, n+1, \dots)$$
$$+ a(\dots n, n-1, \dots) - 2U_0 a(\dots n, n, \dots) = -2Ea(\dots n, n, \dots). \tag{10.103}$$

We first notice that the first and fourth terms written on the left-hand side of (10.103) have their arguments not in Q, but in $Q_{(23)}$. We then subtract from (10.103) the following relation, verified by any amplitude of a given sector, here Q:

$$\dots + a_Q(\dots n+1, n, \dots) + a_Q(\dots n-1, n, \dots) + a_Q(\dots n, n+1, \dots)$$
$$+ a_Q(\dots n, n-1, \dots) = -2Ea_Q(\dots n, n, \dots), \tag{10.104}$$

to obtain

$$a_{Q(23)}(\dots n+1, n, \dots) - a_Q(\dots n+1, n, \dots) + q_{Q(23)}(\dots n, n-1, \dots)$$
$$- a_Q(\dots n, n-1, \dots) = -2Ea_Q(\dots, n, n, \dots). \tag{10.105}$$

Substituting expansions (10.101) in (10.105), we obtain

$$\sum_P (\langle Q(23)||P\rangle - \langle Q||P(23)\rangle) \times (e^{\cdots +ik_{P2}n + ik_{P3}(n+1)+\cdots} + e^{\cdots +ik_{P2}(n-1)+ik_{P3}n+\cdots})$$

$$+2U_0\langle Q||P\rangle e^{\cdots +ik_{P2}n + ik_{P3}n+\cdots} = 0. \tag{10.106}$$

Associating the terms P and $P(23)$ in the above sum, we have

$$(\langle Q(23)||P\rangle - \langle Q||P(23)\rangle) \cdot i(\sin k_{P3} - \sin k_{P2}) + U_0(\langle Q||P\rangle + \langle Q||P(23)\rangle) = 0. \tag{10.107}$$

This is the same relation as for the continuum model (10.11):

$$\langle Q(23)||P(23)\rangle = \langle Q||P\rangle(1 + x_{P2,P3}) + \langle Q||P(23)\rangle x_{P2,P3},$$

with

$$x_{ij} = \frac{iU_0}{\sin k_i - \sin k_j}. \tag{10.108}$$

Bethe's hypothesis will thus be justified in the same manner, and the theory of the Hubbard model then becomes identical in structure to that of the continuum model, to which it tends in the limit $N \to \infty$, $\frac{U}{c} = \frac{L}{N}$, c, L, M finite (Lieb and Wu, 1968).

10.4.3 Lai's model

We shall content ourselves with giving the definition of Lai's model (1974), which relates to both the Hubbard model and the magnetic chain. It is, however, solvable by Bethe–Yang's method only for a specific value of the coupling. The wave equation is written

$$\sum_{j=1}^{M} \left(\Psi(\ldots, n_j + 1, \ldots) + \Psi(\ldots, n_j - 1, \ldots) - 2\Psi(\ldots)\right)$$

$$+ \sum_{j=1}^{M} \pm(1 + (j, j+1))\delta(n_{j+1} - n_1 - j)\Psi = -2E\Psi. \tag{10.109}$$

In addition, ψ vanishes as soon as two space coordinates coincide. The permutation operator $(j, j + 1)$ acts on the spin and space variables. We thus have a Hamiltonian for particles of arbitrary spin whose mutual interaction comprises an infinitely repulsive hard core and a nearest-neighbour interaction whose intensity is equal to that of the kinetic term.

11

Identical particles with δ-interactions: General solution for two internal states

11.1 The spin-1/2 fermion problem

11.1.1 Symmetry conditions

We treat here the more specific problem of spin-1/2 fermions with a δ-interaction, in other words the problem of identical particles with two internal states. This is a useful step on the way to the general solution, which is rather complicated.

The Hamiltonian (10.1) being independent of the internal variables, a wavefunction totally antisymmetric in the N particle indices belonging to total spin $S = S_z = \dfrac{N}{2} - M = \bar{M} - \dfrac{N}{2}$:

$$\Psi_S = \sum_{\pi \in \pi_N} I(\pi) \chi_+(\pi 1) \ldots \chi_+(\pi \bar{M}) \chi_-(\pi(\bar{M}+1)) \ldots$$
$$\chi_-(\pi N) \psi(x_{\pi 1}, x_{\pi 2}, \ldots, x_{\pi N}) \tag{11.1}$$

can be recomposed into a sum of products of a spin function $\chi_{\mathscr{C}}$ with an orbital function $\psi_{\mathscr{C}}$ belonging to conjugate representations of π_N (Hammermesh, 1964). The spin wavefunctions, which cannot be antisymmetric with respect to more than $2S + 1 = 2$ variables, have the symmetry type of the two-row tableau $[\bar{M}, N - \bar{M}]$ in which M is determined by the spin $S = S_z$. The orbital function thus belongs to type

$$\mathscr{C} = [1^{2\bar{M}-N}, 2^{N-\bar{M}}] = [1^{N-2M}, 2^M].$$

It is separately antisymmetric in the $\bar{M}$ arguments of the first column and the $M = N - \bar{M}$ ones of the second, and obeys the *Fock condition*. This condition expresses the fact that the orbital function cannot be antisymmetrized with respect to $M + 1$ arguments:

223

$$F \cdot \psi = 0, \quad F = \sum_{j=1}^{\bar{M}} (jm) - 1 \quad (m > \bar{M}), \quad \psi = \psi(x_1 x_2 \ldots x_{\bar{M}} | x_{\bar{M}+1} \ldots x_N).$$

$$(11.2)$$

Finally, the sum $\Psi = \sum \chi_{\mathscr{C}} \psi_{\mathscr{C}}$ runs over the Young tableaux numbered with $\mathscr{C}$; for example, for $N = 3$, $S = \dfrac{1}{2}$, we have $\bar{M} = 2$ and the Young tableau ⬜⬜/⬜; there are two tableaux $\mathscr{C}$: $\boxed{\begin{array}{cc}1&2\\\hline 3\end{array}}$, $\boxed{\begin{array}{cc}1&3\\\hline 2\end{array}}$.

Let us translate the symmetry conditions into conditions for the amplitudes $\langle Q \| P \rangle$. According to the expansion (10.6) of the real-space wavefunction, the antisymmetry in the arguments of each column is equivalent to the conditions

$$\langle Q(j, j+1) \| P \rangle = -\langle Q \| P \rangle \quad \forall j = 1, 2, \ldots, \bar{M} - 1, \bar{M} + 1, \ldots, N - 1 \quad (11.3)$$

and, using the transition matrices (10.11),

$$\langle Q \| P(j, j+1) \rangle = -\langle Q \| P \rangle, \quad j \neq \bar{M}. \tag{11.4}$$

In other words, in each sector, the antisymmetry in each of the two groups $x_{Q1}, \ldots, x_{Q\bar{M}}$ and $x_{Q(\bar{M}+1)}, \ldots, x_{QN}$.

The Fock condition can also be written explicitly, though this is not useful in Yang's method. It suffices to calculate the transpositions (n, m) $(m = \bar{M} + 1, n \leq \bar{M})$ starting from the Y generators, or T as in (10.21). Let us give the result without proof:

$$\langle F \| P \rangle \equiv \sum_{k=1}^{\bar{M}} y_{Pk, P1} y_{Pk, P2} \cdots y_{Pk, Pm} \langle I \| P(km) \rangle$$

$$- y_{Pm, P1} y_{Pm, P2} \cdots y_{Pm, P\bar{M}} \langle I \| P \rangle = 0, \tag{11.5}$$

$$y_{jl} = 1 + x_{jl}, \quad x_{jl} = \frac{ic}{k_j - k_l}, \tag{11.6}$$

which is a relation between amplitudes in a given sector (Gaudin, 1967a).

These explicit relations are useless in Yang's method, which simply notices that the symmetry conditions can be expressed on the elements $|\bar{P}\rangle$ of the ring of π_N (10.19):

$$(j, j+1) | \bar{P} \rangle = -|\bar{P}\rangle, \quad j \neq \bar{M}, \quad F | \bar{P} \rangle = 0, \tag{11.7}$$

and thus the elements $|\bar{P}\rangle$ belong to the representation $\mathscr{C}$ of the symmetric group π_N. But we know a convenient realization of the conjugate representation $\tilde{\mathscr{C}}$; this is the spin representation already invoked in Chapter 8 for $n = 2$ and in (10.56) for arbitrary n ($n = 2s + 1$). From this, we obviously deduce a realization of $\mathscr{C}$ by using the conjugate basis

$$|P\rangle \rightarrow |\tilde{P}\rangle = \sum_Q I(Q)\bar{Q}\langle Q||P\rangle \quad (I(Q) = \text{parity sign of } Q) \tag{11.8}$$

with

$$(\tilde{0j}) = -(0j) = -\frac{1}{2}(1 + \vec{\tau} \cdot \vec{\sigma}_j), \tag{11.9}$$

which allows us to represent the $\tilde{X}_{jl}$ and $\tilde{Z}$ in the $\mathscr{C}$ representation. However, to lighten the notation, we shall omit the $\tilde{\ }$ sign on the operators.

11.1.2 Reduction to an inhomogeneous six-vertex model

According to the results of Subsections 10.2.4, 10.3.1 and 11.1.1, the periodicity and symmetry conditions lead to the diagonalization of a family of commuting operators $Z(k)$ (10.53) in the $\mathscr{C}$ representation of π_N. According to (10.50), the coupled equations for the spectrum of the fermion system will be written

$$Z_j = Z(k_j) = e^{ik_j L}, \quad j = [1, N], \tag{11.10}$$

in which we find, on the left-hand side, the eigenvalues of the Z_j operator. The matrix $Z(k)$ is formed by using the X_{n0} operators of the conjugate representation $\tilde{\mathscr{C}}$:

$$X_{n0} = \frac{1 + \frac{1}{2}(1 + \tau \cdot \sigma_n)x_n}{1 + x_n},$$

$$x_n \equiv x_{n0} = \frac{ic}{k_n - k}, \quad y_n = 1 + x_n. \tag{11.11}$$

$Z(k)$ thus appears as the transfer matrix of an inhomogeneous six-vertex model

$$A_n = X_{n0} = \begin{pmatrix} a_n\left(\frac{1}{2} + S_n^z\right) + b_n\left(\frac{1}{2} - S_n^z\right) & c_n S_n^- \\ c_n S_n^+ & b_n\left(\frac{1}{2} + S_n^z\right) + a_n\left(\frac{1}{2} - S_n^z\right) \end{pmatrix} \tag{11.12}$$

with weights

$$a_n = 1, \quad b_n = \frac{1}{y_n}, \quad c_n = \frac{x_n}{y_n} \tag{11.13}$$

which are respectively those of type 1, 3 or 5 vertices (or 2, 4 and 6) (Chapter 7, Figure. 7.1).

We are thus facing an analogous problem to that treated in detail in Section 7.2, but the weights now vary with the site and it turns out that this inhomogeneous problem also falls to Baxter's method, generalized by Baxter (1971a). The reason is that the *relative invariants* are equal:

$$\frac{J_2}{J_1} = \frac{J_3}{J_1} = \frac{a_n^2 + b_n^2 - c_n^2}{2a_n b_n} \equiv 1, \tag{11.14}$$

by virtue of (11.13).

The transfer matrix commutes with the total spin component $\tilde{S}^z = \frac{1}{2}\sum_n \sigma_n^z$ associated with the $\tilde{\mathscr{C}}$ representation. More precisely, one has

$$\tilde{S}^z = S^z = \frac{N}{2} - M.$$

The eigenstate $|\tilde{I}\rangle$, using the notation of formula (10.46) with $P = I$, can be expanded in the form

$$|I\rangle \equiv |M\rangle = \sum_{n_1 < n_2 < \ldots} c(n_1 n_2 \ldots n_M) S_{n_1}^- S_{n_2}^- \ldots S_{n_M}^- |\frac{1}{2}N\rangle. \tag{11.15}$$

The matrix elements of Z can be distributed into two classes, as illustrated in Figure 11.1.

By inspection of the figure, we can write the matrix elements in the form

$$\langle m|Z_+|n\rangle = \frac{1}{y} D(0n_1)\tilde{D}(n_1 m_1)D(m_1 n_2)\ldots D(m_M N + 1),$$

$$\langle m|Z_-|n\rangle = \frac{1}{y}\tilde{D}(0m_1)D(m_1 n_1)\tilde{D}(n_1 m_2)\ldots \tilde{D}(n_M N + 1), \tag{11.16}$$

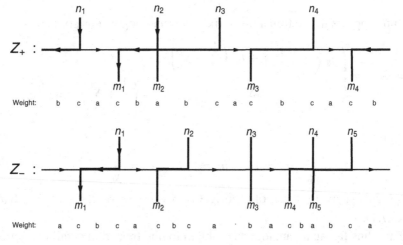

Figure 11.1 Elements of matrix Z.

with the definitions

$$D(mn) = x_n, \qquad\qquad m < n$$
$$D(nn) = \frac{1}{x_n}, \qquad\qquad m = n \qquad\qquad (11.17)$$

$$\tilde{D}(n, m) = y_{n+1}y_{n+2}\cdots y_{m-1}x_m, \quad n < m$$
$$\tilde{D}(n, n) = \frac{y_n}{x_n}, \qquad\qquad n = m$$
$$y = y_1 y_2 \cdots y_N, \qquad\qquad x_0 = x_{N+1} = 1. \qquad (11.18)$$

If we postulate the cyclic invariance of the $c(n)$ coefficients analogously to (7.40):

$$c(m_1 m_2 \ldots m_M) \equiv c(m_2 m_3 \ldots m_{M+1}), \qquad m_{M+1} \equiv m_1 + N, \qquad (11.19)$$

the eigenvalue equation for z,

$$Z|\bar{I}\rangle = z|\bar{I}\rangle, \qquad\qquad (11.20)$$

is simply written in the following explicit manner:

$$\sum_{\{n\}} c(n) \prod_{i=1}^{M} D(m_i n_i)\tilde{D}(n_i m_{i+1}) = zyc(m) \qquad (11.21)$$

in which the sum runs over all sets $\{n\}$ such that

$$m_i \le n_i \le m_{i+1}, \qquad 1 \le n_1 < n_2 < \ldots < n_M \le N. \qquad (11.22)$$

The cyclic invariance of c allows us to regroup the two types of terms, Z_+ and Z_-, into a single notational form.

11.1.3 Solution of the inhomogeneous model

Yang postulates an expansion of Bethe type for the $c(n)$ coefficient:

$$c(n) = \sum_{Q \in \pi_M} B(Q)\varphi(v_{Q1}, n_1)\varphi(v_{Q2}, n_2)\ldots\varphi(v_{QM}, n_M) \qquad (11.23)$$

with

$$\varphi(v, n) = y_1' y_2' \cdots y_{n-1}' x_n',$$
$$x_n' = x_n(v) = \frac{ic}{k_n - v}, \qquad y_n' = y_n(v) = 1 + x_n'. \qquad (11.24)$$

The set $\{v\}$ remains to be determined.

After substitution of the expansion (11.23) on the left-hand side of (11.21), we are led to calculate sums of the form

$$\mathscr{L}_2 = \sum_{n_2=m_2}^{m_3} D(m_2 n_2)\tilde{D}(n_2 m_3)\varphi(v, n_2)$$

$$= \frac{1}{x_{m_2}} y_{m_2+1} \cdots y_{m_3-1} x_{m_3} \underline{\varphi(v, m_2)} + y_{m_3} \underline{\varphi(v, m_3)}$$

$$+ \sum_{n=m_2+1}^{m_3-1} x_n y_{n+1} y_{n+2} \cdots y_{m_3-1} x_{m_3} \varphi(v, n), \qquad (11.25)$$

in which we have underlined the two extremal terms of the sum, called *dangerous terms* in Subsection 7.3.2 since they can contribute to the violation of rule (11.22). Taking (11.24) into account, we must perform the partial sum

$$\mathscr{L} = \sum_{n=m_2+1}^{m_3-1} y_1' y_2' \cdots y_{n-1}' x_n' x_n y_{n+1} \cdots y_{m_3-1} \qquad (11.26)$$

but

$$x_n' x_n = (y_n - y_n')x_0', \qquad x_0' = x_0(v) = \frac{ic}{k - v}, \qquad (11.27)$$

from which we deduce

$$\mathscr{L} = x_0'(y_1' y_2' \cdots y_{m_2}' y_{m_2+1} \cdots y_{m_3-1} - y_1' y_2' \cdots y_{m_3-1}'). \qquad (11.28)$$

We then obtain for $\mathscr{L}_2$:

$$\mathscr{L}_2 = (1 + x_0')\frac{1}{x_{m_2}} y_{m_2} y_{m_2+1} \cdots y_{m_3-1} x_{m_3}\varphi(v, m_2) + (1 - x_0')\varphi(v, m_3). \quad (11.29)$$

After substitution in (11.21), we thus had to sum over the n_i variables subject to restrictions (11.22). If we didn't take into account the fact that the n_i are distinct, it would suffice to independently sum over each n_i in the interval $m_i \le n_i \le m_{i+1}$. The overlap of two n's of successive indices can however occur, for example for $n_2 = m_3$ and $n_3 = m_3$. These terms originate from the product of two successive *dangerous terms*. It is thus necessary to eliminate them from the straight n sum, an operation that we will denote by _____ as in Section 7.2.

If we denote in a schematic way a sum such as $\mathscr{L}_2$ in (11.29) in the form $X_2 + Y_3$, we can write equality (11.21) in the condensed form

$$zyc(m) = \sum_Q B(Q)(X_1 + Y_2)(X_2 + Y_3)(X_3 + Y_4) \quad \cdots \quad (X_M + Y_{M+1}), \quad (11.30)$$

with

$$X_2 = (1 + x_0(v_{Q2}))\frac{1}{x_{m_2}} y_{m_2} \cdots y_{m_3-1} x_{m_3}\varphi(v_{Q2}, m_2) \qquad (11.31)$$

containing the dangerous term, cf. (11.25):

$$\frac{1}{x_{m_2}} y_{m_2+1} \dots x_{m_3} \varphi(v_{Q2}, m_2),$$

and

$$Y_3 = (1 - x_0(v_{Q2}))\varphi(v_{Q2}, m_3) \tag{11.32}$$

containing the dangerous term $y_{m_3}\varphi(v_{Q2}, m_3)$.

Following the reasoning of Section 3.2, we can eliminate from the sum (11.30) all terms containing any factor $Y_i X_i$ by choosing $B(Q)$ appropriately. For example, for the term containing $\varphi(v_{Q1}, m_2) \cdot \varphi(v_{Q2}, m_2)$, it is sufficient to choose

$$B(Q) \underbrace{Y_2 X_2}_{} |_Q + B(Q(12)) \underbrace{Y_2 X_2}_{} |_{Q(12)} = 0 \tag{11.33}$$

or in other words, taking definitions (11.31) and (11.32) into account,

$$B(Q)[(1 - x_0(v_{Q1}))(1 + x_0(v_{Q2})) - 1]$$
$$+ B(Q(12))[(1 - x_0(v_{Q2}))(1 + x_0(v_{Q1})) - 1] = 0. \tag{11.34}$$

One obtains

$$B(Q(12)) = -B(Q)\frac{v_{Q1} - v_{Q2} + ic}{v_{Q1} - v_{Q2} - ic},$$

from which one gets the unique solution for $B(Q)$:

$$B(Q) = \prod_{1 \le j < l \le M} \left(1 - \frac{ic}{v_{Qj} - v_{Ql}}\right). \tag{11.35}$$

11.1.4 Coupled equations for the fermion spectrum

Carrying on with the solution of the problem posed in the preceding subsection, it remains to express the cyclicity condition (11.19), which gives

$$B(QC) = y^{-1}(v_{Q1})B(Q) \tag{11.36}$$

with

$$y(v) = y'_1 y'_2 \dots y'_N. \tag{11.37}$$

From (11.35), we deduce that

$$B(Q)/B(QC) = \prod_{j=1}^{N} \left(1 + \frac{ic}{k_j - v_{Q1}}\right) = y(v_{Q1}). \tag{11.38}$$

This must hold irrespective of Q. We deduce from this that the M quantities V_a, $a = [1, M]$, are linked to k by the M algebraic equations

$$\prod_{b \neq a} \frac{v_a - v_b - ic}{v_a - v_b + ic} = \prod_{j=1}^{N} \left(1 + \frac{ic}{k_j - v_a}\right), \tag{11.39}$$

which determine the eigenvectors of Z.

Returning to the eigenvalues, let us examine (11.30). The cyclicity condition (11.36) has the effect of also eliminating the $X_1 Y_{M+1}$ term in (11.30). After summation over Q, there remains on the right-hand side of (11.30)

$$\sum_Q B(Q)(X_1 X_2 \ldots X_M + Y_2 Y_3 \ldots Y_{M+1}),$$

or in other words, taking (11.31) and (11.32) into account,

$$zyc(m) = \sum_Q B(Q) \left\{ y \sum_{a=1}^{M} (1 + x_0(v_a)) \varphi(v_{Q1}, m_1) \ldots \varphi(v_{QM}, m_M) \right.$$
$$\left. + \prod_{a=1}^{M} (1 - x_0(v_a)) \varphi(v_{Q1}, m_2) \ldots \varphi(v_{QM}, m_{M+1}) \right\}. \tag{11.40}$$

Owing to the cyclicity condition, equality (11.40) becomes an identity provided we choose

$$z = \prod_{a=1}^{M} \left(1 + \frac{ic}{k - v_a}\right) + \frac{1}{y} \prod_{a=1}^{M} \left(1 - \frac{ic}{k - v_a}\right),$$
$$y = \prod_{j=1}^{N} \left(1 + \frac{ic}{k_j - k}\right). \tag{11.41}$$

Equations (11.10), which determine the spectrum of the sets $\{k\}$, are written

$$z(k_j) = \prod_{a=1}^{M} \frac{k_j - v_a + ic}{k_j - v_a} = \exp(ik_j L). \tag{11.42}$$

We have thus obtained the coupled equations for the fermion problem according to Yang's method (1968). We note that the system (11.39) expresses the fact that the rational fraction $z(k)$ defined by (11.41) does not admit poles at $k = v_a$ ($a = 1, 2, \ldots, M$). This reminds us once again of property (8.87).

If we define the two polynomials

$$Q(k) = \prod_{a=1}^{M} (k - v_a) \quad \text{and} \quad \Phi(k) = \prod_{j=1}^{N} (k - k_j) \tag{11.43}$$

and set

$$Z_1(k) = Z(k)\Phi(k - ic) = \mathop{\mathrm{Tr}}_0 \prod_n (k - k_n - ic(n0)), \qquad (11.44)$$

property (11.41) is equivalent to

$$Z_1(k)Q(k) = Q(k + ic)\Phi(k - ic) + Q(k - ic)\Phi(k). \qquad (11.45)$$

There indeed exists a matrix Q commuting with Z and verifying (11.45) identically, but we shall not construct it.

11.2 The operatorial method

The simplest and most recent method to diagonalize the transfer matrix of the fermion model or of the eight-vertex model is certainly Faddeev's operatorial method (1980), which expresses the Bethe wavefunction in a compact form as a product of operators acting on a reference state. This condensed form of the Bethe 'Ansatz' had already been encountered in certain limiting cases, and the associated algebraic method could be used for a family of quadratic spin Hamiltonians deriving from the transfer matrices (Gaudin, 1976). This question will be reprised in Chapter 13.

The method essentially rests on the algebra of transition matrices from Subsection 10.3.4 and in our particular case on formulas (10.97) and (10.98).

The transfer matrix $Z(k)$ for spin-$1/2$ fermions in the conjugate representation $\tilde{\mathscr{C}}$ is nothing but

$$Z(k) = T_{11}(k) + T_{22}(k), \qquad (11.46)$$

in the notation of formula (10.89). Applied onto the ferromagnetic reference state $|N/2\rangle$, the X_{n0} operators (11.12) take the form of triangular matrices:

$$\prod_n X_{n0}|0\rangle = \prod_n \begin{pmatrix} a_n & c_n S_n^- \\ 0 & b_n \end{pmatrix} |0\rangle,$$

from which we deduce

$$T_{11}(k)|0\rangle = \prod_n a_n|0\rangle = |0\rangle,$$

$$T_{22}(k)|0\rangle = \prod_n b_n|0\rangle = y^{-1}|0\rangle \qquad (11.47)$$

(y: notation of (11.41)).

Given a set $\{v\}$ of M indeterminate quantities v_j, one considers the commuting family of operators $T_{12}(v_j) = T_+(v_j)$ and forms the state

$$|M\rangle = T_+(v_1)T_+(v_2)\dots T_+(v_M)|0\rangle. \qquad (11.48)$$

One then calculates $T_{11}|M\rangle$ by commuting T_{11} through the T_+, applying the transition formula (10.98) repeatedly:

$$T_{11}(k)T_+(v_j) = (1 + x_{0j})T_+(v_j)T_{11}(k) - x_{0j}T_+(k)T_{11}(v_j),$$
$$T_{11}(v_2)T_+(v_3) = (1 + x_{23})T_+(v_3)T_{11}(v_2) - x_{23}T_+(v_2)T_{11}(v_3), \quad (11.49)$$

$$x_{01} = \frac{ic}{k - v_1}, \qquad x_{23} = \frac{ic}{v_2 - v_3}, \text{ etc.} \qquad (11.50)$$

The following result is easily proven by recursion:

$$T_{11}(k)|M\rangle = \prod_{j=1}^{M}(1 + x_{0j})T_+(v_1)\ldots T_+(v_M)T_{11}(k)|0\rangle$$

$$+ x_{10}\prod_{j\neq 1}(1 + x_{1j})T_+(k)T_+(v_2)\ldots T_+(v_M)T_{11}(v_1)|0\rangle$$

$$+ x_{20}\prod_{j\neq 2}(1 + x_{2j})T_+(v_1)T_+(k)T_+(v_3)\ldots T_{11}(v_2)|0\rangle + \ldots$$

$$+ x_{M0}\prod_{j\neq M}(1 + x_{Mj})T_+(v_1)\ldots T_+(v_{M-1})T_+(k)T_{11}(v_M)|0\rangle. \, (11.51)$$

This recursive method completely parallels that of Subsection 11.1.1, formula (11.6) (Gaudin, 1967a). One gets an analogous formula for $T_{22}|M\rangle$ by changing $+x$ to $-x$ in (11.51). Summing term by term and using (11.47), one gets

$$T_{11}(v_j)|0\rangle = |0\rangle,$$
$$T_{22}(v_j)|0\rangle = y(v_j)^{-1}|0\rangle. \qquad (11.52)$$

If we can choose $\{v\}$ so as to cancel all coefficients of operators containing $T_+(k)$, we get

$$(T_{11}(k) + T_{22}(k))|M\rangle = \left\{\prod_{j=1}^{M}(1 + x_{0j}) + \frac{1}{y}\prod_{j=1}^{M}(1 - x_{0j})\right\}|M\rangle \qquad (11.53)$$

with the conditions

$$\prod_{j\neq m}(1 + x_{mj}) - \frac{1}{y(v_m)}\prod_{j\neq m}(1 - x_{mj}) = 0, \qquad m = 1, 2, \ldots, M. \qquad (11.54)$$

From this we get the M relations, identical to (11.39):

$$y(v_i) = \prod_{j}{}' \frac{1 - x_{ij}}{1 + x_{ij}} = \prod_{j}{}' \frac{v_i - v_j - ic}{v_i - v_j + ic},$$

while the eigenvalue given by (11.53) coincides with the expression for z in (11.41).

Expression (11.48) is the most condensed operator form of the Bethe expansion (11.15), (11.23).

11.3 Sketch of the original solution of the fermion problem

11.3.1 First step

The solution method due to Yang, exposed in Section 11.1, is certainly the most natural and in line with Lieb's works on the ice models and those of Baxter on the extension of Bethe's method to inhomogeneous models. Moreover, it leads directly to Sutherland's generalization to an arbitrary type of symmetry (see Chapter 12). In contrast, Faddeev's method is the simplest and most direct one to obtain the coupled equations and express the Bethe sum in operator form, which could represent substantial progress for the calculation of norms or correlations. However, the original solution (Gaudin, 1967a) presents itself as a rather intuitive path to follow, which reaches its goal through the use of a certain number of algebraic or geometric lemmas. We shall devote this section to a review of some points of this approach and of an algebraic identity in which the Bethe sum admits an explicit compact form.

The first step consists of writing the periodicity condition in L for the wavefunction (11.2) for a first-column variable, for example $x_{\bar{M}}$. Commuting, with the help of the transition matrix from one sector to its neighbour, particle $x_{\bar{M}}$ through the M opposite spin particles x_n, $\bar{M} < n \leq N$, one obtains, as for the Fock condition (11.6), a homogeneous linear system for the amplitudes of the initial sector:

$$\langle I \| P \rangle, \quad \langle I \| P(\bar{M}, \bar{M}+1) \rangle, \dots, \langle I \| P(\bar{M}, N) \rangle, \tag{11.55}$$

of which the dimension $M + 1$ matrix is

$$M_{\alpha\beta}(\mathscr{E}) = \left(-e^{-ik_\alpha L} \delta_{\alpha\beta} + \mathscr{Y}_\alpha(\mathscr{E}) \frac{x_{\beta\alpha}}{y_{\beta\alpha}} \right), \tag{11.56}$$

$$\alpha, \beta \in \mathscr{E} = \{ P\bar{M}, P(\bar{M}+1), \dots, PN \}, \quad \mathscr{Y}_\alpha(\mathscr{E}) = \prod_{\beta \in \mathscr{E}}' y_{\beta\alpha}. \tag{11.57}$$

It is thus necessary that all determinants $|M(\mathscr{E})|$ vanish $\forall \mathscr{E}$. By virtue of a remarkable identity demonstrated in Appendix J, this is realized if there exists a set of M quantities $\{q\}$:

$$q_a = v_a - ic/2, \quad a = 1, 2, \dots, M,$$

such that the $\{k\}$ solve the transcendental system (11.42):

$$\exp(ik_j L) = \prod_{a=1}^{M} \frac{k_j - q_a + ic/2}{k_j - q_a - ic/2}, \tag{11.58}$$

or further still

$$Lk_j = 2\pi n_j + \sum_{a=1}^{M} \theta_{ja}, \qquad n_j \in \mathbb{Z}$$

$$\cot \frac{\theta_{ja}}{2} = \frac{2}{c}(k_j - q_a). \tag{11.59}$$

The second step consists of constructing the coefficients (11.55):

$$\langle I || P \rangle = I(P)c(\mathscr{F}) = I(P)c(\alpha\beta\gamma \ldots \delta),$$

$$\mathscr{F} = \{\alpha\beta\gamma \ldots \delta\} = \{P(\bar{M}+1), \ldots, PN\}, \tag{11.60}$$

as being proportional to the corresponding minors of the $M+1$-dimensional determinants $M(\mathscr{E})$. One easily finds that these coefficients depend symmetrically on the whole set of indices of $\mathscr{F}$ and only on the $k_\alpha, \alpha \in \mathscr{F}$:

$$c(\mathscr{F}) = c(k_\alpha k_\beta \ldots k_\delta) = \left| -e^{-ik_\alpha L} \delta_{\alpha\beta} + \mathscr{Y}_\alpha(\mathscr{F}) \frac{x_{\beta\alpha}}{y_{\beta\alpha}} \right|_{\mathscr{F}}. \tag{11.61}$$

This order-M determinant has no reason to vanish (cf. identity (J.1)). In contrast, identity (J.2) shows that this coefficient has the structure of a Bethe sum:

$$c(k_1 \ldots k_M) = \sum_{R \in \pi_M} A(R)\varphi_1(k_{R1})\varphi_2(k_{R2}) \ldots \varphi_M(k_{RM}), \tag{11.62}$$

with

$$A(R) = \prod_{\alpha < \beta} \left(1 + \frac{ic}{k_{R\alpha} - k_{R\beta}} \right) \equiv \prod_{\alpha < \beta} y_{R\alpha, R\beta},$$

$$\varphi_\alpha(k_j) = \frac{ic}{k_j - q_\alpha + ic/2} \prod_{b=a+1}^{M} \frac{k_j - q_b - ic/2}{k_j - q_b + ic/2} \equiv x_{ja} \prod_{b>a} y_{jb}. \tag{11.63}$$

One will have noted the close analogy between the real space wavefunction coefficients given by (11.62) and the coefficients $c(n)$ of the eigenvectors of Yang's Z_j matrices (11.23), in which the role of k and q is almost interchanged.

11.3.2 Second step

The third step makes use of the (Fock) symmetry conditions which we have ignored until now. It turns out that, in the case of total spin $S = 0$ ($M = \bar{M}$), the Fock condition can be written not with the coefficients intervening in (11.6) (sum over the indices of the first column), but with those of set (11.55) pertaining to the second column. The compatibility between the Fock relation and the system issued from (11.56) can then be written by setting to zero a bounded determinant of order $M+2$:

$$\left| \begin{array}{ccc|c} & & & 1 \\ & M(\mathscr{E}) & & 1 \\ & & & \vdots \\ \hline \mathscr{Y}_{P(M+1)} & \mathscr{Y}_{P(M+2)} & \cdots \quad \mathscr{Y}_{PN} & 1 \end{array} \right| = 0. \qquad (11.64)$$

One can show that this condition can be rewritten as

$$\sum_{\alpha \in \mathscr{E}} Z_\alpha \frac{\partial M(\mathscr{E})}{\partial Z_\alpha} = 0, \quad Z_\alpha = e^{-ik_\alpha L}, \quad \alpha \in \mathscr{E}, \quad \forall \mathscr{E}. \qquad (11.65)$$

Considering the equation $|M(\mathscr{E})| = 0$ as that of a hypersurface in the cartesian coordinates $\{Z_1 Z_2 \ldots Z_{M+1}\}$, equation (11.65) can be interpreted geometrically and the apparent contour of this hypersurface viewed from the origin of coordinates. There thus exist vectors tangent to this surface, of components δZ_j, such that

$$\frac{\delta Z_1}{Z_1} = \frac{\delta Z_2}{Z_2} = \ldots = \delta t. \qquad (11.66)$$

But the surface Det $|M(\mathscr{E})| = 0$ is uniformized by the M parameters q_a (formula (11.58) by virtue of identity (J.1)). Equations (11.66) can thus be rewritten

$$\sum_{a=1}^{M} \frac{\delta q_a}{(k_j - q_a + ic/2)(k_j - q_a - ic/2)} = \delta t. \qquad (11.67)$$

There thus exist $M + 1$ unknowns $(\delta q_a, \delta t)$ for each set $\mathscr{E}$ such that the rational fraction in k

$$G(k) = \sum_{a=1}^{M} \frac{\delta q_a}{(k - q_a)^2 + c^2/4} - \delta t \qquad (11.68)$$

vanishes $\forall k_j \in \mathscr{E}$.

This being shown, for any given $\mathscr{E}$, $G(k)$ is then the quotient of the two degree $2M = N$ polynomials:

$$\delta t \prod_{j=1}^{N} (k - k_j) / \prod_{a=1}^{M} ((k - q_a)^2 + c^2/4) = -G(k). \qquad (11.69)$$

Moreover, the residues at the associated poles $q_a \pm \dfrac{ic}{2}$ are opposite according to (11.68), which leads in (11.69) to

$$\frac{\prod_j (q_a + ic/2 - k_j)}{\prod_b'(q_a - q_b + ic)} = \frac{\prod_j (q_a - ic/2 - k_j)}{\prod_b'(q_a - q_b - ic)}, \qquad (11.70)$$

in other words exactly to the second set of coupled equations between k and q, equivalent to (11.39). This is the heuristic method which led us to the solution for $S = 0$ and, from there, to the general solution.

By a convenient choice for the phases, we can rewrite system (11.70) in the form

$$\sum_{j=1}^{N} \theta_{ja} - \sum_{b=1}^{M} {}' \psi_{ba} = 0, \qquad a = 1, 2, \ldots, M,$$

$$\cot(\psi_{ab}/2) = (q_a - q_b)/c. \tag{11.71}$$

Relations (11.59) and (11.71) form the basis for the study of the spin-1/2 fermion system summarized in the next section.

11.4 On the thermodynamic limit of the fermion system in the vicinity of its ground state

11.4.1 Quantum numbers

The admissible sets of quantum numbers allowing us to classify the states are simply obtained with the help of the principle of continuity in the coupling constant c starting from either of the two remarkable limits

$c = 0$: free fermion system,

$c^{-1} = 0^{+}$: two mutually impenetrable fermion species.

The $c = 0$ limit

It is possible to choose the branches of the auxiliary phases ψ_{ab} and θ_{ja} such that these tend to zero with c; the integers n_j of relation (11.59) then define the occupation of the orbital states of a non-interacting fermion system. The integers n are thus distinct or at most pairwise equal. We shall call $n_a, a = 1, 2, \ldots, r$, the quantum numbers of the 'paired' states in the formal sense defined above, and n_j, $j > 2r$, the unpaired states. We thus have $M - r$ distinct quantum numbers.

The behaviour of a paired set of momenta in the vicinity of $c = 0$ is as follows. If for example

$$\lim_{c \to 0} k_1 = \lim_{c \to 0} k_2 = 2\pi n_1/L,$$

then there exists a q_1 such that

$$q_1 = \frac{2\pi n_1}{L} + O(c), \qquad k_{1,2} = \frac{2\pi n_1}{L} \pm \left(\frac{c}{L}\right)^{1/2} + O(c). \tag{11.72}$$

We shall write

$$\lim k_j = \frac{2\pi n_j}{L} = k_j, \qquad \lim q_a = q_a. \tag{11.73}$$

One can show, starting from relations (11.71), that the unpaired quasi-momenta k and q verify the algebraic system

$$\sum_{j=2r+1}^{N} \frac{1}{k_j - q_a} = \sum_{b=r+1}^{M-r} \frac{1}{q_b - q_a}, \qquad a > r, \tag{11.74}$$

and that this system, in which the k_j are given, admits exactly

$$g_r = c_{M-r}^{N-2r} - C_{M-r-1}^{N-2r} \tag{11.75}$$

solutions by virtue of a theorem due to Heine (Szegö, 1939, Chapter VI.8), which is indeed the degeneracy of a fermion system with total spin $S = \dfrac{N}{2} - M$, knowing that r pairs are necessarily coupled at zero. One can consider the systems $\{q\}$ as sets of 'quantum numbers' labelling the states by completing the sets of integers $\{n\}$. The limit wavefunctions are obtained starting from expressions (11.62), (11.63); it is necessary to normalize the coefficient $c(\mathscr{F})$ by multiplying it by $c^{r/2}$ to obtain the limit amplitude

$$\lim_{c \to 0} c^{r/2} c(\mathscr{F}) = |\frac{1}{k_j - q_a}|_{M-r}^{+} \tag{11.76}$$

in the form of a *permanent* of order $M - r$, if the set $\mathscr{F} = \{k_1 k_2 \ldots k_M\}$ does not contain any pair index; the limit coefficient vanishes in the other cases.

Let us finally give the expansion of the total energy of a state $\{n, q\}$ in the vicinity of $q = 0$, limiting ourselves to the unpaired case $r = 0$:

$$E = \sum_{j=1}^{N} k_j^2 = E^{(0)} + E^{(1)} + E^{(2)} + O(c^3), \tag{11.77}$$

with

$$E^{(0)} = \sum_j (2\pi n_j / L)^2,$$

$$E^{(1)} = \frac{c}{2L}[N(N+2) - 4S(S+1)],$$

$$E^{(2)} = -c^2 \sum_j \left(\sum_a \frac{1}{k_j - q_a} \right)^2. \tag{11.78}$$

Attractive gas

When $c < 0$, according to (11.72), the paired quasi-momenta become complex conjugate pairs; one can show that in the limit of a large system ($L \gg |c|$), one has the exponential approximation

$$k_{1,2} = q_1 \pm ic/2 + O(\exp(L/c)). \tag{11.79}$$

The asymptotic form of the coupled equations of the attractive case is the following:

$$\sum_{j>2r} \theta_{ja} = \sum_{b>r} \psi_{ba}, \qquad M \ge a > r$$

$$Lq_a = 2\pi n_a + \frac{1}{2}\sum_{b\le r}\psi_{ab} - \frac{1}{2}\sum_{l>2r}\theta_{ja}, \qquad a \le r$$

$$Lk_j = 2\pi n_j + \sum_b \theta_{jb}, \qquad N \ge j > 2r. \tag{11.80}$$

We then notice that the wavefunction describes a system containing r pairs linked in a singlet state. When all these particles are linked pairwise, for example in the $S = 0$ ground state, the coupled equations become identical to those of the bosonic system (Chapter 4, equations (4.60)) whose quasi-momenta are $2q$ and the coupling $2c$. The quantum numbers n_a must, however, remain distinct. The total energy is

$$E = 2\sum_{a=1}^{M} q_a^2 - \frac{1}{4}Nc^2 \quad (N = 2M). \tag{11.81}$$

The limit $c^{-1} = 0$

Yang presents the coupled equations in a form which is adapted to the limit $c = \infty$ and introduces two sets of quantum numbers $\{I\}$ and $\{J\}$:

$$Lk = 2\pi I_k - 2\sum_a \text{atan } 2(k - q)/c,$$

$$\sum_k \text{atan } (q - k)/c = \pi J_q + \sum_{q'} \text{atan } (q - q')/c,$$

$$\left(\sum_k \Leftrightarrow \sum_{j=1}^{N}, \quad \sum_q \Leftrightarrow \sum_{a=1}^{M}\right), \tag{11.82}$$

J is half-integer (integer) if $n - M$ is even (odd);
I is half-integer (integer) if M is even (odd).

The transition formula (10.22) gives us in the $c^{-1} = 0$ limit

$$Y_{j,j+1}|\bar{P}\rangle = -|\bar{P}\rangle = |\bar{P}(j, j + 1)\rangle.$$

We deduce from this that the limit vector $|\bar{P}\rangle$ is proportional to the antisymmetric character $\chi(P)$ and thus that, in each sector Q, the limit wavefunction is proportional to the Slater determinant constructed using the limit wavenumbers

$$\lim_{c^{-1}\to 0} k_j = \hat{k}_j. \tag{11.83}$$

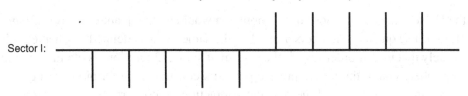

Figure 11.2 Sector *I*.

This is not surprising since the infinitely repulsive potential forces the vanishing of the wavefunction when the coordinates of two fermions of opposite spin coincide.

The proportionality coefficient multiplying the determinant $|\exp i\hat{k}_j x_l|$ can however *a priori* vary from one sector to the other. Because of the necessary antisymmetry between particles of the same spin, this coefficient varies with the spin configuration, which, in fact, describes each sector modulo the antisymmetric exchange between neighbours of the same species.

If we use the most compact expression for the coefficients $c(\mathscr{F})$ of the real-space wavefunction in sector $Q = I$ (Figure 11.2) or (cf. Subsection 11.3.1 and Appendix B)

$$c(\mathscr{F}) = c(k_1 \ldots k_M) = \left|\frac{c^2}{(k_j - q_a)^2 + c^2/4}\right| \bigg/ \left|\frac{ic/2}{k_j - q_a + ic/2}\right|, \qquad (11.84)$$

we obtain, with the normalization chosen above,

$$\lim_{c\to\infty} c(\mathscr{F}) = \prod_{1\le a<b\le M} \frac{1}{2}\left(\hat{q}_a\hat{q}_b - \frac{1}{4}\right) \qquad (11.85)$$

provided we postulate the existence of the finite limit

$$\lim_{c\to\infty} q_a/c = \hat{q}_a. \qquad (11.86)$$

We see in (11.83) that the value of coefficient $c(\mathscr{F})$ does not depend on the set $\mathscr{F}$, which confirms the result established previously in sector I. Under these conditions, the coupled equations (11.82) for $\hat{k}$ and q take the limit form

$$L\hat{k} = 2\pi \left(I_k + N^{-1}\sum_q J_q\right),$$

$$N \operatorname{atan} 2\hat{q} = \pi J_q + \sum_{q'} \operatorname{atan}(\hat{q} - \hat{q}'). \qquad (11.87)$$

Since the $\hat{k}$ must be distinct, we deduce that the numbers I_k are distinct. In contrast, the second equation (11.87) is identical to the periodicity relations for the Heisenberg chain whose quasi-momenta would be the quantities $\hat{q}$ ((1.32) and (1.39)). The quantum numbers J are thus the same as those defined in Chapter 1. It is likely that to first order in c^{-1}, the system of spin-1/2 fermions is identical to the Heisenberg chain: the very repulsive potential constrains the fermions to a quasi-crystalline correlation and the residual interaction, in conformity to the theory of Dirac and Heisenberg cited in Chapter 1, can only be the exchange interaction between neighbours. It would thus suffice to perform the perturbative calculation to first order in c^{-1}.

As a closing to this subsection, let us mention the quantum numbers of the repulsive ground state of spin $S = \dfrac{N}{2} - M$:

$$\{J\} = \left\{ -\frac{M-1}{2}, -\frac{M-3}{2}, \ldots, \frac{M-1}{2} \right\},$$

$$\{I\} = \left\{ -\frac{N-1}{2}, \ldots, \frac{N-1}{2} \right\}. \tag{11.88}$$

11.4.2 The integral equations for the energy of the ground state

The energy per particle in the thermodynamic limit can be obtained starting from the coupled equations by using the same hypotheses as those adopted in Chapter 2 for the spin chain. We shall therefore present, without proof, the integral equations allowing us to calculate the energy as a function of density and magnetization. We shall set

$$\lim \frac{N}{L} = \rho = \frac{2k_0}{\pi}, \quad k_0 - \text{Fermi momentum}$$

$$\lim \frac{2S}{N} = \sigma, \quad \sigma - \text{magnetization per site } (-1 \le \sigma \le 1)$$

$$\lim \frac{E}{N} = \varepsilon, \quad \rho_\pm = \rho(1 \pm \sigma)/2. \tag{11.89}$$

We suppose that the roots k_j and q_a of equations (11.59) and (11.71), or (11.82) associated with the ground state quantum numbers (11.88), become dense on real segments

$$[-k_1, +k_1], \quad [-q_0, +q_0],$$

with the respective densities

$$\frac{1}{2\pi} f_1(k), \quad \frac{1}{2\pi} f(q). \tag{11.90}$$

We obtain:

Repulsive case $c > 0$

$$2\pi \rho_+ = \int_{-k_1}^{k_1} f_1(k)dk - \int_{-q_0}^{q_0} f(q)dq,$$

$$2\pi \rho_- = \int_{-q_0}^{q_0} f(q)dq,$$

$$2\pi \varepsilon = \frac{1}{\rho} \int_{-k_1}^{k_1} k^2 f_1(k)dk. \tag{11.91}$$

The densities are determined from the two relations

$$f_1(k) - \frac{c}{2\pi} \int_{-q_0}^{q_0} \frac{f(q)dq}{(k-q)^2 + c^2/4} = 1,$$

$$f(q) + \frac{c}{\pi} \int_{-q_0}^{q_0} \frac{f(q')dq'}{(q-q')^2 + c^2} = \frac{c}{2\pi} \int_{-k_1}^{k_1} \frac{f_1(k)dk}{(k-q)^2 + c^2/4}. \tag{11.92}$$

Attractive case $c < 0$

$$2\pi \rho_+ = \int_{-k_1}^{k_1} f_1(k)dk + \int_{-q_0}^{q_0} f(q)dq,$$

$$2\pi \rho_- = \int_{-q_0}^{q_0} f(q)dq,$$

$$2\pi \rho \varepsilon = \int_{-k_1}^{k_1} k^2 f_1(k)dk + 2 \int_{-q_0}^{q_0} (q^2 - c^2/4) f(q)dq \tag{11.93}$$

and the integral relations

$$f_1(k) - \frac{c}{2\pi} \int_{-q_0}^{q_0} \frac{f(q)dq}{(k-q)^2 + c^2/4} = 1,$$

$$f(q) - \frac{c}{\pi} \int_{-q_0}^{q_0} \frac{f(q')dq'}{(q-q')^2 + c^2} = 2 + \frac{c}{2\pi} \int_{-k_1}^{k_1} \frac{f_1(k)dk}{(k-q)^2 + c^2/4}. \tag{11.94}$$

One can, incidentally, easily eliminate f_1 and be left with an inhomogeneous equation for $f(q)$ depending on two parameters k_1 and q. The density and spin will then be given parametrically; in the attractive case, we have for example

$$\sigma \rho = \frac{k_1}{\pi} - \frac{1}{\pi^2} \int_{-q_0}^{q_0} f(q) \, \text{atan} \, \frac{2(k_1 - q)}{c} dq,$$

$$\rho(1 - \sigma) = \frac{1}{\pi} \int_{-q_0}^{q_0} f(q)dq. \tag{11.95}$$

Let us point out that the integral equations of the repulsive and attractive cases can be written in a unified way. Let us extend the definition of $f_1(k)$ outside the interval $[-k_1, +k_1]$ with the help of the first relations (11.92) or (11.94). At infinity, we have the asymptotic behaviour

$$f_1(k) = 1 + c\rho_-/k^2 + O(k^{-3}).$$

By integration, one obtains

$$\int_{-\infty}^{\infty} (f_1(k) - 1)dk = \frac{c}{|c|} \int_{-q_0}^{q_0} f(q)dq.$$

In the two cases, we can thus write

$$2\pi\rho_+ = 2k_1 - \int_{|k|>k_1} (f_1(k) - 1)dk,$$

$$2\pi\rho_- = \int_{-q_0}^{q_0} f(q)dq. \tag{11.96}$$

Also, in both cases we have

$$f_1(k) - \frac{c}{2\pi} \int_{-q_0}^{q_0} \frac{f(q)dq}{(k-q)^2 + c^2/4} = 1,$$

$$f(q) + \frac{c}{2\pi} \int_{|k|>k_1} \frac{f_1(k)dk}{(k-q)^2 + c^2/4} = 1. \tag{11.97}$$

11.4.3 The ground state of the attractive case

According to a theorem of Lieb and Mattis (1962), the ground state of the attractive gas has total spin $S = 0$. The fermions are linked pairwise in the singlet state. Relations (11.93) for $\sigma = 0$ give us $\rho_+ = \rho_-$, and the relations become

$$f(q) - \frac{c}{\pi} \int_{-q_0}^{q_0} \frac{f(q')dq'}{(q-q')^2 + c^2} = 2,$$

$$\rho = \frac{2k_0}{\pi} = \frac{1}{\pi} \int_{-q_0}^{q_0} f(q)dq, \quad (k_1 = 0)$$

$$\varepsilon = \varepsilon_0 - c^2/4, \quad \varepsilon_0 = \frac{1}{\pi\rho} \int_{-q_0}^{q_0} q^2 f(q)dq. \tag{11.98}$$

We will have noticed that the integral equation for the paired fermions differs from that obtained by Lieb and Liniger for repulsive bosons, by the sign of the coupling constant. Our paired fermions behave partly as a system of attractive bosons, but which still remember the exclusion principle between the underlying fermions.

Using dimensional reasonings and eliminating q_0 between ρ and ε_0 gives us a relation

$$\varepsilon_0 = c^2 \mathscr{E}(k_0/|c|), \tag{11.99}$$

and the remaining task is to determine the function $\mathscr{E}(y)$. However, if we write $f(q) = F(q/q_0)$, the integral equation of the problem on interval $[-1, +1]$ depends only on parameter $K = |c|/q_0$:

$$F(x) + \frac{K}{\pi} \int_{-1}^{1} \frac{F(y)dy}{(x-y)^2 + K^2} = 2. \tag{11.100}$$

We must now study the integral equation either in the form (11.100), or in the form (11.98) where it suffices to take $|c| = 1$ due to the homogeneity. This equation already has a history and was obtained in 1949 by Love (Sneddon, 1966) to reduce a classical electrostatic problem in three dimensions. This problem is that of two circular and coaxial conducting discs, of unit radius and separated by a distance $|K|$ (or radius q_0 at unit distance). If K is negative, this is the circular capacitor in which the two discs are put at opposite potentials. If K is positive, which is our case here, we are dealing with two discs at the same potential.

This electrostatic analogy is useful for the following reason: for values of K which are not too small, it is straightforward to obtain the function F numerically. However, in the vicinity of $K = 0$, the integral operator which appears in (11.100) tends to the unit operator and the numerical inversion becomes difficult due to the bad convergence of the Neumann series. It is likely that the analytic functions $k_0(K)$ and $\mathscr{E}(K)$ possess a singularity in $K = 0$, given the very different characters of the electrostatic fields depending on the sign of K. The vicinity of $K = 0$, in other words of $c = 0$, corresponds to large values of k_0, or to the high-density region.

The unknown function $F(x)$ is related to the charge density $\sigma(r)$ on one of the conducting discs via the Abel transformation

$$F(x) = \int_{x}^{1} \frac{\sigma(r)rdr}{(r^2 - x^2)^{1/2}}. \tag{11.101}$$

An approximation for the calculation of $\sigma(r)$ due to Kirchhoff (1877) and whose validity was proven by Huston (1963) rests on the separation between the peripheral region $1 - r = O(K)$ and the central part $1 - r \gg K$. It is intuitively clear that in the limit $K \to 0$ or $q_0 \to \infty$, the boundary effect pertains to a two-dimensional electrostatics problem, which was solved by Maxwell. It remains to calculate the reaction of the peripheral distribution to the central part. This is possible in the case of the capacitor ($K < 0$) since there exists in this case a formula due to Kac and Pollard (1956) which allows us to find an approximate solution to (11.100) far from the edges (cf. Subsection 5.1.3). Let us here quote the Kirchhoff approximation for the capacity of the capacitor (or the density of the Bose gas of Chapter 5):

$$\int_{-1}^{1} F(x)dx = \frac{\pi}{2|K|} + \frac{1}{2} \ln \frac{16\pi}{e|K|} + o(1). \tag{11.102}$$

Getting back to our problem $(K > 0)$, we have obtained using a different analysis

$$k_0 = q_0 \left(1 + \frac{K}{2\pi} \ln \frac{e\pi}{K} + o(K)\right),$$

$$\varepsilon_0 = \frac{q_0^3}{k_0} \left(\frac{1}{3} + \frac{K}{2\pi} \ln \frac{\pi}{eK} + o(k)\right), \tag{11.103}$$

which corresponds to the following approximation for $F(x)$ far from the edges:

$$F(x) = 1 + \frac{K}{\pi} \frac{1}{1 - x^2} + O(K), \quad 1 - x \gg K. \tag{11.104}$$

The elimination of $q_0 = K^{-1}$ gives us only the first two terms in the expansion of the energy $\mathscr{E}(k_0)$ (for $c = 1$):

$$\mathscr{E}(k_0) = \frac{k_0^2}{3} - \frac{k_0}{\pi} + o(1), \tag{11.105}$$

which does not differ from the perturbative expansion. The latter gives us for less effort an additional term: $o(1) = 1/6$.

The behaviour thus seems regular around $c = 0$. For a numerical study we refer the reader to the author's thesis (1967a), in which some elementary excitations of the attractive gas are also described.

Appendix J

Let us consider $(p + 1)$ parameters u_j, and $(p + 1)$ variables Z_j; the indices take their value in sets which we denote $\mathscr{E}$, or $\mathscr{E}_{p+1}$ if we wish to indicate their cardinality. We further set

$$y_{ij} = 1 + \frac{1}{u_i - u_j}, \quad \mathscr{Y}_j(\mathscr{E}) = \prod_{i \neq j} y_{ji}. \tag{J.1}$$

We consider the multilinear function of the Z_j variables defined as follows:

$$M(Z_1, Z_2, \dots, Z_{p+1}) = |-Z_j \delta_{ij} + \mathscr{Y}_j(\mathscr{E})/(u_j - u_i + 1)|_{\mathscr{E}_{p+1}}. \tag{J.2}$$

Proposition (J_1): The algebraic surface of degree $p + 1$ defined by

$$M(Z_1, Z_2, \dots, Z_{p+1}) = 0 \tag{J.3}$$

admits the following rational parametrization:

$$Z_i = \prod_{a=1}^{P} \left(1 + \frac{1}{u_i - v_a}\right), \tag{J.4}$$

in terms of p parameters v_a.

We must first prove that the rational fraction in u and v, $M(Z_1, \ldots, Z_{p+1})$ is identically zero.

Lemma 1:

$$M(Z_1, Z_2, \ldots, Z_{p+1}) = 0 \tag{J.5}$$

if

$$Z_j = 1, \quad \forall j.$$

This corresponds to the case in which all parameters v are infinite.

Proof: We must show that

$$|-\delta_{ij} + \mathscr{Y}_j(\mathscr{E})/(u_j - u_i + 1)| = 0, \tag{J.6}$$

which results from the identity

$$\sum_{j \in \mathscr{E}} \mathscr{Y}_j(\mathscr{E})/(u_j - u_i + 1) = 1. \tag{J.7}$$

Indeed, $\mathscr{Y}_i$ is a rational function of variable u_i whose simple poles are u_j, $j \neq i$. Given the asymptotic behaviour of $\mathscr{Y}_i$ at infinity in u_i, we have the decomposition

$$\mathscr{Y}_i = 1 + \sum_{j \neq i} \frac{(\text{Res in } u_j)}{u_i - u_j}.$$

But the definition is

$$\mathscr{Y}_i = \left(1 + \frac{1}{u_i - u_j}\right) \prod_{l \neq j, i} y_{il}.$$

The residue in u_j thus takes the value $\prod_{l \neq j, i} y_{jl}$, which is equal to $\mathscr{Y}_j/y_{ji}$. We thus get

$$\mathscr{Y}_i = 1 + \sum_{j \neq i} \mathscr{Y}_j/y_{ji}(u_i - u_j),$$

which is the announced identity (J.7).

Lemma 2:

$$M(Z_1, \ldots, Z_{p+1}) = 0, \quad \text{if} \quad Z_i = 1 + \frac{1}{u_i - v_1}. \tag{J.8}$$

Proof: M is a rational function in v_1 whose poles are those of the quantities Z_i, in other words $u_1, u_2, \ldots, u_{p+1}$. When v_1 increases indefinitely, we have $Z_i = 1$ and, by virtue of Lemma 1, $M(1, 1, \ldots, 1) = 0$. We thus have the following decomposition:

$$M = \sum_{j \in \mathscr{E}_{p+1}} \frac{R_j}{v_1 - u_j}. \tag{J.9}$$

It is sufficient to calculate the residue R_1 relative to pole u_1; the others can be deduced by symmetry.

According to the definition, the pole of M intervenes via the Z_1 term in M. The residue of the pole is thus the coefficient of $-Z_1$ in the determinant (J.2) for the value $v_1 = u_1$. Z_i ($i \neq 1$) must be replaced everywhere by y_{i1}. The residue in u_1 is thus the order-p determinant

$$R_1 = \begin{vmatrix} \mathscr{Y}_2 - y_{21} & \dfrac{\mathscr{Y}_3}{u_3 - u_2 + 1} & \dfrac{\mathscr{Y}_4}{u_4 - u_2 + 1} & \cdots \\[2ex] \dfrac{\mathscr{Y}_2}{u_2 - u_3 + 1} & \mathscr{Y}_3 - y_{31} & \dfrac{\mathscr{Y}_4}{u_4 - u_3 + 1} & \cdots \\[2ex] \dfrac{\mathscr{Y}_2}{u_2 - u_4 + 1} & \dfrac{\mathscr{Y}_3}{u_3 - u_4 + 1} & \mathscr{Y}_4 - u_{41} & \cdots \\[2ex] \cdots & \cdots & \cdots & \cdots \end{vmatrix}_p . \tag{J.10}$$

Let us call $\mathscr{E}_p^1$ the set $\mathscr{E}_{p+1}$ from which element 1 is excluded. According to the definition of the $\mathscr{Y}(\mathscr{E})$, we have

$$\mathscr{Y}_j(\mathscr{E}) = y_{j1}\mathscr{Y}_j(\mathscr{E}^1).$$

One can then, in the determinant which defines R_1, factorize y_{21} in column number 1, y_{31} in column number 2, etc.; there remains a residue R_1 proportional to the determinant

$$\left| -\delta_{ij} + \frac{\mathscr{Y}_j(\mathscr{E}^1)}{u_j - u_i + 1} \right|_{\mathscr{E}^1} \tag{J.11}$$

constructed on the indices of set $\mathscr{E}_p^1$. But this determinant vanishes according to Lemma 1 applied to the set $\mathscr{E}^1$ of order p. We thus have $R_1 = 0$, and similarly $R_j = 0$, which proves Lemma 2.

Proof of Proposition J_1: This is done using a double recursion on the number p and the number k of effective variables on which the $Z_j\{v\}$ depend.

We call $\mathscr{H}_{k,p}$ the following proposition:

$M(Z_1, \ldots, Z_{p+1})$ vanishes identically in the $p+1$ parameters u_j, $j \in \mathscr{E}_{p+1}$ and in the k variables v_a.

The general proposition to prove is stated as: '$\mathscr{H}_{p,p}$ is true'.

We shall take '$\mathscr{H}_{k,p}$ is true for $k < p$' as recursion hypothesis.

Lemmas 1 and 2 have established that $\mathscr{H}_{0,p}$ and $\mathscr{H}_{1,p}$ are true. We show in Lemma 3 that

$$\mathscr{H}_{k-1,p} \cap \mathscr{H}_{k-1,p-1} \Rightarrow \mathscr{H}_{k,p} \quad \text{if} \quad k \leq p. \tag{J.12}$$

Lemma 3: If

$$M(Z_1, Z_2, \ldots, Z_{p+1}) = 0, \quad \text{for} \quad Z_i = \prod_{a=1}^{k-1}(1 + 1/(u_i - v_a)),$$

then

$$M(\zeta_1, \zeta_2, \ldots, \zeta_{p+1}) = 0, \quad \text{for} \quad \zeta_i = Z_i(1 + 1/(u_i - v_k)).$$

Proof: The determinant $M(\zeta_1, \ldots, \zeta_{p+1})$ is a rational function of v_k, whose simple poles are the u_i, $i \in \mathcal{E}_{p+1}$. It vanishes for $v_k = \infty$ by virtue of the recursion hypothesis $\mathcal{H}_{k-1,p}$, $k < p$. Let us show that the residue of any given pole, for example u_1, is zero. Such a residue is proportional to the principal minor coefficient of Z_1 in the determinant defining $M(\zeta_1, \ldots, \zeta_{p+1})$ for the value $v_k = u_1$. One thus replaces ζ_i everywhere by $Z_i(1 + 1/(u_i - u_1))$ for $i = 2, 3, \ldots, p + 1$. The residue of the pole in u_1 is thus given by a determinant analogous to (J.10) which, after factorizing y_{j1} in each column j, is proportional to

$$R_1' = \left| -Z_i' \delta_{ij} + \frac{\mathcal{Y}_j(\mathcal{E}^1)}{u_j - u_i + 1} \right|_{\mathcal{E}^1}.$$

It turns out that this order-p determinant, built on the indices of $\mathcal{E}^1$, is the function $M(Z_2', Z_3', \ldots, Z_{p+1}')$ in which the Z_i' are a product of $(k - 1)$ *factors* $y_{ia}' = 1 + 1/(u_i - v_a)$. Thus, if proposition $\mathcal{H}_{k-1,p-1}$ ($k \le p$) is true, this implies that all residues vanish. We thus have proven the logical implication (J.12).

The inductive chain which allows us to prove all $\mathcal{H}_{k,p}$ ($k \le p$) has the same structure as the algorithm of Pascal's arithmetic triangle:

$$
\begin{array}{ccccccc}
\mathcal{H}_{0,1} & \to & \mathcal{H}_{1,1} & & & & \\
\downarrow & & \downarrow & & & & \\
\mathcal{H}_{0,2} & \to & \mathcal{H}_{1,2} & \to & \mathcal{H}_{2,2} & & \\
\downarrow & & \downarrow & & \downarrow & & \\
\mathcal{H}_{0,3} & \to & \mathcal{H}_{1,3} & \to & \mathcal{H}_{2,3} & \to & \mathcal{H}_{3,3} \\
\downarrow & & \downarrow & & \downarrow & & \downarrow \\
\cdots & & \cdots & & \cdots & & \cdots
\end{array}
\qquad (J.13)
$$

All propositions in this table are true, thus so are all $\mathcal{H}_{p,p}$.

It remains to show the inverse proposition, namely that any solution to $M(Z_i) = 0$ can be expressed in the form $Z_i = \prod_a y_{ia}'$. We start from the following identity whose easy demonstration we omit:

$$M(Z_1, \ldots, Z_{p+1}) \equiv \left| u_i^n Z_i - (u_i + 1)^n \right|_{p+1} / \prod_{i<j}(u_i - u_j). \qquad (J.14)$$

There thus exist coefficients a_n $(n = 0, 1, \ldots, p)$, in other words a polynomial of degree $p + 1$ such that we have

$$Z_i = \frac{Q(u_i + 1)}{Q(u_i)}, \qquad Q(u) = a_0 + a_1 u + \ldots + a_p u^{p+1}.$$

The roots of $Q(u)$ are the $(p + 1)$ quantities v_a. Proposition J_1 is thus proven. This does not tell us anything concerning the determinant $M(Z_1, \ldots, Z_p)$ when the number of variables Z_i is precisely equal to p. The objective of the following proposition is to give an expanded expression for $M(\mathscr{E}_p)$ providing a remarkable identity between a determinant and a Bethe sum.

Proposition J_2:

$$M(\mathscr{E}_p) \equiv \sum_{R \in \pi_p} A(R) \varphi_1(u_{R1}) \varphi_2(u_{R2}) \ldots \varphi_p(u_{Rp}), \tag{J.15}$$

in which the sum runs over the order-p permutations R, with the definitions

$$\varphi_a(u) = \frac{1}{v_a - u} \prod_{b=a+1}^{p} \left(1 + \frac{1}{u - v_b}\right); \tag{J.16}$$

$$A(R) = \prod_{1 \leq j \leq l \leq p} y_{Rl, Rj}. \tag{J.17}$$

Proof: $M(\mathscr{E}_p)$ is a rational fraction of v_1 whose simple poles are the u_i. According to Proposition J_1, $M(\mathscr{E}_p)$ vanishes when $v_1 = \infty$, and we thus have the expansion

$$M(\mathscr{E}_p) = \sum_{j=1}^{p} \frac{R_j}{v_1 - u_j}. \tag{J.18}$$

The residue R_1 of the $v_1 = u_j$ pole is the principal (j, j) minor of determinant M for $v_1 = u_j$, multiplied by ζ_j:

$$\zeta_j = \prod_{a=2}^{p} y'_{ja}. \tag{J.19}$$

But for $v_1 = u_j$ we have $(l \neq j)$

$$Z_l = y_{lj} \zeta_l; \qquad \mathscr{Y}_l(\mathscr{E}_p) = y_{lj} \mathscr{Y}_l(\mathscr{E}_{p-1}^j), \tag{J.20}$$

with

$$\{\mathscr{E}_{p-1}^j\} = \{\mathscr{E}\} - \{j\}.$$

In each column l of this minor (j, j), the quantity y_{lj} is a factor and we obtain

$$R_j = \prod_{a=2}^{p} y'_{ja} \prod_{l \neq j} y_{lj} M(\mathscr{E}_{p-1}^j)(\zeta_2, \zeta_3, \ldots, \zeta_p), \tag{J.21}$$

in which the order-$(p-1)$ determinants of type $M(\mathscr{E}_{p-1})$ relative to each set $\mathscr{E}_{p-1}^{j}$ intervene.

Substituting expression (J.21) of the residue in the sum (J.18), we have

$$M(\mathscr{E}_p)(Z_1, \ldots, Z_p) = \sum_j \frac{1}{v_1 - u_j} \prod_{a>1} y'_{ja} \prod_{l \neq j} y_{lj} M(\mathscr{E}_{p-1}^{j})(\zeta_2, \ldots, \zeta_p). \quad \text{(J.22)}$$

The repeated application of the recursion formula (J.22) gives the aforementioned result

$$M(\mathscr{E}_p)(Z_1, Z_2, \ldots, Z_p) = \sum_R \frac{1}{v_1 - u_{R1}} \frac{1}{v_2 - u_{R2}} \cdots \frac{1}{v_p - u_{Rp}}$$

$$\left(\prod_{a=2}^{p} y'_{R1,a} \prod_{b=3}^{p} y'_{R2,b} \cdots \prod_{l>j} y_{Rl,Rj} \right), \quad \text{(J.23)}$$

expressed by formulas (J.15), (J.16), (J.17).

Let us finally mention an extension of this result to the case $k > p$: one can easily show, using the same method as above, the summation property

$$\sum_{\{n\}} a(n_1, n_2, \ldots, n_p) \equiv M(\mathscr{E}) \quad \text{(J.24)}$$

with

$$a\{n\} = \sum_{R \in \pi_p} A(R) \varphi_{n_1}(u_{R1}) \varphi_{n_2}(u_{R2}) \ldots \varphi_{n_p}(u_{Rp}). \quad \text{(J.25)}$$

Appendix K

We have already obtained in Appendix J the expansion for the amplitudes $M(\mathscr{E}_p)$, formula (J.15), as a Bethe sum identical to (B.6) up to notation. These amplitudes were defined as determinants (J.2). From this analogy comes a third identity for the quantity $\Delta_N \begin{pmatrix} k \\ k' \end{pmatrix}$ defined in (B.3).

With the change of notation

$$\gamma u_i = -k'_i, \quad \gamma v_i = -k_i, \quad i = [1, N] \quad \text{(K.1)}$$

the sums (B.6) and (J.15) are identical:

$$(-1)^N M_N = \gamma^N \Delta_N \begin{pmatrix} k_1 & \cdots & k_N \\ k'_1 & \cdots & k'_N \end{pmatrix}. \quad \text{(K.2)}$$

But we had

$$(-1)^N M_N = \left| Z'_j \delta_{ij} - \frac{\mathcal{Y}'_j}{u_j - u_i + 1} \right| \tag{K.3}$$

with the definitions

$$Z'_j = \prod_l \left(1 + \frac{1}{u_j - v_l} \right) = \prod_l \left(1 + \frac{\gamma}{k_l - k'_j} \right),$$

$$\mathcal{Y}'_j = \prod_l \left(1 + \frac{1}{u_j - u_l} \right) = \prod_{l \neq j} \left(1 + \frac{\gamma}{k'_l - k'_j} \right). \tag{K.4}$$

From this one can deduce the *identity*

$$\Delta_N \begin{pmatrix} k_1 & \cdots & k_N \\ k'_1 & \cdots & k'_N \end{pmatrix} \equiv G(k') \left| \frac{Z'_j}{\gamma \mathcal{Y}'_j} \delta_{ij} - \frac{1}{k'_i - k'_j + \gamma} \right|_N \tag{K.5}$$

with

$$G(k') = \prod_{i < j} \left(1 + \frac{c^2}{(k'_i - k'_j)^2} \right) \qquad (\gamma = ic). \tag{K.6}$$

A direct proof of (K.5) can be given; in the notation of Appendix B, we have

$$\frac{Z_j}{\gamma \mathcal{Y}_j} = \frac{1}{e_j f'_j}, \qquad \frac{1}{e_i} = 1 - \sum_j \frac{f'_j}{k_j - k_i - \gamma}. \tag{K.7}$$

One must then prove the identity

$$\left| \frac{f'_j}{(k_j - k'_l)(k_j - k'_l + \gamma)} \right| \cdot \left| \frac{1}{k_j - k_i - \gamma} \right| = \left| \frac{\delta_{ij}}{e_j} + \frac{f'_j}{k_j - k_i - \gamma} \right| \cdot \left| \frac{1}{k_j - k'_l + \gamma} \right|. \tag{K.8}$$

It is sufficient to perform the matrix multiplications intervening on both sides of (K.8) (summation over indices j) and use the definitions of quantities e_j and e'_j (equations (B.12) and (B.13)) to prove that both sides have the common value

$$\left| \frac{1}{e_i} \left(\frac{1}{k_i - k'_l + \gamma} \right) - \left(\frac{1}{e_i} - e'_l \right) \frac{1}{k_i - k'_l + 2\gamma} \right|_N. \tag{K.9}$$

Appendix L

Operator form of the eigenstates of the anisotropic chain

The method we have just described in Section 11.2 also allows us to easily construct the eigenstates of Hamiltonian H_Δ (1.3) or, more generally, those of the transfer matrix of the inhomogeneous six-vertex model. The transfer micromatrix

$$X(\varphi - \varphi') = (W(\varphi - \varphi')\tau \cdot \tau') \tag{L.1}$$

admits the 4×4 representation

$$X(\varphi) = \begin{pmatrix} a & d & \cdot & \cdot \\ d & a & \cdot & \cdot \\ \cdot & \cdot & b & c \\ \cdot & \cdot & c & b \end{pmatrix}, \tag{L.2}$$

and the ternary relation gives us the commutation of the transition matrices

$$(\tau' \cdot A'(\varphi'))(\tau \cdot A(\varphi)) = X(\varphi - \varphi')(\tau \cdot A(\varphi))(\tau' \cdot A'(\varphi'))X(\varphi' - \varphi). \tag{L.3}$$

The product $(\tau \cdot A)(\tau' \cdot A')$ admits the representation

$$(\tau \cdot A)(\tau' \cdot A') = \begin{pmatrix} A_{11}A'_{11} & A_{12}A'_{12} & A_{11}A'_{12} & A_{12}A'_{11} \\ A_{21}A'_{21} & A_{22}A'_{22} & A_{21}A'_{22} & A_{22}A'_{21} \\ A_{11}A'_{21} & A_{12}A'_{22} & A_{11}A'_{22} & A_{12}A'_{21} \\ A_{21}A'_{11} & A_{22}A'_{12} & A_{21}A'_{12} & A_{22}A'_{11} \end{pmatrix}. \tag{L.4}$$

In the chosen particular case $d = 0$ ($J_1 = J_2 = 1$, $J_3 = \Delta = \cosh \Phi$),

$$b = \frac{\sin \frac{\varphi}{2}}{\sin \left(\frac{\varphi}{2} + i\Phi\right)}, \qquad c = \frac{i \sinh \Phi}{\sin \left(\frac{\varphi}{2} + i\Phi\right)}, \qquad a = 1. \tag{L.5}$$

Relations (L.3) give us

$$A'_{12}A_{11} = \left[\sin \left(\frac{\varphi - \varphi'}{2} - i\Phi\right)\right]^{-1} \left(\sin \frac{\varphi - \varphi'}{2} A_{11}A'_{12} - i \sinh \Phi A_{12}A'_{11}\right),$$

$$A'_{11}A_{12} = \left[\sin \left(\frac{\varphi - \varphi'}{2} - i\Phi\right)\right]^{-1} \left(-i \sinh \Phi A_{11}A'_{12} + \sin \frac{\varphi - \varphi'}{2} A_{12}A'_{11}\right), \text{ etc.}$$

$$A_{11}A'_{11} = A'_{11}A_{11}, \qquad A_{12}A'_{12} = A'_{12}A_{12}, \dots \tag{L.6}$$

From this we deduce

$$A_{11}A'_{12} = i \frac{\sinh \Phi}{\sin \frac{\varphi - \varphi'}{2}} A_{12}A'_{11} + \frac{\sin \left(\frac{\varphi - \varphi'}{2} - i\Phi\right)}{\sin \frac{\varphi - \varphi'}{2}} A'_{12}A_{11},$$

$$A_{22}A'_{12} = -i \frac{\sinh \Phi}{\sin \frac{\varphi - \varphi'}{2}} A_{12}A'_{22} + \frac{\sin \left(\frac{\varphi - \varphi'}{2} + i\Phi\right)}{\sin \frac{\varphi - \varphi'}{2}} A'_{12}A_{22}, \tag{L.7}$$

which are the exact analogues of (10.98), to which they moreover tend in the isotropic limit ($\Phi \to 0$, $\varphi/2\Phi = k/c$).

The transfer matrix to be diagonalized is defined as

$$T(\varphi) = \text{Tr } A(\varphi) = A_{11} + A_{22} \tag{L.8}$$

with

$$A(\varphi) = X(\varphi - \psi_1)X(\varphi - \psi_2)\dots X(\varphi - \psi_N) \tag{L.9}$$

depending on N arbitrary parameters ψ_α.

We notice directly that the ferromagnetic reference state $|S^z = \frac{N}{2}\rangle$ is a simultaneous eigenstate of A_{11} and A_{22}:

$$(A_{11}(\varphi) - 1)|0\rangle = 0, \quad |0\rangle \equiv |S^z = \frac{N}{2}\rangle,$$

$$\left(A_{22}(\varphi) - \prod_{\alpha=1}^{N} \frac{\sin\left(\frac{\varphi - \psi_\alpha}{2}\right)}{\sin\left(\frac{\varphi - \psi_\alpha}{2} + i\Phi\right)}\right)|0\rangle = 0. \tag{L.10}$$

The eigenstate $|M\rangle$ of spin $S^z = \frac{N}{2} - M$ is simply obtained by applying on $|0\rangle$ the set of M commuting operators

$$|M\rangle = A_{12}(\varphi_1)A_{12}(\varphi_2)\dots A_{12}(\varphi_M)|0\rangle. \tag{L.11}$$

Owing to an earlier remark concerning the isotropic limit, formula (11.57) can be transposed directly without needing to redo the calculation, since it is sufficient to consider the corresponding periodic rational fractions. By repeated application of (L.7), one obtains

$$T(\varphi)|M\rangle = t(\varphi)|M\rangle \tag{L.12}$$

with the eigenvalue

$$t(\varphi) = \prod_{j=1}^{M} \frac{\sin\left(\frac{\varphi - \varphi_j}{2} - i\Phi\right)}{\sin\frac{\varphi - \varphi_j}{2}} + \prod_{\alpha=1}^{N} \frac{\sin\frac{\varphi - \psi_\alpha}{2}}{\sin\left(\frac{\varphi - \psi_\alpha}{2} + i\Phi\right)} \prod_{j=1}^{M} \frac{\sin\left(\frac{\varphi - \varphi_j}{2} + i\Phi\right)}{\sin\frac{\varphi - \varphi_j}{2}}. \tag{L.13}$$

The M parameters φ_j are determined by the condition that $t(\varphi)$ is regular in $\varphi = \varphi_j$, $j = 1, 2, \dots, M$. We fall back on our coupled equations; for example, using $\psi_\alpha = i\Phi$ for the homogeneous case, in the form (1.40).

12

Identical particles with δ-interactions: General solution for n components and limiting cases

12.1 The transfer matrix $Z(k)$ in a symmetry-adapted basis

12.1.1 Recursive definition of a basis for the representation $\mathscr{C}_N$ of π_N

Yang's method for spin-1/2 fermions described in Chapter 11 was generalized by Sutherland (1975, 1980) to the case of arbitrary spin, in other words to identical particles with $n = 2s + 1$ internal states. The task is then to diagonalize the matrix $Z(k)$ introduced to express the periodicity conditions, in a subring of π_N belonging to a given irreducible representation. Indeed, according to the conclusions of Subsection 10.4.2, the eigenvector $|\bar{I}\rangle$ common to all operators Z_j, and thus also $Z(k)$, is an element of π_N having the symmetry of the real-space wavefunction associated with a certain given Young tableau. As for the $n = 2$ case, we shall introduce the conjugate representation $(0j) = -\sigma_j \cdot \tau$, and we shall diagonalize $Z(k)$ in the representation

$$\mathscr{T}_N = [M_1, M_2, \ldots, M_n], \qquad N = M_1 + M_2 + \ldots + M_n. \qquad (12.1)$$

The difficulty with this problem consists in finding a convenient basis for the representation space adapted to $\mathscr{C}_N$.

Let us consider the tableau $\mathscr{C}_M$, $M = N - M_1$, obtained by suppressing the first row of tableau $\mathscr{C}_N$. In order to obtain a basis of $\mathscr{C}_N$, we can first introduce scalar functions of a running permutation π of order N, taken to be right-invariant under permutations of the first M_1 indices, and which constitute a representation basis for $\mathscr{C}_M$ in the permutation of the following M indices. Following this we can eventually, in a further step, operate the reduction $\mathscr{C}_N \subset \mathscr{C}_M \otimes [M_1]$.

The simplest means to realize the symmetry in the first M_1 indices of the set

$$\pi 1, \pi 2, \ldots, \pi M_1, \ldots, \pi N$$

is to take basis functions which are *independent* of these and functions only of the M following:

$$\pi(M_1 + 1), \ldots, \pi N. \tag{12.2}$$

The latter set constitutes a set of distinct integers

$$n_1, n_2, \ldots, n_M, \quad 1 \le n_a \le N, \tag{12.3}$$

which if one wants represent the coordinates of the M distinguishable particles occupying M sites of a linear chain of length N. The unoccupied sites (vacancies) play a symmetrical role, in contrast the M occupied sites are occupied by distinguishable particles.

A given set $\{n\}$ defines a permutation $P \in \pi_M$ via the inequalities

$$1 \le n_{P1} < n_{P2} < \ldots < n_{PM} \le N. \tag{12.4}$$

The set $\{n\}$ thus defines a basis 'vector' of $\mathscr{C}_M \otimes M_1$ which we shall denote $|\bar{P}n\rangle$.

Action of π on the basis $|\bar{P}n\rangle$

A running permutation $\pi \in \pi_N$ operates in a natural way on the vectors $|\bar{P}n\rangle$: for example, the transposition $\pi = (4, 9)$ exchanges the occupation of sites 4 and 9, that is according to Figure 12.1, puts particle number 1 on the site with coordinate 4. This induces the permutation $P_{12} \in \pi_M$ of the order of the particles along the chain. In general, it is sufficient to define the action of the generators of π_N, in other words of the particle or vacancy transpositions on neighbouring sites to obtain the representation

- $(n, n + 1)|\bar{P}n\rangle = |\bar{P}n'\rangle$

if there exists an index i such that

$$n'_{Pi} = n_{Pi} + 1, \quad n'_{Pj} = n_{Pj}, \quad j \ne i,$$
$$n_{Pi} = n, \quad n_{P(i+1)} > n + 1, \quad \text{etc.}$$

- $(n, n + 1)|\bar{P}n\rangle = |P_{i,i+1}\bar{P}n\rangle$

if there exists an index i such that

$$n_{Pi} = n, \quad n_{P(i+1)} = n + 1. \tag{12.5}$$

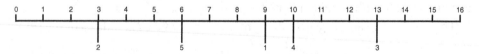

Figure 12.1 Set $\{n\} \equiv \{9, 3, 13, 10, 6\}$ defining the permutation $P = \downarrow$
$\begin{smallmatrix} 1 & 2 & 3 & 4 & 5 \\ 2 & 5 & 1 & 4 & 3 \end{smallmatrix}$, $N = 16, M = 5$.

12.1.2 The transfer matrix Z in the $|\bar{P}n\rangle$ basis

The matrix representation of the transpositions according to (10.56) allows us to interpret the matrix Z as a transfer matrix for a vertex model on a square lattice, each link of the lattice being open to n states labelled by an index α taking values from 1 to n. (For $n = 2$, one specifies two states of orientation $\downarrow$, $\uparrow$.) Figure 12.2 represents the two types of vertices and the relative weights associated with the transfer micromatrix:

$$X_{j0} = (1 + x_{j0}\sigma \cdot \tau)/(1 + x_{j0}).$$

We can also consider X_{j0} as a two-body scattering matrix, with the only two elastic processes being the direct and exchange ones, and consider a two-dimensional configuration of the vertex model as spatio-temporal trajectories along the two diagonal (time) and antidiagonal (space) axes according to a well-known analogy.

The graphical representation allows us to easily express the matrix elements of Z in the basis $|\bar{P}n\rangle$ adapted to $\mathscr{C}_M \otimes [M_1]$. Indeed, with the variables of each row of tableau $\mathscr{C}_N$ we can associate a state with the same spin. The first row of variables (unoccupied sites of the chain) will carry the first spin index ($\alpha = 1$) and the variables of $\mathscr{C}_M$, of coordinates $\{n\}$ are partially distinguished by the spin indices $\alpha > 1$, each coordinate n_j of given rank j being associated with the same spin α_j.

The graphical representation of the transfer matrix elements according to the convention of Figure 12.2 with the additional prescription of representing the same (greater than one) spin index states with a continuous thick line, leads to the same structure for the Z matrix elements as that described in Subsection 11.1.2, up to the fact that the initial and final states $\{n\}$ and $\{m\}$ which refer to the second spin component ($\downarrow$) are here specified by $\{\bar{Q}n\}$ and $\{\bar{P}m\}$ describing 'particles' discerned by $(n - 1)$ 'spin states'.

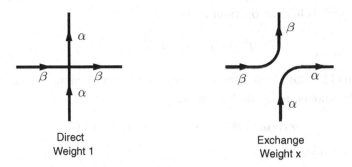

Direct
Weight 1

Exchange
Weight x

Figure 12.2 The two types of vertices and their weights.

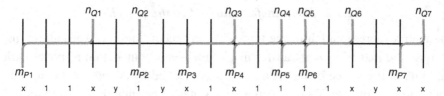

$P1 = Q1, \quad P2 = Q2, \quad P3 = Q3, \quad P4 = Q6, \quad P5 = Q4, \quad P6 = Q5, \ldots \Rightarrow Q = PQ_{45}Q_{56}.$

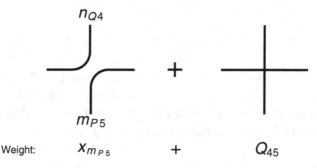

Weight: $\qquad X_{mP5} \qquad + \qquad Q_{45}$

Figure 12.3 Top: graphical representation of the matrix element $\langle \bar{P}m|Z|\bar{Q}n \rangle$; the lines are thick or thin depending on whether the index α is greater than or equal to 1.

It remains to determine, for a given graph, which transition $Q \to P$ it is associated with. It is sufficient for this purpose to determine the transpositions operated by the crossing of continuous lines on a typical graph (Figure 12.3). This corresponds to giving an operatorial weight operating in $\mathscr{C}_M$ to each crossing: that is $Q_{j,j+1}$, when $n_{Qj} = m_{P(j+1)}$. The structure of the matrix element can further be written in the general form

$$y\langle \bar{P}m|Z|\bar{Q}n \rangle \equiv \prod_{j=1}^{M} D(m_{Pj}n_{Qj})\bar{D}_j(n_{Qj}m_{P(j+1)}), \tag{12.6}$$

in which the functions $D(n, m)$ have the same definition as in (11.17), except for $\bar{D}(n, m)$ which is here an operator in π_N:

$$\bar{D}_j(m, m) = 1 + \frac{1}{x_m}Q_{j,j+1}. \tag{12.7}$$

Expression (12.6) takes into account the cyclic invariance along the fictitious chain and must be understood with the notation

$$m_{P(M+1)} \equiv m_{P1} + N, \qquad Q_{M,M+1} \equiv Q_{M1} \tag{12.8}$$

and the restrictions

$$\begin{cases} m_{Pj} \le n_{Qj} \le m_{P(j+1)}, \\ n_{Qj} < n_{Q(j+1)}, \qquad j = 1, 2, \ldots, M. \end{cases} \tag{12.9}$$

The cyclic invariance of our basis $|\bar{P}m\rangle$ is expressed by the relations

$$|m_{P1}m_{P2}\ldots m_{PM}\rangle \equiv |m_{P2}, m_{P3}, \ldots, m_{P1} + N\rangle. \tag{12.10}$$

12.2 Recursive diagonalization of matrix Z

12.2.1 Adaptation of Lieb's method

The eigenvalue (z) equation for the transfer matrix Z is written

$$\sum_{\bar{Q}n} \langle \bar{P}m|Z|\bar{Q}n\rangle c(\bar{Q}n) = zc(\bar{P}m), \tag{12.11}$$

in which $c(\bar{P}m) = \langle \bar{P}m|\bar{I}_M\rangle$ are the unknown amplitude functions of $\pi \in \pi_N$ having symmetry $\mathscr{C}_M \otimes [M_1]$.

Sutherland tries the Bethe sum

$$c(\bar{Q}n) = \sum_{R \in \pi_M} \langle Q|R\rangle \varphi(v_{R1}, n_{Q1})\varphi(v_{R2}, n_{Q2})\ldots\varphi(v_{RM}, n_{QM}), \tag{12.12}$$

in which the functions $\varphi(V, n)$ have the same definition as in (11.24). The $\langle Q|R\rangle$ coefficients must not be confused with those of Chapter 11, but are of similar nature. We shall introduce for convenience the elements of the ring of π_M

$$|\bar{R}\rangle = \sum_Q \bar{Q}\langle Q|R\rangle, \tag{12.13}$$

which allow us to write (12.11) in the form

$$\sum_R \sum_{\{n\}} \prod_{j=1}^{M} D(m_jn_j)\bar{D}(n_jm_{j+1})\varphi(v_{Rj}, n_j)|\bar{R}\rangle = zy \sum_R \prod_{j=1}^{M} \varphi(v_{Rj}, m_j)|\bar{R}\rangle. \tag{12.14}$$

The summation over $\{n\}$ in the left-hand side of (12.14) and the elimination of the 'off-diagonal terms' following Lieb's method are performed in an analogous way to that described in Subsection 11.1.3, but now being careful with the order of operations since the $\bar{D}$ are operators in π_M. The sum over n gives us

$$\sum_{n=m_3}^{m_4} D(m_3n)\bar{D}_3(n, m_4)\varphi(v_{R3}, n) \equiv X_3 + Y_4, \tag{12.15}$$

with

$$\begin{cases} X_3 \equiv (1 + x_0(v_{R3}))\dfrac{y_{m_3}}{x_{m_3}}y_{m_3+1}\cdots y_{m_4-1}x_{m_4}\varphi(v_{R3}, m_3), \\[2mm] Y_4 \equiv (Q_{34} - x_0(v_{R3}))\varphi(v_{R3}, m_4). \end{cases} \tag{12.16}$$

X_3 contains the *dangerous term* (using the terminology of Chapter 7) corresponding to $n = m_3$ in sum (12.15):

$$\frac{1}{x_3} y_{m_3+1} \cdots y_{m_4-1} x_{m_4} \varphi(v_{R3}, m_3), \tag{12.17}$$

and Y_4 contains the *dangerous term* $n = m_4$:

$$(x_{m_1} + Q_{34}) \varphi(v_{R3}, m_4). \tag{12.18}$$

The left-hand side of the $\{n\}$ and R sums in (12.14) will be reduced to the sum over the only two *diagonal terms*

$$\sum_R (X_1 X_2 \ldots X_M + Y_2 Y_3 \ldots Y_{M+1}) | \bar{R} \rangle, \tag{12.19}$$

if we ensure for each 'off-diagonal' term the equalities

$$\sum_R (\ldots Y_i X_i \ldots Y_j X_j \ldots) | \bar{R} \rangle \equiv 0. \tag{12.20}$$

We have the restriction $j - 1 > 1$, which ensures that the operators appearing in the various $\underline{YX}$ commute among themselves. Condition (12.20) is then written, by virtue of (12.16) and expressions (12.17) and (12.18) for the dangerous terms:

$$[(1 + x_0(v_{R4}))(Q_{34} - x_0(v_{R3})) y_{m_4} - (Q_{34} + x_{m_4})] | \bar{R} \rangle$$
$$+ [(1 + x_0(v_{R3}))(Q_{34} - x_0(v_{R4})) y_{m_4} - (Q_{34} + x_{m_4})] | Q_{34} \bar{R} \rangle = 0. \tag{12.21}$$

In order to be valid for any m_4, this requires the two relations

$$(Q_{34} - 1)(|\bar{R}\rangle + |Q_{34}\bar{R}\rangle) = 0 \tag{12.22}$$

and

$$[x_0(v_{R4}) Q_{34} - x_0(v_{R3}) - x_0(v_{R3}) x_0(v_{R4})] | \bar{R} \rangle$$
$$+ [x_0(v_{R3}) Q_{34} - x_0(v_{R4}) - x_0(v_{R4}) x_0(v_{R3})] | Q_{34} \bar{R} \rangle = 0. \tag{12.23}$$

The first condition expresses the 'continuity' of the coefficient $c(Qn)$ extrapolated to the boundary of two adjacent sectors (cf. Subsection 10.4.2):

$$c(\bar{Q} n) \equiv c(Q_{34} \bar{Q} n), \quad \text{if} \quad n_{Q3} = n_{Q4}. \tag{12.24}$$

The second condition (12.23) can be rewritten, taking (12.22) into account, as

$$|Q_{34} \bar{R}\rangle = \frac{Q_{34} + x_{R3,R4}}{1 - x_{R3,R4}} |\bar{R}\rangle, \tag{12.25}$$

with

$$x_{ij} = \frac{ic}{v_i - v_j}. \tag{12.26}$$

Relations (12.25) lead to (12.22) and are therefore sufficient to ensure the decomposition (12.19) into two terms, under the condition that the cyclic invariance is taken into account. It is remarkable that we fall back on the coherence conditions for $|\bar{R}\rangle$ whose algebraic structure is identical to that obtained in (10.22). The problem for the $\langle Q|R\rangle$ amplitudes in $\mathscr{C}_M$ is thus analogous to that for $\langle I\|\rangle$ in $\mathscr{C}_N$.

12.2.2 The cyclicity condition in $\mathscr{C}_M$

The cyclicity condition (12.10) on the expansion (12.12) is expressed by the relations

$$\langle QC_M|RC_M\rangle\varphi(v_{R1}, n+N) \equiv \langle Q|R\rangle\varphi(v_{R1}, n), \tag{12.27}$$

in which C_M is the circular permutation in $\pi_M : (12\ldots M)$. Taking (12.13) into account we can write it as

$$\bar{C}_M|\bar{R}\rangle = y(v_{R1})|\bar{C}_M\bar{R}\rangle, \tag{12.28}$$

with

$$y(v) = \prod_{n=1}^{N} y_n(v), \quad y_n(v) = 1 + \frac{ic}{k_n - v} \tag{12.29}$$

following the notation (11.24).

Since the representation (12.25) of π_M is the exact analogue of (10.22) with the correspondence $\{k\} \leftrightarrow \{-v\}$, we shall set

$$\tilde{X}_{ij} = \frac{1 + x_{ij}Q_{ji}}{1 - x_{ij}}, \tag{12.30}$$

and since the transpositions Q_{ij} operate in $\mathscr{C}_M$ or in other words already in the conjugate representation, we shall also define

$$X'_{ij} = \frac{1 - x_{ij}}{1 + x_{ij}}\tilde{X}_{ij} = \frac{1 + x_{ij}Q_{ij}}{1 + x_{ij}}, \tag{12.31}$$

which are the analogues of operators (10.54) in the conjugate representation.

Equation (12.28) is analogous to the cyclicity condition (10.26). From this we deduce that $R|\bar{R}\rangle$ is an eigenvector of the operator

$$\tilde{Z}_{R1} = \tilde{X}_{R2,R1}\tilde{X}_{R3,R1}\ldots\tilde{X}_{RM,R1} \tag{12.32}$$

for the eigenvalue $y(V_{R1})$. It is thus also an eigenvector of the operator $\tilde{Z}_R(k_0)$ or

$$Z'_R(k_0) = \mathop{\mathrm{Tr}}_{0} X'_{R1,0}X'_{R2,0}\ldots X'_{RM,0}, \tag{12.33}$$

$$Z'_R(k_0)R|\bar{R}\rangle = z'(k_0)R|\bar{R}\rangle, \tag{12.34}$$

where

$$Z'_R = \bar{R} Z' R. \tag{12.35}$$

Z' is the analogue in $\mathscr{C}_M$ of operator $Z(k_0)$ in $\mathscr{C}_N$ and the eigenvalue $z'(k_0)$ does not depend on R. Since we have the identity $\bar{Z}_{R1} \equiv Z'_R(k_0 = v_{R1})$, we deduce the relations

$$z'(v_i) = y(v_i) \prod_{\substack{j=1 \\ j \neq i}}^{M} \frac{1 + x_{ij}}{1 - x_{ij}}, \qquad i = 1, 2, \dots, M, \tag{12.36}$$

equivalent to our cyclicity conditions, which are constraints on the eigenvalue of Z'.

12.2.3 Relation between Z and Z'

The transition formulas (12.25) and the cyclicity conditions (12.36) thus allow us to reduce to form (12.19) the left-hand side of the eigenvalue equation (12.11) or (12.14), which becomes

$$\sum_R (X_1 X_2 \dots X_M + Y_2 Y_3 \dots Y_{M+1}) | \bar{R} \rangle = zy \sum_R \prod_{j=1}^{M} \varphi(v_{Rj}, m_j) | \bar{R} \rangle. \tag{12.37}$$

On the left-hand side, there only subsist the 'diagonal' terms; according to (12.16), the product of X symbols in (12.37) is a number, but the product of the Y is an operator of π_M. Let us examine the latter, which is written as

$$\prod_{j=1}^{M+1} Y_j \equiv (Q_{12} - x_0(v_{R1}))(Q_{23} - x_0(v_{R2})) \dots (Q_{M1} - x_0(v_{RM})) \prod_{j=1}^{M} \varphi(v_{Rj}, m_{j+1}). \tag{12.38}$$

If we introduce the spin representation for the transpositions of π_M, we can introduce an $(M+1)$th spin operator τ (of dimension $n-1$) and write

$$\prod_{j=1}^{M} (Q_{j,j+1} - x_0(v_{Rj})) = C_M \mathop{\mathrm{Tr}}_0 (1 - x_0(v_{R1}) Q_{10})(1 - x_0(v_{R2}) Q_{20}) \dots$$
$$\dots (1 - x_0(v_{RM}) Q_{M0}). \tag{12.39}$$

Let us substitute on the left-hand side of (12.37) the product of X according to (12.16) and the product of Y according to (12.39). For this last term, let us perform the translation $R \to RC_M$ in the sum over R and let us commute C_M through the trace. Taking the cyclicity condition (12.28) into account, we obtain for equality (12.37)

$$\sum_R \left\{ \underset{0}{\text{Tr}} \prod_{j=1}^{M}(1 - x_0(v_{Rj})Q_{j0}) - y \prod_{j=1}^{M}(1 + x_0(v_j)) - yz \right\}$$

$$\times \prod_{j=1}^{M} \varphi(v_{Rj}, m_j) | \bar{R} \rangle = 0, \tag{12.40}$$

but according to definition (12.31) of the X'_{j0} operators in $\mathscr{C}_M$, we have

$$\underset{0}{\text{Tr}} \prod_j (1 - x_0(v_{Rj})Q_{j0}) = \bar{R} \, \underset{0}{\text{Tr}} \prod_j (1 - x_0(v_{Rj})Q_{Rj0})R$$

$$= \prod_j (1 - x_0(v_j)) \bar{R} Z_R(k_0) R. \tag{12.41}$$

According to lemma (12.34), $R|\bar{R}\rangle$ is an eigenvector of Z'_R with eigenvalue $z'(k_0)$ (independent of R). Relation (12.40) is thus an identity in $|\bar{R}\rangle$ and $\{m\}$ provided we have

$$z = \prod_{j=1}^{M}(1 + x_0(v_j)) + z'y^{-1} \prod_{j=1}^{M}(1 - x_0(v_j)). \tag{12.42}$$

This is the relation between the spectrum of Z and that of Z'. We note that conditions (12.36), which determine the V_j, express the fact that the rational fraction $z(k_0)$ does not have a pole in $k_0 = v_j$: *the residue of the apparent pole vanishes.*

12.2.4 Sutherland's coupled equations

We have shown how to reduce the diagonalization of matrix Z belonging to the ring of π_N in the $\mathscr{C}_N = [M_1, M_2, \ldots, M_n]$ representation to that of a strictly analogous matrix $Z' \in \pi_M$ in the $\mathscr{C}_M = [M_2, \ldots, M_n]$ representation obtained from $\mathscr{C}_N$ by suppressing the first row of the tableau; $M = N - M_1$. We have constructed the components of the standard eigenvector as functions $\langle |\bar{Q}n \rangle = c(n_{Q1}, n_{Q2}, \ldots, n_{QM})$ of the variables of tableau $\mathscr{C}_N$, thereby automatically guaranteeing the symmetry in the group of the first M_1 variables of $\mathscr{C}_N$. This condition is, however, not sufficient to ensure that the amplitudes $c(\bar{Q}n)$ have symmetry type $\mathscr{C}_N$. We must show that $c(\bar{Q}n)$, viewed as a function of $\pi \in \pi_N$, cannot be symmetrized with respect to the $(M_1 + 1)$ variables containing those of the first row $\pi 1, \ldots, \pi M_1$ and an arbitrary variable of $\mathscr{C}_M$, n_{Q1} for example. This is the *Fock condition*, which is here written as

$$\left\{ \sum_{n \neq n_{Q1}} (n, n_{Q1}) + I \right\} c(\bar{Q}n) \equiv 0, \tag{12.43}$$

or further still

$$\sum_{n=1}^{n_{Q2}-1} c(n\, n_{Q2}\ldots n_{QM}) + \sum_{n=n_{Q2}+1}^{n_{Q3}-1} c(n_{Q2}\, n\, n_{Q3}\ldots n_{QM}) + \ldots \equiv 0.$$

We leave it to the reader to perform the summations and show identity (12.43) starting from the transition relations (12.25) and the cyclicity condition.

The *reduction process* $Z \in \mathscr{C}_N \to Z' \in \mathscr{C}_M$ can be iterated up to the point where all rows of tableau $\mathscr{C}_N$ have run out, and gives finally the equations for the spectrum of Z. Each reduction brings a sub-tableau into play together with the associated parameters $\{k\}$, $\{v\}$, etc. In order to present the result in a systematic manner, we perform the following change of notation: let us call $\{A\}$ the set $\{k\}$, $\{B\}$ the set $\{v\}$, etc. such that

N quantities A are associated with tableau $[M_1, M_2, \ldots, M_n]$

$N - M_1$ quantities B are associated with tableau $[M_2, M_3, \ldots, M_n]$

$N - M_1 - M_2$ quantities C are associated with tableau $[M_3, \ldots, M_n]$

$\ldots$

M_n quantities F are associated with tableau $[M_n]$.

We shall set

$$y_A \equiv y = \prod_{n=1}^{N}\left(1 + \frac{ic}{k_n - k_0}\right) \equiv \prod_{A}\left(1 + \frac{ic}{A - k_0}\right),$$

$$y_B = \prod_{n=1}^{M}\left(1 + \frac{ic}{v_n - k_0}\right) \equiv \prod_{B}\left(1 + \frac{ic}{B - k_0}\right),$$

$$y_C = \prod_{C}\left(1 + \frac{ic}{C - k_0}\right),$$

$$\ldots$$

$$y_F = \prod_{F}\left(1 + \frac{ic}{F - k_0}\right). \tag{12.44}$$

The fundamental relation (12.42) can be written as

$$z y_A = y_A \bar{y}_B + z' y_B, \tag{12.45}$$

in which $\bar{y}_B$ originates from the substitution $c \to -c$ in y_B. Iterating formula (12.45) gives us

$$z y_A = y_A \bar{y}_B + y_B \bar{y}_C + \ldots + y_E \bar{y}_F + y_F. \tag{12.46}$$

According to the remark following (12.42), the formulas analogous to (12.36) express the fact that the rational fraction $z(k_0)$ does not admit any pole at points $A, B, C, \ldots, F$ or in other words

$$\operatorname*{Res}_{k_0=B}(y_A \bar{y}_B + y_B \bar{y}_C) = 0,$$

$$\operatorname*{Res}_{k_0=C}(y_B \bar{y}_C + y_C \bar{y}_D) = 0,$$

$$\cdots$$

$$\operatorname*{Res}_{k_0=F}(y_E \bar{y}_F + y_F) = 0. \tag{12.47}$$

We hereby obtain Sutherland's equations:

$$\prod_A \left(1 + \frac{ic}{A-B}\right) = \prod_C \left(1 + \frac{ic}{C-B}\right) \prod_{B'}' \frac{B'-B+ic}{B'-B-ic},$$

$$\prod_B \left(1 + \frac{ic}{B-C}\right) = \prod_D \left(1 + \frac{ic}{D-C}\right) \prod_{C'}' \frac{C'-C+ic}{C'-C-ic},$$

$$\cdots$$

$$\prod_E \left(1 + \frac{ic}{E-F}\right) = \prod_{F'}' \frac{F'-F+ic}{F'-F-ic}, \tag{12.48}$$

which determine the sets $\{B\}, \{C\}, \ldots, \{F\}$ associated with the sub-tableaux $\mathscr{C}_{N-M_1}, \mathscr{C}_{N-M_1-M_2}, \ldots, \mathscr{C}_{M_n}$ as a function of $\{A\} = \{k\}$. This solves the problem of the spectrum of Z.

If we go back to the original problem of the determination of the energy spectrum of N identical particles with periodicity conditions in L, we must add to the symmetry conditions $\mathscr{C}$ (12.48) the transcendental relations originating from (10.50):

$$Z(k_j) = e^{ik_j L}, \quad j = 1, 2, \ldots, N,$$

or in other words according to (12.46):

$$\exp(ik_j L) = \prod_B \left(1 + \frac{ic}{k_j - B}\right), \tag{12.49}$$

which completes the determination of admissible sets $\{k\}$ such that $E = \sum_j k_j^2$.

Corollary: The spectrum of the generalized exchange Hamiltonian (10.71)

$$H = \sum_{n=1}^{N}(n, n+1) = \sum_n \sum_{\alpha,\beta} \sigma_n^{\alpha\beta} \sigma_{n+1}^{\beta\alpha}$$

is determined by the coupled equations (12.48) in which we have set $A_n \equiv 0$.

$$H = N - \frac{\partial \ln z}{\partial k_0}\Big|_{k_0=0} \simeq N - \sum_B \frac{c^2}{B(B-ic)}. \tag{12.50}$$

If we set

$$B_\alpha = \frac{c}{2} \tan \frac{k_\alpha}{2} + \frac{ic}{2}, \quad \alpha = 1, 2, \ldots, N - M_1,$$

we have

$$H = N - 2 \sum_{\alpha=1}^{N-M_1} (\cos k_\alpha + 1), \tag{12.51}$$

and the left-hand side of the first equation (12.48) can be written as e^{iNk_α}.

12.3 Zero coupling limit

12.3.1 Limit spectrum

The study of the vicinity of $c = 0$ for the system of commuting operators Z'_j or $Z'(k_0)$ leads to various corollaries which are both interesting and nontrivial, but somewhat simpler to prove than the complete results of the preceding sections. To first order in the coupling constant c, we obtain from the definition of the Z'_j operators

$$Z'_j = I_N + ic \sum_l \frac{(jl) - I_N}{k_l - k_j} + O(c^2), \tag{12.52}$$

in which I_N designates the identity in π_N and (jl) a running transposition. From this we deduce the following.

Proposition: The N elements of the ring of π_N

$$\Xi_j = \sum_l{}' \frac{(lj)}{k_l - k_j} \quad \left(\sum_j \Xi_j \equiv 0 \right) \tag{12.53}$$

commute among each other.

The direct verification of this is very easy. We have thus constructed a commuting sub-algebra of π_N depending simply on $N - 1$ parameters. From Sutherland's equations (12.48), we easily deduce the equations for the eigenvalues ξ_j of operators Ξ_j. Using again the notation of Subsection 12.2.4, we shall write $\{A\} \equiv \{k\}$, $\{B\} = \lim_{c \to 0}\{B\}$, etc. We shall also set

$$a = N, \quad b = N - M_1, \quad c = N - M_1 - M_2, \quad \ldots, \quad f = M_n$$

in such a way that the quantities $\{A\}, \{B\}, \ldots, \{F\}$ are associated with the Young tableau

$$[a - b, b - c, \ldots, f]. \tag{12.54}$$

The limit equations take the form

$$\sum_A \frac{1}{A-B} - \sum_C \frac{1}{C-B} = 2\sum_{B'} \frac{1}{B'-B},$$

$$\sum_B \frac{1}{B-C} - \sum_D \frac{1}{D-C} = 2\sum_{C'} \frac{1}{C'-C},$$

$$\cdots$$

$$\sum_E \frac{1}{E-F} = 2\sum_{F'} \frac{1}{F'-F}, \tag{12.55}$$

and we have the spectrum

$$\xi_j = {\sum_l}' \frac{1}{k_l - k_j} - \sum_B \frac{1}{B - k_j}. \tag{12.56}$$

12.3.2 Limit basis

The simultaneous eigenvector of the set Z'_j, denoted $|I_N\rangle$, was constructed using recursion (cf. Subsection 12.2.4) and, at each step of the reduction, the components are given by *Bethe sums*. The limit basis of the eigenvectors is rather simple and can be given explicitly. We had decomposed the standard eigenvector $|I_N\rangle$ on the basis $|\bar{Q}n\rangle$; our amplitudes $\langle\bar{Q}n|I_N\rangle = c(\bar{Q}n)$ are functions of $P \in \pi_N$ (or π_a) ($N \equiv a$), the permutation $Q \in \pi_M$ (or π_b) is defined by

$$n_{Q1} < n_{Q2} < \ldots < n_{QM}, \quad n_1 = P1, \; n_2 = P2, \; \ldots, \; n_M = PM. \tag{12.57}$$

According to (12.12) and definition (11.24) for $\varphi(B, n)$, we have

$$\lim_{c\to 0} c^{-1}\varphi(B, n) = \frac{i}{B - A_n}, \quad A_n \equiv k_n.$$

Consequently, the limit amplitudes, after a convenient normalization, take the form (in the new notation $a, b, \ldots$)

$$\langle\bar{P}_a||I_a\rangle_a = \lim_{c\to 0}(ic)^{-b}\langle\bar{Q}n|I_n\rangle$$

$$= \sum_{R\in\pi_b} \langle Q||R\rangle_b \frac{1}{B_{R1} - A_{n_{Q1}}} \frac{1}{B_{R2} - A_{n_{Q2}}} \cdots \frac{1}{B_{Rb} - A_{n_{Qb}}}. \tag{12.58}$$

But in the $c = 0$ limit, the vector or element of the ring of π_b,

$$|\bar{R}\rangle = \sum_{Q\in\pi_b} \bar{Q}\langle Q||R\rangle_b, \tag{12.59}$$

is easily determined as a function of R. Indeed, the limit of the transition formulas can be written

$$|Q_{34}\bar{R}\rangle = Q_{34}|\bar{R}\rangle, \tag{12.60}$$

which leads to

$$|\bar{R}\rangle = R|I_b\rangle, \qquad \forall R \in \pi_b \qquad (12.61)$$

and, according to (12.59),

$$\langle Q||R\rangle_b \equiv \langle Q\bar{R}||I_b\rangle_b. \qquad (12.62)$$

We notice that the amplitudes $\langle R||I\rangle_b$ are the exact analogues for tableau $\mathscr{C}_b$ of the initial amplitudes $\langle P||I\rangle_a$ defined for $\mathscr{C}_a$. Taking (12.62) into account as well as the fact that $n_{Qi} = PQi$, according to (12.57), the expansion (12.58) for the limit amplitude gives us

$$\langle \bar{P}||I_a\rangle = \sum_{R \in \pi_b} \langle Q\bar{R}||I_b\rangle \frac{1}{B_{R1} - A_{PQ1}} \dots \qquad (12.63)$$

After a translation $R \to RQ$ and change of variables $\bar{P} \to P$, $\bar{R} \to Q$, we finally obtain

$$\langle P||I_a\rangle = \sum_{Q \in \pi_b} \langle Q||I_b\rangle (B_{Q1} - A_{P1})^{-1}(B_{Q2} - A_{P2})^{-1} \dots (B_{Qb} - A_{Pb})^{-1}. \quad (12.64)$$

The repeated application of the recursion formula (12.64) leads to an obvious expression which there is no need to write since it will not be clearer than the relation above.

The amplitude displays itself as a sum over the subgroups of nested permutations

$$\pi_a \supset \pi_b \supset \pi_c \dots$$

associated with the sequence of Young tableaux

$$\mathscr{C}_a = [a - b, b - c, \dots, f], \qquad \mathscr{C}_b = [b - c, \dots, b], \qquad \dots, \qquad \mathscr{T}_f = [f],$$

with

$$\langle S||I\rangle_f \equiv 1, \qquad \forall S \in \pi_f. \qquad (12.65)$$

12.3.3 Representation basis for π_N

We have thus obtained the elements of the ring of π_a,

$$|I_a\rangle = \sum_{P \in \pi_a} \bar{P}\langle P||I_a\rangle,$$

which are eigenvectors of the set Ξ_j, provided the auxiliary quantities $\{B\}$, $\{C\}$, ... which enter in the definition of $|I_a\rangle$ verify the algebraic system (12.55). We could check that $|I_a\rangle$ satisfies all Fock conditions of the type

$$((b, b+1) + (b, b+2) + \ldots + (b, a) + I)|I_a\rangle = 0,$$

$$((c, b+1) + (c, b+2) + \ldots + (c, a) + I)|I_a\rangle = 0, \ldots \quad (12.66)$$

which lead to the fact that $|I_a\rangle$ is indeed of symmetry type $\mathscr{C}_a$.

The question of the number of linearly independent vectors thus obtained is not generally solved by direct counting. In the case of tableau $\mathscr{C}_N = [N - M, M]$ with two rows, we could count the solutions of system

$$\sum_A \frac{1}{A - B} = 2 \sum_{B'} \frac{1}{B' - B}, \quad \forall B. \quad (12.67)$$

This is equivalent to counting the number of polynomials $P_b(Z)$ of degree b which satisfy the identity

$$P_a(Z) P_b''(Z) - P_a'(Z) P_b'(Z) - R_{a-2}(Z) P_b(Z) \equiv 0, \quad (12.68)$$

with

$$P_a(Z) = \prod_A (Z - A), \quad P_b(Z) = \prod_B (Z - B), \quad (12.69)$$

in which R is an undetermined polynomial. For a generic position of the A, we find exactly $C_b^a - C_{b-1}^a$ distinct solutions, which is indeed the dimension of the $[a - b, b]$ representation of π_a.

The limit basis for the fermion wavefunctions with spatial symmetry $\bar{\mathscr{C}}$ conjugate to $\mathscr{C}$ can be written

$$\psi(x_1, x_2, \ldots, x_N) = \sum_{P \in \pi_N} \chi(P) \langle P || I_N \rangle \exp(i(\bar{P}k, x)). \quad (12.70)$$

13

Various corollaries and extensions

13.1 A class of completely integrable spin Hamiltonians

13.1.1 Presentation of some results

We here present some consequences of the theory in Chapter 12, concerning the diagonalization of a certain class of Hamiltonians which are quadratic in spin variables, these Hamiltonians being related to various models of interacting quantum systems. The results concerned are sufficiently interesting to be considered independently from their origin (see Section 12.3) and are sufficiently simple to allow the presentation of a direct proof.

One can thus calculate the energy levels of the following Hamiltonian:

$$H_0 = \sum_n v_n \vec{S}_n \cdot \vec{S} \tag{13.1}$$

describing the magnetic interaction of a distinct spin $\vec{S}$ with a collection of environmental spins $\vec{S}_n$ supposedly without mutual interaction, the intensity of the magnetic moments v_n being arbitrary.

Another example is obtained starting from an anisotropic spin Hamiltonian taken in a certain limit. The Hamiltonian

$$\tilde{H} = \omega a^\dagger a + \sum_n (a S_n^+ + a^\dagger S_n^- + \varepsilon_n S_n^z) \tag{13.2}$$

is integrable and describes the interaction of an oscillator mode of a field of phonons or photons with a set of two-level systems whose energy spacings are arbitrary. In the two quoted cases, the calculation of the spectrum and of the states is brought back to the solution of coupled algebraic equations.

Let us indeed apply the results of Section 12.3 concerning the commuting set $\{\Xi_j\}$ by using the spin-1/2 representation of π_N defined by the Young tableau with two rows $\mathscr{C} = [N - M, M]$. With the help of spin-1/2 matrices, we have the representation of the transposition $(lj) = (1 + \vec{\sigma}_l \cdot \vec{\sigma}_j)/2$ and as a consequence, the operators

$$H_j = \sum_{l=1}^{N}{}' \frac{\sigma_j \cdot \sigma_l}{\varepsilon_j - \varepsilon_l}, \tag{13.3}$$

in which the $\{\varepsilon_j\}$ are a given set, are simultaneously diagonalizable. For the total spin $S = N/2 - M$, the eigenvalues are given by the expression

$$h_j = \sum_{l=1}^{N}{}' \frac{1}{\varepsilon_l - \varepsilon_j} - 2 \sum_{\alpha=1}^{M} \frac{1}{E_\alpha - \varepsilon_j} \tag{13.4}$$

provided we have the M equations defining the $\{E_\alpha\}$

$$\sum_{l=1}^{N} \frac{1}{\varepsilon_l - E_\alpha} - 2 \sum_{\beta=1}^{M}{}' \frac{1}{E_\beta - E_\alpha} = 0. \tag{13.5}$$

It is thus clear that the class of Heisenberg Hamiltonians

$$\mathcal{H} = \sum_{j<l} v_{jl} \sigma_j \cdot \sigma_l \equiv \sum_{j} \eta_j H_j, \tag{13.6}$$

depending on $2N - 2$ effective parameters, is completely integrable. It is sufficient for the coefficients to have the form $v_{jl} = (\eta_j - \eta_l)/(\varepsilon_j - \varepsilon_l)$.

It is almost obvious that the result above does not depend on the size of the spin; in a first instance, we are dealing with an assembly of spin-1/2 but it is sufficient to bring the ε together in packets to obtain operators analogous to (13.3) or (13.6), the size of each spin being arbitrary. Let us give, for example, the equations for the spectrum of (13.1) with

$$\vec{S}_n^2 = s_n(s_n + 1). \tag{13.7}$$

For total spin $S = S^z$, and the spin deviation

$$M = s_0 + \sum_{j} s_j - S^z, \tag{13.8}$$

the spectrum is given by the formula

$$H_0 = s_0 \sum_{j=1}^{N} s_j v_j - \sum_{\alpha=1}^{M} \omega_\alpha, \tag{13.9}$$

$$\frac{S^z + 1}{\omega_\alpha} + \sum_{j}^{N} \frac{s_j}{v_j - \omega_\alpha} - \sum_{\beta}^{M}{}' \frac{1}{\omega_\beta - \omega_\alpha} = 0, \quad \forall \alpha. \tag{13.10}$$

Going back to the more general Hamiltonian (13.6), it is curious that for $N = 4$ it is sufficient to have a single relation between the six couplings

v_{jl} to ensure integrability. Here the existence of quantities η_j and ε_j such that $v_{ij} = (\eta_i - \eta_j)/(\varepsilon_i - \varepsilon_j)$ is equivalent to the single condition

$$\frac{(v_{14} - v_{12})(v_{24} - v_{23})(v_{34} - v_{13})}{(v_{24} - v_{12})(v_{34} - v_{23})(v_{14} - v_{13})} = 1, \tag{13.11}$$

which can also be written in a symmetric fashion as

$$\sum_{circ.} v_{12} v_{34} (v_{14} + v_{23} - v_{13} - v_{24}) = 0. \tag{13.12}$$

The integrability of these systems is, of course, valid in the classical limit.

13.1.2 Direct proof

We consider a set $\{\varepsilon\}$ and a set of independent spin operators $\vec{S}_j$, basis elements of the $su(2)$ Lie algebra (more generally, basis elements of an arbitrary Lie algebra $\mathcal{G}$). We define the meromorphic in E operator

$$\vec{S}(E) = \sum_j \frac{\vec{S}_j}{\varepsilon_j - E}. \tag{13.13}$$

We have the commutation relations

$$\left[S^{\alpha}(E), S^{\beta}(E') \right] = \frac{c_{\gamma}^{\alpha\beta}}{E - E'} (S^{\gamma}(E) - S^{\gamma}(E')), \tag{13.14}$$

in which the coefficients $c_{\gamma}^{\alpha\beta}$ are the structure constants of $\mathcal{G}$.

Proposition:

$$\left[\vec{S}^2(E), \vec{S}^2(E') \right] = 0, \qquad \forall E, E' \tag{13.15}$$
$$\vec{S}^2(E) = \sum_{\alpha\beta} g_{\alpha\beta} S^{\alpha}(E) S^{\beta}(E)$$

in which $g_{\alpha\beta}$ is the metric of $\mathcal{G}$.

For the proof, we can make direct use of (13.14). One can also write

$$\vec{S}^2(E) = \sum_j \frac{\vec{S}_j^2}{(\varepsilon_j - E)^2} - 2 \sum_j \frac{H_j}{\varepsilon_j - E} \tag{13.16}$$

with

$$H_j = \sum_k{}' \frac{\vec{S}_j \cdot \vec{S}_k}{\varepsilon_j - \varepsilon_k}. \tag{13.17}$$

It is sufficient to prove that

$$[H_j, H_k] = 0, \tag{13.18}$$

which results from the Jacobi identity. For each distinct triplet of indices, we have

$$\sum_{circ.} \frac{[\vec{S}_j \cdot \vec{S}_l, \vec{S}_k \cdot \vec{S}_l]}{(\varepsilon_j - \varepsilon_l)(\varepsilon_k - \varepsilon_l)} = \sum_{circ.} c_{\alpha\beta\gamma} S_j^\alpha S_k^\beta S_l^\gamma \frac{1}{(\varepsilon_j - \varepsilon_l)(\varepsilon_k - \varepsilon_l)} \equiv 0, \tag{13.19}$$

from the total antisymmetry of $c_{\alpha\beta\gamma}$. $\square$

Going back to spin-1/2, in other words to $su(2)$, we construct the eigenstates of $\vec{S}^2(E)$ in the form (Gaudin, 1976)

$$|M\rangle = S^-(E_1)S^-(E_2)\dots S^-(E_M)|0\rangle \tag{13.20}$$

in which the set $\{E\}$ is not determined. The reference state $|0\rangle$ is the completely aligned state defined by

$$S^z|0\rangle = \left(\sum_j s_j\right)|0\rangle, \tag{13.21}$$

which leads to

$$S^z|M\rangle = \left(\sum_j s_j - M\right)|M\rangle,$$

$$S^z(E)|M\rangle = \sum_j \frac{s_j}{\varepsilon_j - E}|M\rangle. \tag{13.22}$$

With the help of the commutation relations (13.14), we have

$$[S^z(E), S^-(E')] = \frac{-1}{E - E'}(S^-(E) - S^-(E')),$$

$$[\vec{S}^2(E), S^-(E')] = \frac{2}{E - E'}(S^-(E')S^z(E) - S^-(E)S^z(E')) \tag{13.23}$$

and as a consequence,

$$\vec{S}^2(E)|M\rangle = \left\{\left(\sum_j \frac{s_j}{\varepsilon_j - E}\right)^2 + \sum_j \frac{s_j}{(\varepsilon_j - E)^2}\right\}|M\rangle$$

$$+ \sum_{\alpha=1}^M \frac{2}{E - E_\alpha} S^-(E_1)\dots S^-(E_{\alpha-1})\left\{S^-(E_\alpha)S^z(E) - S^-(E)S^z(E_\alpha)\right\}\dots$$

$$\dots S^-(E_M)|0\rangle. \tag{13.24}$$

Commuting the S^z operators towards the right, we obtain

$$
\vec{S}^2(E)|M\rangle = \left\{ \left(\sum_j \frac{s_j}{\varepsilon_j - E} \right)^2 + \sum_j \frac{s_j}{(\varepsilon_j - E)^2} \right\} |M\rangle
$$

$$
+ \sum_{\alpha=1}^{M} \left(\sum_j \frac{s_j}{\varepsilon_j - E} \right) \frac{2}{E - E_\alpha} S^-(E_1) \ldots S^-(E_\alpha) \ldots S^-(E_M)|0\rangle
$$

$$
- \sum_\alpha \left(\sum_j \frac{s_j}{\varepsilon_j - E_\alpha} \right) \frac{2}{E - E_\alpha} S^-(E_1) \ldots S^-(E) \ldots S^-(E_M)|0\rangle
$$

$$
+ \sum_\alpha \sum_{\beta>\alpha} \frac{-1}{E - E_\beta} \frac{2}{E - E_\alpha} S^-(E_1) \ldots S^-(E_\alpha) \ldots \left\{ S^-(E) - S^-(E_\beta) \right\} \ldots |0\rangle
$$

$$
+ \sum_\alpha \sum_{\beta>\alpha} \frac{1}{E_\alpha - E_\beta} \frac{2}{E - E_\alpha} S^-(E_1) \ldots S^-(E) \ldots \left\{ S^-(E_\alpha) - S^-(E_\beta) \right\} \ldots |0\rangle.
\underset{\alpha}{\downarrow} \qquad\qquad \underset{\beta}{\downarrow}
$$

$$(13.25)$$

Let us collect on the right-hand side of (13.25) the coefficients of vector

$$
S^-(E_1) \ldots S^-(E) \ldots S^-(E_M)|0\rangle;
\underset{\alpha}{\downarrow}
$$

we find

$$
- \left(\sum_j \frac{s_j}{\varepsilon_j - E_\alpha} \right) \frac{2}{E - E_\alpha} - \sum_{\beta<\alpha} \frac{2}{E - E_\alpha} \frac{1}{E - E_\beta}
$$

$$
- \sum_{\beta>\alpha} \frac{2}{E - E_\alpha} \frac{1}{E_\alpha - E_\beta} + \sum_{\beta<\alpha} \frac{2}{E - E_\beta} \frac{1}{E_\beta - E_\alpha}
$$

$$
\equiv \frac{2}{E - E_\alpha} \left\{ \sum_j \frac{-s_j}{\varepsilon_j - E_\alpha} + \sum_{\beta<\alpha} \frac{1}{E_\beta - E_\alpha} - \sum_{\beta>\alpha} \frac{1}{E_\alpha - E_\beta} \right\}. \quad (13.26)
$$

Thus, if we impose the M relations

$$
\sideset{}{'}\sum_{\beta=1}^{M} \frac{1}{E_\beta - E_\alpha} - \sum_{j=1}^{N} \frac{s_j}{\varepsilon_j - E_\alpha} = 0, \qquad \forall \alpha = 1, \ldots, M, \qquad (13.27)
$$

the operator products containing $S^-(E)$ disappear on the right-hand side of (13.25). We are then left with a vector proportional to $|M\rangle$. Collecting the factors, one obtains for the corresponding eigenvalue

$$\vec{S}^2(E) = \sum_j \frac{s_j}{(\varepsilon_j - E)^2} + \left(\sum_j \frac{s_j}{\varepsilon_j - E} + \sum_\alpha \frac{1}{E - E_\alpha} \right)^2 - \sum_\alpha \frac{1}{(E - E_\alpha)^2}.$$

(13.28)

The algebraic relations (13.27) determine the sets $\{E_\alpha\}$ and lead to the fact that the residues of the apparent simple poles of function $\vec{S}^2(E)$ at points E_α vanish. The 'residue' of the double pole in ε_j is precisely $s_j(s_j + 1) = \vec{S}_j^2$, in accordance with (13.16). The residue of the simple pole in ε_j is, according to (13.17),

$$H_j = -s_j \left(\sum_k{}' \frac{s_k}{\varepsilon_k - \varepsilon_j} + \sum_\alpha \frac{1}{\varepsilon_j - E_\alpha} \right).$$

(13.29)

13.1.3 Remarks on the spectrum, the norm and the correlation

The coupled equations (13.27) to determine the spectrum of $\vec{S}^2(E)$ or of the operators H_j lead to an elimination problem which has been well known since Lamé's theory of functions (Whittaker and Watson, 1927, p. 536). Indeed, the polynomial $P(E)$ which admits $\{E\}$ as a set of zeroes solves a second-order differential equation with rational coefficients, analogous to Lamé's:

$$P''(E) + \left(\sum_j \frac{2s_j}{\varepsilon_j - E} \right) P'(E) + \left(\sum_j \frac{\alpha_j}{\varepsilon_j - E} \right) P(E) = 0,$$

(13.30)

in which the parameters a_j are to be determined such that the solution $P(E)$ is a polynomial. Lamé's case corresponds to $s_j = -\frac{1}{2}$, $\forall j$. This problem is related to that of the separation of Laplace's equation in N dimensions in ellipsoidal coordinates, the coordinated surfaces being the homofocal quadrics

$$(Q_j): \quad \sum_{i=1}^N \frac{x_i^2}{E_j - \varepsilon_i} = 1.$$

(13.31)

But the separation method can be extended to differential operators of the type $\sum_i \Delta_{\nu_i}$ with

$$\Delta_{\nu_j} = \frac{\partial^2}{\partial x_j^2} - \frac{\nu_i^2 - \frac{1}{4}}{x_i^2},$$

(13.32)

operators which can be considered as resulting from the separation between angular and radial (x_i) variables on multidimensional Laplacians. The values of ν_i can, however, be arbitrary and the associated dimension is not necessarily integer.

If we perform the separation of the equation $\sum_i \Delta_{\nu_i} \varphi = 0$ in ellipsoidal coordinates, we fall exactly on the system (13.27) to determine the roots of the $P(E)$ polynomials which constitute the factors of the ellipsoidal harmonics. One must choose $\nu_i = 2s_i + 1$. We note that the values $\nu_i = 0, 1/2$ corresponding to the disappearance of the 'centrifugal potential' in the radial Laplacian (13.32) precisely define Lamé's case ($s_i = -1/2, -1/4$) in which the roots of P are all real. It is interesting to notice the fact that the problem of determining the spectrum of our *spin operators* is identical to that of the construction of the harmonics of a certain generalized *radial Laplacian*.

Let us finally mention the electrostatic analogy of Stieltjes (Szegö, 1939, p. 151, eq. (6.8)), which interprets the roots E_α as the complex coordinates of M mobile charges $+1$ in (unstable) equilibrium in the field of N fixed charges $-s_j$ in ε_j. In our problem the roots can be complex due to the fact that the s_j are positive.

Let us now tackle the question of the spin–spin correlation in an eigenstate. This can be expressed in an elegant way. It is sufficient to apply a transformation to the $\{\varepsilon_j\}$ parameters

$$\frac{\partial H_j}{\partial \varepsilon_k} = -\frac{\langle \vec{S}_j \cdot \vec{S}_k \rangle}{(\varepsilon_j - \varepsilon_k)^2}, \quad j \neq k. \tag{13.33}$$

By derivation of the eigenvalue (13.29), we obtain

$$\langle \vec{S}_j \cdot \vec{S}_k \rangle = s_j s_k - s_j \sum_{\alpha=1}^{M} \frac{(\varepsilon_j - \varepsilon_k)^2}{(\varepsilon_j - E_\alpha)^2} \frac{\partial E_\alpha}{\partial \varepsilon_k}. \tag{13.34}$$

If we introduce the Jacobian matrix of (13.27):

$$\Delta_{\alpha\beta} = \left(\sum_j \frac{s_j}{(\varepsilon_j - E_\alpha)^2} - \sum_\gamma \frac{1}{(E_\alpha - E_\gamma)^2} \right) \delta_{\alpha\beta} + \frac{1}{(E_\alpha - E_\beta)^2}, \tag{13.35}$$

we have

$$\sum_\beta \Delta_{\alpha\beta} \frac{\partial E_\beta}{\partial \varepsilon_k} = \frac{s_k}{(E_k - E_\alpha)^2},$$

and consequently the average value

$$\langle \vec{S}_j \cdot \vec{S}_k \rangle = s_j s_k + \frac{(\varepsilon_j - \varepsilon_k)^2}{|\Delta|} \begin{vmatrix} 0 & \cdots & \dfrac{s_k}{(\varepsilon_k - E_\alpha)^2} \\ \vdots & & \\ \dfrac{s_j}{(\varepsilon_j - E_\beta)^2} & & \Delta_{\beta\alpha} \end{vmatrix}, \tag{13.36}$$

in the form of a quotient of two determinants. The idea of this method goes back to Richardson (1965), who obtained the eigenvectors of the pairing Hamiltonian

$$H = \sum_l \varepsilon_l S_l^z - G \sum_{j,k} S_j^+ S_k^-$$
(13.37)

in the form (13.20), the equations for the spectrum being similar to (13.27). Richardson's method also allows the computation of the norm of state $|M\rangle$. One finds

$$\langle M|M\rangle = M! |\Delta|.$$
(13.38)

13.2 Other examples of integrable systems

13.2.1 Anisotropic extension

It is natural to try to extend the family of integrable quadratic spin Hamiltonians by considering an anisotropic spin–spin interaction

$$\tilde{H}_j = \sum_{l=j}^{N}{}' \sum_\alpha \omega_{jl}^\alpha \sigma_j^\alpha \sigma_l^\alpha,$$
(13.39)

$$\left[\tilde{H}_j, \tilde{H}_k\right] = 0.$$
(13.40)

The σ^α are elements of a Lie algebra. In this diagonal form, we have found a solution to (13.40) only for $so(3)$ in which we have the commutation

$$\left[\sigma^\alpha, \sigma^\beta\right] = 2\varepsilon_{\alpha\beta\gamma}\sigma^\gamma.$$
(13.41)

The operators $\tilde{H}_j$ can be obtained, if one wants, by derivation of the inhomogeneous eight-vertex transfer matrix considered in Chapter 8. Let us indeed show that in the notation of that chapter, we have

$$T_1 = T(u_1) = X_{21}X_{31}\ldots X_{N1},$$
(13.42)

in which X_{21} is defined in Subsection 8.5.1. In the vicinity of $\eta = 0$, formulas (8.57) and (8.64) give us the expansion

$$X_{12} = 1 + \eta\Theta_1'(0) \sum_{\alpha=1}^{3} \frac{\Theta_{\alpha+1}(u_{12})}{\Theta_{\alpha+1}(0)\Theta_1(u_{12})}\sigma_1^\alpha\sigma_2^\alpha + O(\eta^2).$$
(13.43)

If we set

$$\tilde{H}_1 = \frac{\partial T_1}{\partial \eta}\Big|_{\eta=0},$$
(13.44)

we indeed obtain structure (13.39) with the antisymmetric coefficients

$$\omega_{12}^{\alpha} = \frac{\Theta_1'(0)}{\Theta_1(u_{12})} \frac{\Theta_{\alpha+1}(u_{12})}{\Theta_{\alpha+1}(0)}, \qquad u_{12} = u_1 - u_2. \tag{13.45}$$

Instead of proceeding via the results of Chapter 8, it is obviously simpler to search directly for the conditions on the ω_{ij}^{α} such that commutation (13.40) is ensured. One immediately finds the constraints

$$\omega_{ij}^1 \omega_{jk}^3 + \omega_{jk}^2 \omega_{ki}^1 + \omega_{ki}^3 \omega_{ij}^2 = 0, \qquad \forall i, j, k, \tag{13.46}$$

whose general solution is precisely given by the elliptic uniformization (13.45), relations (13.46) being identical to the Jacobi relations provided we have $u_{ij} + u_{jk} + u_{ki} = 0$. We also have the following algebraic representation:

$$\omega_{ij}^1 : \omega_{ij}^2 : \omega_{ij}^3 = \mathrm{cs}\,(u_{ij}) : \mathrm{ds}\,(u_{ij}) : \mathrm{ns}\,(u_{ij})$$
$$= \frac{s_i c_i d_j + s_j c_j d_i}{s_i^2 - s_j^2} : \frac{s_i d_i c_j + s_j d_j c_i}{s_i^2 - s_j^2} : \frac{c_i d_i s_j + c_j d_j s_i}{s_i^2 - s_j^2},$$

with the relations $s_i = \mathrm{sn}\,(u_i; k)$ (k modulus)

$$s_i^2 + c_i^2 = 1, \qquad k^2 s_i^2 + d_i^2 = 1. \tag{13.47}$$

In the limit $k = q = 0$, we have a supplementary integral of motion: the magnetic component of total spin $S^z = \frac{1}{2} \sum_j \sigma_j^z$. The operators $\tilde{H}$ then take the form

$$\tilde{H}_j = \sum_{l=1}^{N}{}' \frac{1}{\sin(u_j - u_l)} \left\{ \sigma_j^x \sigma_l^x + \sigma_j^y \sigma_l^y + \sigma_j^z \sigma_l^z \cos(u_j - u_l) \right\}. \tag{13.48}$$

Here again, the size of one of the spins can be arbitrarily chosen by letting u_j parameters tend to each other in packets. Hamiltonian (13.2), which is a completely integrable example, is precisely a weak limit of the anisotropic Hamiltonian (13.48) when the size of each spin increases indefinitely. Let us give the results pertaining to this Hamiltonian. We have the constant

$$M = a^\dagger a + \sum_l \left(S_l^z + \frac{1}{2} \right). \tag{13.49}$$

The spectrum is given by the formula

$$\tilde{H} = -\frac{1}{2} \sum_l \varepsilon_l + \sum_{\alpha=1}^{M} \tilde{E}_\alpha \tag{13.50}$$

with the equations

$$\sum_j \frac{1}{\tilde{E}_\alpha - \varepsilon_j} - 2 \sum_\beta{}' \frac{1}{\tilde{E}_\alpha - \tilde{E}_\beta} - \tilde{E}_\alpha = 0. \tag{13.51}$$

The eigenstates are

$$|M\rangle = \prod_{\alpha=1}^{M} \left(a^\dagger - \frac{1}{2} \sum_j \frac{S_j^+}{\varepsilon_j - \tilde{E}_\alpha} \right) |0\rangle. \tag{13.52}$$

13.2.2 Extension to a Lie algebra

Let us simply state that the definition of the commuting family H_j is valid for an arbitrary Lie algebra $\mathscr{G}$:

$$\mathscr{G} : \left[X^\lambda, X^\mu \right] = c_\nu^{\lambda\mu} X^\nu, \tag{13.54}$$

$$X_j = 1 \otimes 1 \otimes \ldots \otimes \underset{\underset{j}{\downarrow}}{X} \otimes \ldots \otimes 1, \tag{13.55}$$

$$H_j = \sum_k{}' \frac{\vec{X}_j \cdot \vec{X}_k}{\varepsilon_j - \varepsilon_k}. \tag{13.56}$$

The $u(n)$ algebra gives us

$$H_j|_{u(n)} = \sum_k{}' \sum_{\alpha\beta} \frac{\sigma_j^{\alpha\beta} \sigma_k^{\beta\alpha}}{\varepsilon_j - \varepsilon_k} \equiv \Xi_j, \tag{13.57}$$

in which we recognize in the ξ_j (formula (12.53)) the derivatives of the Yang matrices Z_j for the problem of identical particles interacting with a δ potential. Their spectrum is given in Section 12.3. We remind the reader of the matrix representation of the transpositions in (10.70).

The $o(n)$ algebra gives us

$$H_j|_{o(n)} = \sum_l{}' \sum_{\alpha<\beta} \frac{\hat{\sigma}_j^{\alpha\beta} \hat{\sigma}_l^{\alpha\beta}}{\varepsilon_j - \varepsilon_l}, \tag{13.58}$$

with

$$\hat{\sigma}_j^{\alpha\beta} = \sigma_j^{\alpha\beta} - \sigma_j^{\beta\alpha}. \tag{13.59}$$

The spectrum has not yet been determined. We do not know of any anisotropic extension for $n > 2$.

The problem of the ternary relations pertaining to a given symmetry group has been treated by Zamolodchikov in the context of the S-matrix formalism (cf. Subsection 10.3.2). The question is, for example, solved for $u(n)$ with Yang's X_{ij}

matrices; for $\mathbb{Z}_2 + \mathbb{Z}_2$, with those of Baxter. Let us give a calculation which leads to the solution for $o(n)$.

The ternary relation $X_{12}X_{13}X_{23} = X_{23}X_{13}X_{12}$ can be written explicitly as

$$\sum_{\sigma\tau\rho}(\beta\gamma|A|\tau\rho)(\alpha\rho|A'|\sigma v)(\sigma\tau|A''|\lambda\mu) = \sum_{\sigma\tau\rho}(\alpha\beta|A''|\sigma\tau)(\sigma\gamma|A'|\lambda\rho)(\tau\rho|A|\mu v)$$

(13.60)

in the notation $X_{12} = A''$, $X_{13} = A'$, $X_{23} = A$, according to the sketch of Figure 10.2. We look for a solution among symmetric matrices, and (13.60) expresses the fact that the product matrix $AA'A''$ is also symmetrical. We shall define the $n \times n$ matrices

$$(\beta\gamma|A|\tau\rho) = (A_{\beta\gamma})_{\tau\rho} \equiv (A_{\tau\rho})_{\beta\gamma},$$
$$(\alpha\rho|A'|\sigma v) = (A'_{v\alpha})_{\rho\sigma} \equiv (A'_{\rho\sigma})_{v\alpha}, \quad \text{etc.}$$

(13.61)

such that (13.60) can be written

$$\text{Tr } A_{\beta\gamma}A'_{v\alpha}A''_{\lambda\mu} = \text{Tr } A_{\mu v}A'_{\gamma\lambda}A''_{\alpha\beta}.$$

(13.62)

Let us look for a solution in the form

$$A_{\beta\gamma} = \delta_{\beta\gamma} \cdot 1 + a\sigma_{\beta\gamma} + b\sigma_{\gamma\beta}, \quad \text{etc.}$$

(13.63)

in which a and b are undetermined parameters and $\sigma_{\beta\gamma}$ is the $n \times n$ matrix $(\sigma_{\beta\gamma})_{\mu v} = \delta_{\beta\mu}\delta_{\gamma v}$. We obtain

$$\begin{aligned}
A_{\beta\gamma}A'_{v\alpha}A''_{\lambda\mu} &\equiv \delta_{\beta\gamma}\delta_{v\alpha}\delta_{\lambda\mu}(\text{Tr } 1 + \sum a + \sum b) \\
&+ \delta_{\beta\gamma}\delta_{\alpha\lambda}\delta_{\mu v}(a'a'' + b'b'') + \ldots + \delta_{\beta\gamma}\delta_{\lambda v}\delta_{\alpha\mu}(a'b'' + b'a'') + \ldots \\
&+ \underline{\delta_{\gamma v}\delta_{\alpha\lambda}\delta_{\mu\beta}aa'a''} + \underline{\delta_{\alpha\beta}\delta_{\mu v}\delta_{\lambda\gamma}bb'b''} + \underline{\delta_{\beta v}\delta_{\alpha\lambda}\delta_{\mu\gamma}ba'a''} + \ldots \\
&+ \delta_{\alpha\gamma}\delta_{\mu v}\delta_{\lambda\beta}ab'b'' + \ldots
\end{aligned}$$

(13.64)

The underlined terms are symmetric under the exchange $\alpha\beta\gamma \leftrightarrow \lambda\mu v$; only the others contribute to the following relations:

$$\begin{cases} bb'b'' = a + a' + a'' + b + b' + b'' + n, \\ ab'b'' = a'b'' + b'a'', \quad \text{and cyclic perm.} \end{cases}$$

(13.65)

From this we deduce the parametrization

$$a = \frac{1}{3}\left(1 - \frac{n}{2}\right)(2 - u_{23}),$$
$$b = \frac{2 - u_{23}}{1 + u_{23}}, \quad \ldots, \quad u_{23} = u_2 - u_3.$$

(13.66)

According to (13.61) and (13.63), we can write

$$A = a\mathbf{1} + \sigma^{\alpha\beta} \otimes \sigma^{\alpha\beta} + b\sigma^{\alpha\beta} \otimes \sigma^{\beta\alpha}. \tag{13.67}$$

Going back to a previous notation (Chapter 8, $X_{12} = W_{12}\sigma_1\sigma_2$), we obtain after a convenient normalization

$$X_{12} = \frac{1}{3}\left(1 - \frac{n}{2}\right) \cdot \mathbf{1} + \sum_{\alpha\beta}\left(\frac{\sigma_1^{\alpha\beta}\sigma_2^{\alpha\beta}}{2 - u_{12}} + \frac{\sigma_1^{\alpha\beta}\sigma_2^{\beta\alpha}}{1 + u_{12}}\right). \tag{13.68}$$

The orthogonal invariance is manifest.

13.3 Ternary relation and star–triangle relation

13.3.1 A model on a planar graph

Baxter (1978) considers a general model defined on a non-periodic planar graph. On an oriented surface, one considers a family of oriented lines (δ_j), all secant to each other, defining a planar graph G. One assigns a statistical weight to each intersection or vertex element of a transfer micromatrix; with each oriented segment or link of G between two consecutive intersections on a given line (δ) one associates a state index, such that the matrix element attached to the vertex configuration at summit A is written $(\beta\gamma|A|\tau\rho)$ according to Figure 13.1. It is necessary to define matrix A on each summit. We suppose that a parameter u_j is associated with each line (δ_j). With the angle between the oriented lines (δ_1, δ_2) is associated the parameter

$$u_{12} = u_1 - u_2. \tag{13.69}$$

We then take as transfer matrix the general matrix $X_{12}(u_{12})$, verifying the ternary relations (cf. Chapter 10 or 13, formula (13.60)). In this way we have defined a vertex model depending on a family of intersecting lines and on as many given parameters. As a simple consequence of the ternary relations and of the graphical representation (10.2), we can formulate the following.

Proposition: The partition function Z of the vertex model on G is independent of the graph G in the sense that Z does not depend on the relative positions of the lines provided that these remain in pairwise intersection in general position.

Figure 13.1 Representation of the matrix elements.

Now that we have mentioned this result, which shall not be exploited further here, let us move on to the definition of the spin model of Ising type corresponding to the vertex model on graph G defined above. With each face of the planar graph, in other words to each summit of the dual planar graph G^*, one associates a spin state indexed by $a, b, c, \ldots$ Let us suppose that it is possible to establish a correspondence between any 4-spin configuration around a summit and the vertex configuration itself. If this correspondence is globally single-valued, or can be made such, the vertex model is then equivalent to an Ising-type model with 4-spin interaction. As an example, consider a spin index which is an integer variable modulo n ($a \in \mathbb{Z}_n$) associated with the circuit limiting each face of G. The index of an edge is the difference between the variables of the two adjacent faces. One has a global multivalued correspondence $1 \to n$ having no importance.

13.3.2 Star–triangle relation in the eight-vertex model

It is sufficient to transcribe the ternary relation associated with the elementary graph of three lines according to the sketch of Figure 10.2 in terms of spins on the dual graph. Figure 13.2 illustrates the correspondence. In this model the variables

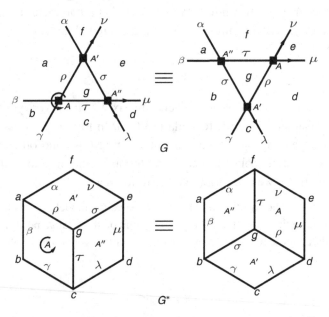

Figure 13.2 Graphs of the ternary relation between A, A', A'' on G or W, W', W'' on the dual G^*.

associated with the edges of G $(\alpha, \beta, \ldots, \sigma, \tau, \ldots)$ or to the faces $(a, b, \ldots, g)$ take the values ± 1. We have the two conservation properties

$$(\beta\gamma|A|\tau\rho) \neq 0, \text{ if } \beta\gamma\tau\rho = 1, \tag{13.70}$$

and symmetry

$$(\beta\gamma|A|\tau\rho) = (\tau\rho|A|\beta\gamma). \tag{13.71}$$

In terms of the spin variables, one defines the coefficients W:

$$(\beta\gamma|A|\tau\rho) = W(a, b, c, g), \tag{13.72}$$

with

$$\beta = a/b, \quad \gamma = b/c, \quad \tau = c/g, \quad \rho = g/a, \tag{13.73}$$

which leads to (13.70). The correspondence between A and W is single-valued by postulating the self-conjugation

$$W(a, b, c, g) = W(\bar{a}, \bar{b}, \bar{c}, \bar{g}). \tag{13.74}$$

The symmetry (13.71) is equivalent to the relation

$$W(a, b, c, g) = W(c, g, a, b). \tag{13.75}$$

It is realized in the self-conjugate eight-vertex model in which the weight W depends on two variables only:

$$W(a, b, c, g) = W(ac, bg). \tag{13.76}$$

In the notation of (8.14):

$$W(\varepsilon, \varepsilon') = \begin{pmatrix} a & d \\ c & b \end{pmatrix}, \tag{13.77}$$

or further still, with the help of the parameters K_3, K_4, K defined in (8.6):

$$W(\varepsilon, \varepsilon') = \exp\left\{K_3\varepsilon' + K_4\varepsilon + K\varepsilon\varepsilon'\right\} = \exp\left\{K_3bg + K_4ac + Kabcg\right\}. \tag{13.78}$$

According to (8.34) and (13.77), the two absolute invariants are K and

$$\cosh K_3 \cosh K_4 \sinh 2K + \sinh K_3 \sinh K_4 \cosh 2K. \tag{13.79}$$

If we modify the definition of W' slightly by taking

$$(\alpha\rho|A'|\sigma\nu) \equiv W'(e, f, a, g),$$

the ternary relation in W displays the ternary symmetry and can be written as

$$\sum_g W(a, b, c, g) W'(e, f, a, g) W''(c, d, e, g)$$

$$= \sum_g W(d, e, f, g) W'(b, c, d, g) W''(f, a, b, g). \qquad (13.80)$$

This identity expresses the fact that the left-hand side of (13.80), which is a function of the six variables $a, b, \ldots, f$, is invariant under the simultaneous exchange

$$a \leftrightarrow d, \qquad b \leftrightarrow e, \qquad c \leftrightarrow f,$$

which is a central symmetry operation. One will have noted that the triple sum over σ, τ, ρ in (13.60) is replaced by a single sum over g, which is a variable associated with the face $(\sigma\tau\rho)$ due to the conservation relations. Relation (13.80) is equivalent to the following identity:

$$\exp(K_4 ca + K'_4 ae + K''_4 ec) \cosh[K_3 b + K'_3 f + K''_3 d + K(abc + efa + cde)]$$
$$\equiv (a, b, c \leftrightarrow d, e, f). \qquad (13.81)$$

In the particular case in which the 4-spin interaction is absent, $K = 0$, one can recognize in (13.81) the simplest star–triangle relation in the form

$$\cosh(K_3 e + K'_3 c + K''_3 a) = \exp(K_4 ca + K'_4 ae + K''_4 ec), \qquad (13.82)$$

which gives us three relations instead of the six in (13.81). The problem of their uniformization has been solved in Chapter 8.

13.4 Ternary relation with $\mathbb{Z}_5$ symmetry

13.4.1 Star–triangle relation without self-conjugation

Let us consider the star–triangle relations in their general form (13.80) in which the W coefficients are subjected only to the symmetry relation (13.75). By losing the self-conjugation relation (13.74), the correspondence between the A and the W is no longer bijective, but one can construct A starting from W. Let us, for example, take one conservation law (which we shall write in an additive rather than a multiplicative form); the analogue of (13.75) will be

$$\beta = \frac{1}{2}(a - b), \qquad \gamma = \frac{1}{2}(b - c), \qquad \tau = \frac{1}{2}(g - c), \qquad \rho = \frac{1}{2}(a - g), \quad (13.83)$$

giving rise to a dimension 3 ($\beta = \pm 1, 0$).

To ease the writing of all relations between the three sets of 10 coefficients, let us classify them according to the value of the three products ad, be, cf, according to the following table:

ad	be	cf	No. of classes of relations
+	+	+	0
+	+	−	3
−	−	+	3
−	−	−	1

We obtain three families of relations in the three spin variables

$$\sum_g W(a, b, c, g) W'(b, \bar{c}, a, g) W''(c, a, b, g) = (c \to \bar{c}), \qquad (13.84)$$

and those permuted cyclically in the W. In the same way,

$$\sum_g W(a, b, c, g) W'(\bar{b}, c, a, g) W''(c, \bar{a}, \bar{b}, g) = (a \to \bar{a}, b \to \bar{b}) \qquad (13.85)$$

and the permuted ones. Finally,

$$\sum_g W(a, \bar{b}, c, g) W'(b, \bar{c}, a, g) W''(c, \bar{a}, b, g) = (a \to \bar{a}, b \to \bar{b}, c \to \bar{c}).$$

$$(13.86)$$

13.4.2 Solution of the star–triangle relations for the discrete gas model with neighbour exclusion

Let us give in this subsection the particular solution of the star–triangle relations recently obtained by Baxter (1980), a solution with $\mathbb{Z}_5$ symmetry which led him to exactly solve the problem of hard hexagons on the triangular lattice. By adapting the notation to the problem of a discrete gas on a square lattice, the spin variables are replaced by site occupation variables taking the value 0 or 1 $(a \to (1 + a)/2)$.

The exclusion of nearest neighbours is translated on the weights W into the property

$$W(a, b, c, g) \neq 0 \Rightarrow ab = bc = cg = ga = 0. \qquad (13.87)$$

We thus have five independent W coefficients:

$$W(0, 0, 0, 0) = \omega_1, \quad W(1, 0, 0, 0) = \omega_2, \quad W(0, 1, 0, 0) = \omega_3,$$
$$W(1, 0, 1, 0) = \omega_4, \quad W(0, 1, 0, 1) = \omega_5. \qquad (13.88)$$

According to Subsection 13.4.1, we have the seven relations

$$\omega_3\omega_3'\omega_1'' = \omega_1\omega_1'\omega_2'' + \omega_2\omega_2'\omega_4'',$$

$$\omega_1\omega_3'\omega_3'' = \omega_2\omega_1'\omega_1'' + \omega_4\omega_2'\omega_2'',$$

$$\omega_3\omega_1'\omega_3'' = \omega_1\omega_2'\omega_1'' + \omega_2\omega_4'\omega_2'',$$

$$\omega_5\omega_3'\omega_3'' = \omega_1\omega_2'\omega_2'' + \omega_2\omega_4'\omega_4'',$$

$$\omega_3\omega_5'\omega_3'' = \omega_2\omega_1'\omega_2'' + \omega_4\omega_2'\omega_4'',$$

$$\omega_3\omega_3'\omega_5'' = \omega_2\omega_2'\omega_1'' + \omega_4\omega_4'\omega_2'',$$

$$\omega_5\omega_5'\omega_5'' = \omega_2\omega_2'\omega_2'' + \omega_4\omega_4'\omega_4''. \tag{13.89}$$

The linear system in $\omega_{1,2,3,4}''$ is solved using relations 2, 3, 4, 5 to obtain the ratios

$$\frac{\omega_1''}{\begin{vmatrix} \omega_3\omega_4 & \omega_1\omega_2 \\ \omega_3'\omega_4' & \omega_1'\omega_2' \end{vmatrix}} = \frac{\omega_2''}{\begin{vmatrix} \omega_1^2 & \omega_2\omega_3 \\ \omega_1'^2 & \omega_2'\omega_3' \end{vmatrix}} = \frac{\omega_3''}{\begin{vmatrix} \omega_1\omega_4 & \omega_2^2 \\ \omega_1'\omega_4' & \omega_2'^2 \end{vmatrix}} = \frac{\omega_4''}{\begin{vmatrix} \omega_1\omega_3 & \omega_2\omega_5 \\ \omega_1'\omega_3' & \omega_2'\omega_5' \end{vmatrix}}$$

$$= \frac{\omega_2''}{\begin{vmatrix} \omega_4\omega_5 & \omega_2\omega_3 \\ \omega_4'\omega_5' & \omega_2'\omega_3' \end{vmatrix}}. \tag{13.90}$$

From this we deduce the existence of a first invariant r_1 such that

$$\omega_1^2 = \omega_4\omega_5 + r_1\omega_2\omega_3. \tag{13.91}$$

Substituting $\omega_{1,2,4}''$ from (13.90) in the first of (13.89), we deduce a second invariant r_2 such that

$$\omega_1^3 = \omega_4\omega_3^2 + \omega_5\omega_2^2 + r_2\omega_1\omega_2\omega_3. \tag{13.92}$$

If we finally use the last two relations, we obtain a third invariant r_3 such that

$$r_3\omega_1^2 = (r_1 - r_2)\omega_4\omega_5 + \omega_2\omega_3. \tag{13.93}$$

We must then distinguish two cases according to whether (13.93) is distinct from (13.91) or not.

If the invariants are such that the relations are distinct, we can choose as new invariants the ratios $\dfrac{\omega_1\omega_4}{\omega_2^2}$, $\dfrac{\omega_4\omega_5}{\omega_1^2}$ and $\dfrac{\omega_2\omega_3}{\omega_1^2}$. This leads to the parametrization

$$\omega_2/\omega_1 = \omega, \quad \omega_4/\omega_1 = C_4\omega^2, \quad \omega_3/\omega_1 = C_3/\omega, \quad \omega_5/\omega_1 = C_5/\omega^2, \tag{13.94}$$

in which $C_{3,4,5}$ are invariants and ω is a parameter. We then notice that the relations (13.89) are such that only the product $\omega\omega'\omega''$ appears:

$$C_3^2 = w(1 + C_4 w),$$
$$C_5 C_3^2 = w^2(1 + C_4^2), \qquad w \equiv \omega\omega'\omega'',$$
$$C_5^3 = w^3(1 + C_4^3 w), \tag{13.95}$$

which leads to a single arbitrary invariant. The point to notice is that the three objects $W(\omega)$, $W(\omega')$ and $W(\omega'')$ can be permuted, which leads to the commutativity of the associated matrices and the trivial character of this solution.

We are thus led to consider the second case in which relations (13.93) and (13.91) coincide for the choices of invariants

$$r_3 = r_1^{-1}, \qquad r_2 = r_1 - r_1^{-1}. \tag{13.96}$$

Relation (13.92) combined with (13.91) gives us

$$\omega_4(\omega_3^2 - \omega_1\omega_5) = \omega_2(r_1^{-1}\omega_1\omega_3 - \omega_2\omega_5). \tag{13.97}$$

We then perform an apparently unnecessary decoupling of the above relation into two relations having a similar structure to the first (13.91), which gives the complete system

$$\omega_3^2 = \omega_1\omega_5 - r^{-1}\omega_2\omega_4,$$
$$\omega_4^2 = r\omega_2\omega_5 - rr_1^{-1}\omega_1\omega_3,$$
$$\omega_1^2 = r_1\omega_2\omega_3 + \omega_4\omega_5, \tag{13.98}$$

for which we will provide a one-parameter solution. By a change of scale

$$\omega_j = \alpha_j \theta_j, \tag{13.99}$$

the three relations (13.98) take a form which is globally invariant under circular permutations of order 5: $(1\ 2\ 3\ 4\ 5)$"

$$\theta_3^2 = \delta\theta_1\theta_5 - \delta^{-1}\theta_2\theta_4,$$
$$\theta_4^2 = \delta\theta_2\theta_5 - \delta^{-1}\theta_1\theta_3,$$
$$\theta_1^2 = \delta\theta_2\theta_3 - \delta^{-1}\theta_4\theta_5, \tag{13.100}$$

in which δ refers to an invariant. It suffices to choose the α_j such that we have

$$\frac{\alpha_1\alpha_4}{\alpha_2^2} = \frac{\alpha_1\alpha_5}{\alpha_3^2} = \delta, \qquad \frac{\alpha_1^4}{\alpha_2^2\alpha_3^2} = -\delta^3. \tag{13.101}$$

Relations (13.100) remind us of the quadratic relations between θ-functions. The $\mathbb{Z}_5$ cyclic invariance leads us to take

$$
\theta_1 = \theta_1(u), \quad
\begin{cases}
\theta_2 = \theta_1\left(u - \dfrac{4\pi}{5}\right), & \theta_3 = \theta_1\left(u + \dfrac{4\pi}{5}\right), \\[2mm]
\theta_4 = \theta_4\left(u + \dfrac{2\pi}{5}\right), & \theta_5 = \theta_1\left(u - \dfrac{2\pi}{5}\right),
\end{cases}
\tag{13.102}
$$

in which $\theta_1(u; q)$ is the odd Jacobi function with 2π period in u and of parameter q, which remains undetermined at this stage. Relations (13.100) become identities in u provided we choose

$$
\delta = \frac{\theta_1\left(\frac{2\pi}{5}\right)}{\theta_1\left(\frac{\pi}{5}\right)}.
\tag{13.103}
$$

In order to show this, it is sufficient to eliminate $\theta_4^2(u)$ between the two quadratic relations

$$
\theta_1\left(u + \frac{\pi}{5}\right)\theta_1\left(u - \frac{\pi}{5}\right)\theta_4^2(0) = \theta_1^2(u)\theta_4^2\left(\frac{\pi}{5}\right) - \theta_4^2(u)\theta_1^2\left(\frac{\pi}{5}\right),
$$

$$
\theta_1\left(u + \frac{2\pi}{5}\right)\theta_1\left(u - \frac{2\pi}{5}\right)\theta_4^2(0) = \theta_1^2(u)\theta_4^2\left(\frac{2\pi}{5}\right) - \theta_4^2(u)\theta_1^2\left(\frac{2\pi}{5}\right). \tag{13.104}
$$

Finally, relations (13.90), which are equivalent to the original ternary relations, can easily be verified with the help of the quartic relation (I.13). It is enough for the three arguments u, u', u'' to verify

$$
u + u' + u'' + \frac{\pi}{5} \equiv 0 \quad (\mathrm{mod}\ \pi).
\tag{13.105}
$$

We shall take

$$
u = u_2 - u_3 - \frac{2\pi}{5}, \quad u' = u_3 - u_1 - \frac{2\pi}{5}, \quad u'' = u_1 - u_2 - \frac{2\pi}{5}. \tag{13.106}
$$

From this, we can write the solution of our star–triangle relation in the form

$$
\omega_1 = \alpha_1\theta_1\left(u - \frac{2\pi}{5}\right), \quad \omega_4 = \alpha_4\theta_1(u),
$$

$$
\omega_2 = -\alpha_2\theta_1\left(u - \frac{\pi}{5}\right), \quad \omega_5 = -\alpha_5\theta_1\left(u + \frac{\pi}{5}\right).
$$

$$
\omega_3 = \alpha_3\theta_1\left(u + \frac{2\pi}{5}\right),
\tag{13.107}
$$

The hard hexagon model, which translates into neighbour exclusion on the triangular lattice, corresponds to a particular choice of the u parameter. Indeed, the exclusion of two neighbours on one (or the other) of the two diagonals of the given square lattice is expressed by the condition

$$\omega_4 = 0 \Leftrightarrow u = 0 \quad \text{or} \quad \omega_5 = 0 \Leftrightarrow u = -\frac{\pi}{5}.$$

In contrast, the constants α_j can be chosen such that we can define a fugacity Z obviously verifying

$$Z = \left(\frac{\omega_2}{\omega_1}\right)^4 = \left(\frac{\omega_3}{\omega_1}\right)^4, \quad Z^2 = \left(\frac{\omega_5}{\omega_1}\right)^4, \quad \omega_4 = 0. \tag{13.108}$$

We thus have

$$\left(\frac{\alpha_2}{\alpha_3}\right)^4 = \delta^4, \quad Z = \left(\frac{\alpha_2}{\delta}\right)^4 = \alpha_3^4, \quad Z^2 = \left(\frac{\alpha_4}{\delta}\right)^4. \tag{13.109}$$

Taking (13.101) into account, it suffices to take

$$\alpha_1 = 1, \quad \alpha_2^4 = \delta^{-1}, \quad \alpha_3^4 = \delta^{-5} = Z, \tag{13.110}$$

from which we obtain the relation between the fugacity Z and the modulus q:

$$Z^{-1} = \left(\theta_1\left(\frac{2\pi}{5}; q\right) \Big/ \theta_1\left(\frac{\pi}{5}; q\right)\right)^5. \tag{13.111}$$

Our objective being simply to mention this recent realization of the ternary relations, we shall not develop Baxter's solution for hard hexagons any further, also considering that the Bethe sum (provided it exists) remains to be found. Baxter is, however, able to calculate the free energy without going through the spectrum of the usual transfer matrix. A singularity is noticed in $q = 0$, which provides a transition point for the critical value of the fugacity, which is

$$Z_c^{-1} = \left(2\cos\frac{\pi}{5}\right)^5 = \left(\frac{\sqrt{5}+1}{2}\right)^5. \tag{13.112}$$

13.5 Ternary relations with $\mathbb{Z}_2^8$ symmetry

13.5.1 Invariance of a ternary product under Fourier transformation

In this subsection, we show how the existence of a ternary relation between symmetric matrices, under the hypothesis analogous to (13.76) of them depending only on two variables, is equivalent to the invariance of a certain ternary product under Fourier transformation on $\mathbb{Z}_R$, where R is the dimension of the space of individual states.

The weights on G^* depend on two variables

$$W(a, b, c, g) = W(a - c, b - g), \tag{13.113}$$

which translates in G to

$$A(\beta, \gamma, \tau, \rho) = W(\beta + \gamma, \beta - \rho)\delta(\beta + \gamma - \tau - \rho).$$

All arguments belong to $\mathbb{Z}_R$ (class of residues modulo R).

The ternary relation (13.80) can thus be written

$$\sum_g W(a - c, b - g)W'(e - a, f - g)W''(c - e, d - g)$$

$$= \sum_g W(d - f, e - g)W'(b - d, c - g)W''(f - b, a - g). \qquad (13.114)$$

The convolution form of these relations suggests the introduction of the Fourier transform on one of the arguments, for example the second:

$$P(d, k) = R^{-1} \sum_{g \in \mathbb{Z}_R} W(d, g)e(kg/R). \qquad (13.115)$$

Setting

$$d - f = p, \quad b - d = p', \quad f - b = -p - p',$$
$$a - c = r, \quad e - a = r', \quad c - e = -r - r', \qquad (13.116)$$

we easily obtain the transformed relation

$$P(r, q)P(r', q')P''(-r - r', -q - q')$$
$$= R^{-2} \sum_{p,k,p',k'} e\left(\frac{qp' - pq' + rk' - r'k}{R}\right)$$
$$P(p, k)P'(p', k')P''(-p - p', -k - k'). \qquad (13.117)$$

The symmetrical role played by the two arguments of the new weight P leads us to define a vector of dimension 2:

$$x = (x_1, x_2) \equiv (r, q) \in Z_R^2 \qquad (13.118)$$

and the scalar product $x \cdot \xi = x_1\xi_2 + x_2\xi_1$.

Relation (13.117) becomes

$$P(x)P'(y)P''(-x - y) = R^{-2} \sum_{\xi, \eta} e\left(\frac{x \cdot \eta - y \cdot \xi}{R}\right) P(\xi)P'(\eta)P''(-\xi - \eta)$$

$$(x, y, \xi, \eta \in Z_R^2) \qquad (13.119)$$

or further still, in symmetric form,

$$P(y-z)P'(z-x)P''(x-y) = R^{-2} \sum_{\substack{\xi,\eta \\ \xi+\eta+\zeta=0}} e\left(\frac{x\cdot\xi + y\cdot\eta + x\cdot\zeta}{R}\right) P(\xi)P'(\eta)P''(\zeta).$$

(13.120)

A further equivalent expression to (13.119) is the following. If the function $Q(x, \xi)$ is defined according to the equality

$$\sum_{\eta} e\left(\frac{\eta\cdot x}{R}\right) P'(\eta)P''(\xi - \eta) = P(x)Q(x, \xi),$$

(13.121)

then the ternary relation is equivalent to the symmetry condition

$$Q(x, \xi) = Q(\xi, x).$$

(13.122)

The proof is obvious.

Let us give the analytical form of the weights P for the model with $\mathbb{Z}_2^2$ symmetry. According to (13.77),

$$W(\alpha, \beta) = \begin{pmatrix} b & c \\ d & a \end{pmatrix}, \quad \alpha, \beta = 0, 1; \quad \alpha = (1 - \varepsilon)/2$$

and according to our definition (13.115),

$$2P(\alpha, \beta) = \begin{pmatrix} b+c & b-c \\ d+a & d-a \end{pmatrix} = \begin{pmatrix} w_0 + w_1 + w_2 - w_3 & w_0 - w_1 - w_2 - w_3 \\ w_0 + w_1 + w_3 - w_2 & w_0 + w_2 + w_3 - w_1 \end{pmatrix}.$$

Using normalization (8.61), we obtain (after the translation $v \rightarrow v + \eta$)

$$2P(0, 1) = \frac{\Theta_1(v/2 - \eta|\tau/2)}{\Theta_1(v/2 + \eta|\tau/2)} \frac{\Theta_1(v + 2\eta|\tau/2)}{\Theta_1(2\eta|\tau/2)},$$

$$2P(0, 0) = \frac{\Theta_2(v/2 - \eta|\tau/2)}{\Theta_2(v/2 + \eta|\tau/2)} \frac{\Theta_1(v + 2\eta|\tau/2)}{\Theta_1(2\eta|\tau/2)},$$

$$2P(1, 0) = \frac{\Theta_3(v/2 - \eta|\tau/2)}{\Theta_3(v/2 + \eta|\tau/2)} \frac{\Theta_1(v + 2\eta|\tau/2)}{\Theta_1(2\eta|\tau/2)},$$

$$2P(1, 1) = \frac{\Theta_4(v/2 - \eta|\tau/2)}{\Theta_4(v/2 + \eta|\tau/2)} \frac{\Theta_1(v + 2\eta|\tau/2)}{\Theta_1(2\eta|\tau/2)}.$$

(13.123)

Using notation (M.7) for θ-functions, we can write

$$P_{\alpha\beta} = (-1)^{\alpha\beta} \frac{\theta_{\beta\bar{\alpha}}(v/2 - \eta|\tau/2)}{\theta_{\beta\bar{\alpha}}(v/2 + \eta|\tau/2)}, \quad \bar{\alpha} = 1 - \alpha.$$

(13.124)

The analytical form of the weights P in terms of θ-quotients is almost as simple as those of Baxter's w. The direct verification of identity (13.119), however, seems much more difficult than that of (8.44) for the w, which immediately followed from the Jacobi relations. It would be interesting to calculate the symmetric quantity $Q(x, \xi)$ which appears on the right-hand side of (13.121) for the already solved cases.

13.5.2 Abelian uniformization of the ternary relations: $\mathbb{Z}_2^g$ symmetry

The uniformizing functions for the ternary relations which we have encountered up to now are rational meromorphic functions, circular or elliptic, of the spectral parameter. The quotients of the weights are rational for the $su(n)$ or $so(n)$ symmetries, circular for $so(2)$, and elliptic for $\mathbb{Z}_2 + \mathbb{Z}_2$. The existence of abelian solutions generalizing the elliptic case to the case of type greater than 1 has been conjectured (Chudnovsky and Chudnovsky, 1980) starting from Frobenius's quadratic formula for multivariable theta functions. The abelian functions, which are multiply periodic functions, can indeed be constructed as quotients of sums of theta products.

An elementary proof of Frobenius's formula (Frobenius, 1968) is given in Appendix N. A reminder of the definition and of some useful properties of the multivariable theta functions is given at the beginning of Appendix M. Formula (N.20) can be written

$$\Theta_\alpha(u_1 + \eta)\Theta_\beta(u_2) = \sum_{\lambda,\mu} S_{\alpha\beta,\lambda\mu}(u_{12}, \eta)\Theta_\mu(u_2 + \eta)\Theta_\lambda(u_1), \qquad (13.125)$$

in which $\Theta_\alpha(u|\tau)$ is the general function defined in (N.6), with argument $u \in \mathbb{C}^g$, characteristic $\alpha \in \mathbb{Z}_2^g$, associated with a symmetric matrix τ with positive imaginary part. We can see how the matrix $S(u_{12})$ performs the transposition of two arguments u_1 and u_2 on the product of the theta functions $\Theta(u_1 + \eta)\Theta(u_2)$. We know how to deduce that $S_{\alpha\beta,\gamma\mu}$ verifies the ternary relations for the case $g = 1$, and we would like to extend this proof to the case of arbitrary type. Let us consider the product

$$\Theta_\alpha(u_1 + \eta)\Theta_\beta(u_2)\Theta_\gamma(u_3 - \eta) \qquad (13.126)$$

and apply the product of transpositions of arguments $(12)(23)(12)$ to (13.126) in representation (13.125). We obtain

$$\Theta_\alpha(u_1 + \eta)\Theta_\beta(u_2)\Theta_\gamma(u_3 - \eta)$$
$$= \sum_{\lambda,\mu,\nu} S^{(3)}_{\alpha\beta\gamma,\lambda\mu\nu}(u_{12}, u_{23}; \eta)\Theta_\nu(u_3 + \eta)\Theta_\mu(u_2)\Theta_\lambda(u_1 - \eta), \quad (13.127)$$

in which $S^{(3)}$ coincides for example with the first ternary product of formula (10.66). If we now apply the product of transpositions $(23)(12)(23)$ on (13.126), we obtain a relation analogous to (13.127) in which $S^{(3)}$ will be replaced by $^tS^{(3)}$, the second ternary product in relation (10.66). Setting $T^{(3)} = S^{(3)} - {}^tS^{(3)}$, we have

$$\sum_{\lambda,\mu,\nu} T^{(3)}_{\alpha\beta\gamma,\lambda\mu\nu}(u_{12}, u_{23}, \eta)\Theta_\lambda(u_1 - \eta)\Theta_\mu(u_2)\Theta_\nu(u_3 + \eta) = 0. \tag{13.128}$$

From (13.128), can we deduce the equalities $T^{(3)}_{\alpha\beta\gamma,\lambda\mu\nu}(\ldots) = 0$, $\forall \alpha, \beta, \gamma, \lambda, \mu, \nu$, u_{12}, u_{13}, η? With this in mind, let us perform the translation $u_j \to u_j + u$. $T^{(3)}$ remains unchanged, and (13.128) provides us with a homogeneous cubic relation between theta functions of the same period for argument u. But there does not exist such a relation for the $g = 1$ case. More precisely, the identity

$$\sum_{\lambda\mu\nu} t_{\lambda\mu\nu}\Theta_\lambda(u + v_1)\Theta_\mu(u + v_2)\Theta_\nu(u + v_3) = 0 \tag{13.129}$$

leads to $t_{\lambda\mu\nu} = 0$ if v_1, v_2 and v_3 are distinct modulo any half-period. The linear independence of functions Θ_λ of type g is known (Siegel, 1973, Vol. 3, p. 72), but we would need more in the general case, more precisely the linear independence of a class of theta products of third degree.

This being clear, let us give the explicit form of the S coefficients as they emerge from Frobenius's formula. We have (Appendix N)

$$S_{\alpha\beta,\lambda\mu}(u_{12}; \eta) = C^{\alpha+\beta}_{\beta+\mu}(y, z), \quad 2y = \eta + u_{12}, \quad 2z = \eta - u_{12}. \tag{13.130}$$

Formula (N.17) involves directly the Fourier transform on $\mathbb{Z}_2^g$ of the coefficients C or S, which is precisely the weight P defined in (13.115):

$$C^\alpha_\beta(y, z) = \sum_\lambda (-1)^{\beta\lambda} \frac{\sum_\mu (-1)^{\mu\lambda}\Theta_\mu(2y + \tau \cdot \alpha/2|2\tau)}{\sum_\mu (-1)^{\mu\lambda}\Theta_\mu(2z + \tau \cdot \alpha/2|2\tau)}. \tag{13.131}$$

But according to (13.113), we have

$$S_{\alpha\beta,\lambda\mu} \equiv A(\alpha, \beta, \lambda, \mu) = W(\alpha + \beta, \alpha + \mu) = C^{\alpha+\beta}_{(\beta+\mu)+(\alpha+\beta)}. \tag{13.132}$$

From this, we deduce the quotient form of weight $P_{\alpha\beta}$:

$$P_{\alpha\beta}(u) = (-1)^{\alpha\beta} \frac{\sum_\mu (-1)^{\beta\mu}\Theta_\mu(\eta + u + \tau \cdot \alpha/2|2\tau)}{\sum_\mu (-1)^{\beta\mu}\Theta_\mu(\eta - u + \tau \cdot \alpha/2|2\tau)}, \tag{13.133}$$

$\alpha, \beta, \mu \in \mathbb{Z}_2^g$.

The application to the $g = 1$ case, in the notation of (M.7), gives us

$$P_{0\alpha} = \frac{\Theta_{00}(2y|2\tau) \pm \Theta_{01}(2y|2\tau)}{\Theta_{00}(2z|2\tau) \pm \Theta_{01}(2y|2\tau)} = \frac{\Theta_{\alpha0}(y|\tau/2)}{\Theta_{\alpha0}(z|\tau/2)},$$

$$P_{1\alpha} = \frac{\Theta_{00}(2y + \tau/2|2\tau) \pm \Theta_{01}(2y + \tau/2|2\tau)}{\Theta_{00}(2z + \tau/2|2\tau) \pm \Theta_{01}(2z + \tau/2|2\tau)} = \pm\frac{\Theta_{\alpha1}(y|\tau/2)}{\Theta_{\alpha1}(z|\tau/2)},$$

$$\pm \leftrightarrow \alpha = 0, 1. \tag{13.134}$$

It is then sufficient to perform the slight change of parameters

$$\eta \to \eta + \frac{\tau}{2}, \qquad \frac{v}{2} \to v,$$

to exactly fall back on the known results (13.124).

Let us finally mention that the general Jacobi formula diplayed in Appendix M also allows us to conjecture the existence of ternary relations of symmetry $\mathbb{Z}_2^g + \mathbb{Z}_2^g$, for the case $r = 2$.

13.6 Notes on a system of distinguishable particles

Other finite quantum integrable systems are known, but it does not seem that they relate in any way to Bethe's method; for example, the Toda chain treated in Chapter 14. Let us also mention the one-dimensional systems with two-body repulsive potential of the form x^{-2} and the corresponding singly and doubly periodic potentials, which were studied by Calogero (1971), Sutherland (1971) and Adler (1977). Besides the ground state, the wavefunctions are not always known or explicitly obtained, except for exceptional values of the coupling constant. It is known that the eigenvalue distribution of a Brownian Hermitian matrix is precisely identical to the particle distribution in the ground state of a one-dimensional gas with x^{-2} potential, for a specific value of the coupling. One would have other coupling values corresponding to symmetric or self-dual random matrices. The associated quantum Hamiltonian is integrable; it originates from the invariant separation of a Fokker–Planck equation in the space of matrices (Dyson, 1962; Mehta, 1980).

Let us mention another example, built from δ potentials. The following Hamiltonian for particles of unequal masses:

$$H = -\sum_j \frac{1}{m_j} \frac{\partial^2}{\partial x_j^2} - g \sum_{i<j} (m_i + m_j) \delta(x_i - x_j) \tag{13.135}$$

admits the bound ground state $(g > 0)$:

$$\psi(x) = \exp\left\{ -\frac{g}{2} \sum_{i<j} m_i m_j |x_i - x_j| \right\}, \tag{13.136}$$

with energy

$$E = -\frac{g^2}{8} \left\{ \sum_{i<j} m_i m_j (m_i + m_j) + 2 \sum_{i<j<k} m_i m_j m_k \right\}. \tag{13.137}$$

The purely exponential form of the ground state, in many ways analogous to that for the case of equal masses, could lead one to think that all diffraction effects vanish for this particular choice of the coupling constants. In each sector D_Q : $x_{Q1} < x_{Q2} < \ldots$, the bound state wavefunction can indeed be written

$$\psi = \exp\left(-\sum_j q_j x_j\right), \tag{13.138}$$

with

$$q_j = \frac{g}{4} m_j \sum_i m_i \varepsilon(\bar{Q}j - \bar{Q}i). \tag{13.139}$$

The sets of quasi-momenta $\{q\}$ vary here from one sector to another. This structure, however, does not seem to originate from a finite Bethe sum. Indeed, we already know that in the limit $g = \infty$, the system of three impenetrable particles of different masses gives rise to diffractive scattering (McGuire and Hurst, 1972). One can in any case see this quite easily by studying the kinematics of energy- and momentum-preserving collisions. Contrary to what was observed for the case of equal masses, the successive two-body collisions almost always lead to an infinite family of sets of quasi-momenta. The $m_1 \leftrightarrow m_2$ collision indeed induces the change from a set $\{k\}$ to a set $\{k'\}$ such that

$$k' = X_{12}k, \qquad X_{12}^2 = 1, \tag{13.140}$$

with

$$k_1 + k_2 = k'_1 + k'_2, \qquad \frac{k_1^2}{m_1} + \frac{k_2^2}{m_2} = \frac{k_1'^2}{m_1} + \frac{k_2'^2}{m_2}. \tag{13.141}$$

We thus have the representation

$$X_{12} = \begin{pmatrix} \dfrac{m_1 - m_2}{m_1 + m_2} & \dfrac{2m_1}{m_1 + m_2} & \cdots & \cdots \\[2mm] \dfrac{2m_2}{m_1 + m_2} & \dfrac{m_2 - m_1}{m_1 + m_2} & \cdots & \cdots \\[2mm] \vdots & \vdots & 1 & \cdots \\[1mm] \vdots & \vdots & \vdots & \ddots \end{pmatrix}, \tag{13.142}$$

in which the X_{ij} are not the elements of a finite group. In the impenetrable case, we could consider $X_{12}, X_{23}, \ldots, X_{N-1,N}$ as the involutive generators of an infinite free group, and pose the problem of the extension of Bethe's method leading to infinite sums over proper discontinuous groups. Essential and difficult convergence problems, however, then manifest themselves. We have already encountered, in an attempt to generalize Sommerfeld's method to diffraction problems of dimension higher than two, such infinite discrete groups as the modular groups, which certainly are of a nature corresponding to the problems at hand and essentially determine the analytical structure of the Fourier transforms of the wavefunctions. This question remains largely unexplored due to difficulties in manipulating automorphous functions (Gaudin, 1976).

Going back to the scattering operators X_{ij} defined by (13.142), one can verify the remarkable relations

$$X_{12}X_{13}X_{23} = X_{23}X_{13}X_{12}, \quad X_{12}^2 = 1 \quad \text{(not to be confused with } X_{12}X_{21} = 1\text{).}$$
$$(13.143)$$

Let us show that they allow us to ensure a simple connection of the wavefunctions from one sector to the other, even if these do not have the structure of a finite Bethe sum. The existence of X_{ij} verifying (13.143) indeed leads to that of a representation for permutations with generators

$$Y_{34} = X_{34}P_{34}, \tag{13.144}$$

in which the P act on the k. With a given set $\{k\}$, we can associate $N!$ sets denoted $\{\bar{P}k\}$ such that

$$\bar{P}k = Y_{\bar{p}}k, \tag{13.145}$$

all wavefunctions

$$\exp i(\bar{P}k, \bar{P}x), \quad (\bar{P}x)_j = x_{Pj}$$

having the same energy

$$E = \sum_j \frac{1}{m_{Pj}}(Pk)_j^2. \tag{13.146}$$

The essential property of the wavefunctions is their *continuity* from one sector to the adjacent one, in the following sense:

$$\exp i(\overline{P_{12}Pk}, \overline{P_{12}Px}) = \exp i(\overline{Pk}, \overline{Px})|_{x_{P1}=x_{P2}}.$$

From this point onwards, the problem remains open.

Appendix M

Extension of the Jacobi relation between theta functions

The entire function $\theta(z)$, $z \in \mathbb{C}^g$, depending on a symmetric matrix τ of order g, is defined by the series

$$\theta(z|\tau) = \sum_{n \in \mathbb{Z}^g} e\left(\frac{1}{2}n \cdot \tau \cdot n + n \cdot z\right), \tag{M.1}$$

normally convergent for $\Im\tau > 0$. We have denoted

$$e(x) = e^{2\pi i x}, \qquad n \cdot z = \sum_{j=1}^{g} n_j z_j.$$

The theta function thus defined has the following multiplicative property under the period group:

$$\theta(z + \Lambda + \tau \cdot L) = e(-L \cdot z - L \cdot \Lambda - \frac{1}{2} L \cdot \tau \cdot L)\theta(z), \qquad \forall \Lambda \text{ and } L \in \mathbb{Z}^g. \quad \text{(M.2)}$$

Given the integer number r, one considers the family of theta functions characterized by a translation of the argument

$$z \to z + \frac{\Lambda + \tau \cdot L}{r}.$$

The pair $(M) = \Lambda, L$ defines a *characteristic* of the theta function

$$\theta_{(M)}(z) = \theta_{\Lambda,L}(z) = \theta_L\left(z + \frac{\Lambda}{r}\right) \qquad \text{(M.3)}$$

with the notation

$$\theta_L(z) = e\left(\frac{L \cdot z}{r} + \frac{L \cdot \tau \cdot L}{2r^2}\right)\theta\left(z + \frac{\tau \cdot L}{r}\right) \equiv \sum_{m - \frac{L}{r} \in \mathbb{Z}^g} e\left(\frac{1}{2} m \cdot \tau \cdot m + m \cdot z\right). \qquad \text{(M.4)}$$

From form (M.4), one can deduce the properties

$$\theta_{L+rL'}(z) = \theta_L(z), \qquad \theta_{\Lambda+r\Lambda',L+rL'}(z) = e\left(\frac{L \cdot \Lambda'}{r}\right)\theta_{\Lambda,L}(z). \qquad \text{(M.5)}$$

It is sufficient to only consider the functions whose characteristics are the classes of residues modulo r, in other words

$$(M) = \Lambda, L \in \mathbb{Z}_r^g \times \mathbb{Z}_r^g \quad (\mathbb{Z}_r = \mathbb{Z}/r\mathbb{Z}). \qquad \text{(M.6)}$$

From this comes the corollary that a product of r functions $\theta_{(M)}$, with arbitrary arguments but having the same characteristic $(M) = \Lambda, L$, is doubly r-periodic in (M).

In the case $r = 2$, $g = 1$, we thus obtain four theta functions of a complex variable

$$\theta_{00} = \theta_3, \qquad \theta_{11} = -\theta_1, \qquad \theta_{01} = \theta_2, \qquad \theta_{10} = \theta_4, \qquad \text{(M.7)}$$

in which the Jacobi functions θ_α, defined in Appendix I, appear on the right-hand side.

The expression for the Jacobi formula can be written

$$\prod_{\alpha=1}^{2r} \theta_{(M+M_\alpha)}(z_\alpha) = r^{-g} \sum_{(M')} e\left(\frac{M \times M'}{r}\right) \cdot \prod_{\alpha=1}^{2r} \theta_{(M'+M_\alpha)}(x_\alpha) \qquad \text{(M.8)}$$

with

$$x_\alpha = z_\alpha - \frac{1}{r}\sum_\beta z_\beta, \qquad z_\alpha \in \mathbb{C}^g \qquad \text{(M.9)}$$

$$(M_\alpha) = (\Lambda_\alpha, L_\alpha) \in \mathbb{Z}^g \times \mathbb{Z}^g, \qquad \sum_\alpha (M_\alpha) = 0 \qquad \text{(M.10)}$$

$$(M), (M') \in \mathbb{Z}_r^g \times \mathbb{Z}_r^g, \qquad M \times M' \equiv \Lambda \cdot L' - L \cdot \Lambda'. \qquad \text{(M.11)}$$

This relation expresses a homomorphism between the involution $(z) \to (x)$ and the Fourier transformation in $\mathbb{Z}_r^g \times \mathbb{Z}_r^g$.

Proof: The involution $z_\alpha \Rightarrow x_\alpha = z_\alpha - \frac{1}{r}\bar{z}, \bar{z} = \sum_\alpha z_\alpha,$

$$z_\alpha = x_\alpha - \frac{1}{r}\bar{x}, \qquad \bar{x} = \sum x_\alpha = -\bar{z}, \qquad \text{(M.12)}$$

in an orthogonal transformation of $\mathbb{R}_{2r}$. Let us consider the product of $2r$ theta functions of the same period $\prod_\alpha \theta(z_\alpha|\tau)$. Let us define r-fractional summation variables

$$m_\alpha = n_\alpha - \frac{1}{r}\bar{n}, \qquad \bar{n} = \sum n_\alpha, \qquad n_\alpha \in \mathbb{Z}^g. \qquad \text{(M.13)}$$

According to (M.1), we have

$$\prod_\alpha \theta(z_\alpha) = \sum_{\{n_\alpha \in \mathbb{Z}^g\}} e\left(\sum_\alpha \frac{1}{2}m_\alpha \tau m_\alpha + m_\alpha x_\alpha\right) \qquad \text{(M.14)}$$

in which we have simply used the orthogonality of the involution.

We can perform the sum on the right-hand side of (M.14) by distinguishing the different classes of $\bar{m} = -\bar{n}$ modulo r. We set

$$\bar{m} \equiv L \ (\text{mod } r), \qquad L \in \mathbb{Z}_r^g. \qquad \text{(M.15)}$$

We thus have, according to (M.13),

$$m_\alpha - \frac{L}{r} = n_\alpha \in \mathbb{Z}^g. \qquad \text{(M.16)}$$

The sum (M.14) can thus be written

$$\prod_\alpha \theta(z_\alpha) = \sum_L \sum_{\{m_\alpha - \frac{L}{r} \in \mathbb{Z}^8\}} e\left(\sum_\alpha \frac{1}{2} m_\alpha \cdot \tau \cdot m_\alpha + m_\alpha \cdot x_\alpha\right), \quad \left(\sum_\alpha m_\alpha \equiv L\right).$$
(M.17)

The summation constraint $\sum_\alpha \equiv L \pmod r$ can be expressed by introducing the Fourier expansion of the delta function on $\mathbb{Z}_r$:

$$\delta(m) = r^{-1} \sum_{k \in \mathbb{Z}_r} e\left(\frac{km}{r}\right),$$

and thus on $\mathbb{Z}_r^8$:

$$\delta(\bar{m} - L) = r^{-8} \sum_{\Lambda \in \mathbb{Z}_r^8} e\left(\frac{\Lambda \cdot (\bar{m} - L)}{r}\right).$$
(M.18)

Substituting this in the right-hand side of (M.17), the summation over the m_α can then be performed by introducing the definitions (M.4); the sum over L remains, giving

$$\prod_\alpha \theta = \sum_{\Lambda, L} e\left(-\frac{L \cdot \Lambda}{r}\right) \prod_\alpha \theta_L\left(x_\alpha + \frac{\Lambda}{r}\right),$$
(M.19)

which is nothing other than formula (M.8) for the case $M_\alpha = M = 0$. We must now simply perform the following translation on the z arguments:

$$z_\alpha \to z_\alpha + \frac{\Lambda_\alpha + \tau \cdot L_\alpha}{r}, \quad \bar{\Lambda} = \sum_\alpha \Lambda_\alpha = 2r \Lambda', \quad \bar{L} = 2r L'.$$
(M.20)

Repeated use of formula (M.3)

$$\theta\left(z + \frac{\Lambda + \tau \cdot L}{r}\right) = \theta_{\Lambda, L}(z)\, e\left(-\frac{z \cdot L}{r} - \frac{\Lambda \cdot L}{r^2} - \frac{L \cdot \tau \cdot L}{2r^2}\right)$$
(M.21)

and a careful calculation of the exponential factors lead to the identity

$$\prod_{\alpha=1}^{2r} \theta_{\Lambda_\alpha, L_\alpha}(z_\alpha) = r^{-8} \sum_{L, \Lambda} e\left(\frac{\Lambda \cdot L' - L \cdot \Lambda'}{r}\right) \prod_{\alpha=1}^{2r} \theta_{\Lambda_\alpha + \Lambda - \Lambda', L_\alpha + L - L'}(x_\alpha).$$
(M.22)

The translation $(M_\alpha) \to (M_\alpha + M')$ on the characteristics gives the formula announced in (M.8). For $r = 2$, we fall back on the usual quartic relations

$$\prod_{\alpha=1}^{4} \theta_{(M+M_\alpha)}(z_\alpha) = 2^{-8} \sum_{(M')} (-1)^{\Lambda \cdot L' - L \cdot \Lambda'} \prod_{\alpha=1}^{4} \theta_{(M'+M_\alpha)}(x_\alpha)$$
(M.23)

whose one-variable version is recalled in (I.12).

Appendix N

The Frobenius formula for theta functions

This consists of a quadratic formula between theta functions of identical period. To demonstrate it, we must go via a preliminary quadratic relation between theta functions of periods τ and 2τ. We indeed have the identity

$$\theta(x|\tau)\theta(y|\tau) = \sum_L \theta_L(x + y|2\tau)\theta_L(x - y|2\tau), \qquad L \in \mathbb{Z}_2^g \qquad (N.1)$$

where the definition of θ_L is given in Appendix M (M.4):

$$\theta_L(x) = \sum_{m-\frac{L}{2}\in\mathbb{Z}^g} e\left(\frac{1}{2}m\cdot\tau\cdot m + m\cdot x\right). \qquad (N.2)$$

The proof of (N.1) is immediate. Let us perform the series product

$$\theta(x)\theta(y) = \sum_{n,m\in\mathbb{Z}^g} e\left(\frac{1}{2}(n\cdot\tau\cdot n + m\cdot\tau\cdot m) + n\cdot x + m\cdot y\right). \qquad (N.3)$$

Let us set

$$n + m = 2p, \qquad n - m = 2q. \qquad (N.4)$$

There exists an $L \in \mathbb{Z}_2^g$ such that $p - (L/2)$ and $q - (L/2)$ belong to $\mathbb{Z}^g$, since each component of the pair p, q is simultaneously integer or half-integer according to (N.4). We deduce that

$$\theta(x|\tau)\theta(y|\tau) = \sum_L \sum_{q-\frac{L}{2}\in\mathbb{Z}^g} \sum_{q-\frac{L}{2}\in\mathbb{Z}^g}$$
$$e\left(\frac{1}{2}(p\cdot2\tau\cdot p + q\cdot2\tau\cdot q) + p\cdot(x+y) + q\cdot(x-y)\right) \qquad (N.5)$$

from which (N.1) follows. The sum runs over 2^g terms.

We shall need to work with functions that are strictly invariant under the translations $z \to z + \tau\cdot L$. For this reason we modify the definition of the theta functions slightly by taking

$$\Theta_L(z|\tau) = e\left(\frac{1}{2}z\cdot\tau^{-1}\cdot z\right)\cdot\theta_L(z|\tau) \qquad (N.6)$$

(do not confuse this Θ notation with that of Chapter 8). By virtue of the definition (M.2), we have

$$\Theta\left(z + \frac{\tau\cdot L}{2}\right) = e\left(\frac{1}{2}z\cdot\tau^{-1}\cdot z + \frac{1}{2}z\cdot L + \frac{1}{8}L\cdot\tau\cdot L\right)\theta\left(z + \frac{\tau\cdot L}{2}\right) = \Theta_L(z).$$
$$(N.7)$$

Identity (N.1) can further be presented in the form

$$\Theta(x + y|\tau)\Theta(x - y|\tau) = \sum_{L \in \mathbb{Z}_2^g} \Theta(2x + \tau \cdot L|2\tau)\Theta(2y + \tau \cdot L|2\tau). \quad \text{(N.8)}$$

Performing the translation $x \to x + \frac{1}{2}\tau \cdot L'$ and applying property (N.7), we obtain

$$\Theta_{L'}(x + y|\tau)\Theta_{L'}(x - y|\tau) = \sum_{L \in \mathbb{Z}_2^g} \Theta_{L+L'}(2x|2\tau)\Theta_L(2y|2\tau). \quad \text{(N.9)}$$

Since $\Theta_L(x)$ has $\mathbb{Z}_2^g$ periodicity in the characteristic, the convolutive form of the right-hand side leads to the introduction of the Fourier transform on $\mathbb{Z}_2^g$. We define

$$\Phi_\Lambda(2x) = \sum_{L \in \mathbb{Z}_2^g} (-1)^{\Lambda \cdot L} \Theta_L(2x), \qquad \Lambda \in \mathbb{Z}_2^g \quad \text{(N.10)}$$

and by inverting (N.9), we obtain

$$\Theta_L(x + y)\Theta_L(x - y) = 2^{-g} \sum_\Lambda (-1)^{L \cdot \Lambda} \Phi_\Lambda(2x)\Phi_\Lambda(2y). \quad \text{(N.11)}$$

Let us now apply the direct formula for arguments $x + z$ and $x - z$:

$$\sum_{L'} (-1)^{L' \cdot \Lambda} \Theta_{L'}(x + z)\Theta_{L'}(x - z) = \Phi_\Lambda(2x)\Phi_\Lambda(2z). \quad \text{(N.12)}$$

Let us now eliminate $\Phi_\Lambda(2x)$ between (N.11) and (N.12) to obtain

$$\Theta_L(x + y|\tau)\Theta_L(x - y|\tau) = \sum_{L \in \mathbb{Z}_2^g} C_{L+L'}(y, z)\Theta_{L'}(x + z|\tau)\Theta_{L'}(x - z|\tau) \quad \text{(N.13)}$$

with the following expression for the coefficient:

$$C_L(y, z) = \sum_{\Lambda \in \mathbb{Z}_2^g} (-1)^{L \cdot \Lambda} \frac{\Phi_\Lambda(2y)}{\Phi_\Lambda(2z)}. \quad \text{(N.14)}$$

Relation (N.13), due to Frobenius, is a quadratic relation in Θ in terms of the x dependence. The C_L coefficients have a rather simple Fourier transformation:

$$C_L(y, z) = \sum_\Lambda (-1)^{L \cdot \Lambda} \frac{\sum_{L'}(-1)^{\Lambda \cdot L'}\Theta_{L'}(2y|2\tau)}{\sum_{L'}(-1)^{\Lambda \cdot L'}\Theta_{L'}(2z|2\tau)}. \quad \text{(N.15)}$$

We could replace Θ by θ in (N.13), provided the same is done in (N.15).

Let us give the formula a more general appearance as a function of the characteristic, by performing the translation $\tau \cdot \tilde{L}/2$ on x, y and z. We finally obtain

$$\Theta_L(x - y|\tau)\Theta_{L+\tilde{L}}(x + y|\tau) = \sum_{L' \in \mathbb{Z}_2^g} C_{L-L'}^{\tilde{L}}(y, z)\Theta_{L'}(x - z|\tau)\Theta_{L'+\tilde{L}}(x + z|\tau),$$

$$\text{(N.16)}$$

with

$$C_L^{\tilde{L}}(y, z) = \sum_{\Lambda}(-1)^{L\cdot\Lambda}\frac{\sum_{L'}(-1)^{\Lambda\cdot L'}\Theta_{L'}\left(2y + \frac{\tau\cdot\tilde{L}}{2}|2\tau\right)}{\sum_{L'}(-1)^{\Lambda\cdot L'}\Theta_{L'}\left(2z + \frac{\tau\cdot\tilde{L}}{2}|2\tau\right)}, \qquad L, \tilde{L} \in \mathbb{Z}_2^g. \quad \text{(N.17)}$$

Let us also present the Frobenius formula in a form very suggestive of Zamolodchikov's algebra (cf. (10.63)). We set

$$x + y = u_1 + \eta, \qquad x + z = u_2 + \eta,$$
$$x - y = u_2, \qquad x - z = u_1,$$
$$2y = u_{12} + \eta, \qquad 2z = \eta - u_{12}. \quad \text{(N.18)}$$

Let us note the characteristics in $\mathbb{Z}_2^g$:

$$\alpha = L + \tilde{L}, \quad j = L' + \tilde{L}, \quad \alpha + \beta \equiv \lambda + \mu, \quad \beta = L, \quad \mu = L'. \quad \text{(N.19)}$$

We have

$$\Theta_\alpha(u_1 + \eta)\Theta_\beta(u_2) = \sum_{\lambda,\mu} S_{\alpha\beta,\lambda\mu}(u_{12}, \eta)\Theta_\mu(u_2 + \eta)\Theta_\lambda(u_1), \quad \text{(N.20)}$$

where $S_{\alpha\beta,\lambda\mu} = C_{\alpha+\lambda}^{\alpha+\beta}(y, z)$ is given by (N.17).

14

On the Toda chain

14.1 Definition

We consider a cyclic chain of identical masses, transversely mobile and coupled between neighbours. Neglecting all longitudinal movement, the state of the chain is described at time t by the transversal coordinates and momenta

$$q_{2n}, \; p_{2n}, \quad n = 1, 2, \ldots, N \tag{14.1}$$

of the mobile masses. The Toda Hamiltonian is the following:

$$H = \frac{1}{2} \sum_{n=1}^{N} p_{2n}^2 + \sum_{n=1}^{N} e^{q_{2n} - q_{2n+2}}. \tag{14.2}$$

The chain being closed, we identify q_{2N+2} with q_2. Figure 14.1 illustrates the corresponding mechanical system.

If the transversal amplitudes q_{2n} remain very small during motion, we have a system of harmonically coupled masses. One should note that the total potential of the cyclic chain presents an absolute minimum defining the equilibrium state $q_{2n} = 0$, $\forall n$. This would not be the case for an open chain. For finite displacements, the effect of the potential transmitted by a link $(q_4 - q_2)$ is somewhat analogous to that of a joint stop on an articulation: the force (or moment) is negligible in one direction $(q_4 > q_2)$ but increases very rapidly in the other $(q_4 < q_2)$ with an intermediate elastic zone between blocking and liberty. Other analogies are possible, for example the electric analogy (Hirota, 1973).

14.2 Bäcklund transformation

Let us consider the classical Hamiltonian system defined by $H(q_{2n}, p_{2n})$ in which q_{2n} and p_{2n} are the conjugate coordinates and momenta. There exists a remarkable canonical transformation

On the Toda chain

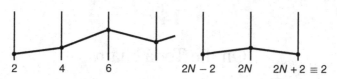

Figure 14.1 Mechanical system corresponding to the Toda chain.

$$(B): \quad q_{2n}, p_{2n} \to q_{2n+1}, p_{2n+1}, \tag{14.3}$$

which leaves the form of H invariant. This transformation is generated by the function $W(q_n)$ such that

$$dW = \sum_{n=1}^{N} p_{2n} dq_{2n} - p_{2n+1} dq_{2n+1} = \sum_{n=1}^{2N} (-1)^n p_n dq_n. \tag{14.4}$$

The expression for W is the following:

$$W(q_n) = \sum_{n=1}^{2N} (-1)^n (e^{q_n - q_{n+1}} + c q_n), \tag{14.5}$$

in which c is an undetermined constant. We thus have, according to (14.4) and (14.5),

$$p_n = e^{q_n - q_{n+1}} + e^{q_{n-1} - q_n} + c, \quad \forall n \tag{14.6}$$

which in principle permits us to pass from the system with variables $\{q_{2n}, p_{2n}\}$ to that with $\{q_{2n+1}, p_{2n+1}\}$. We shall come back to this point at the end of Section 14.4. Let us calculate, with the help of (14.6), the sum

$$\sum_{n=1}^{2N} (-1)^n p_n^2 = \sum_n (-1)^n (e^{q_n - q_{n+1}} + e^{q_{n-1} - q_n} + c)^2$$

$$= \sum_n (-1)^n (e^{2(q_n - q_{n+1})} + e^{2(q_{n-1} - q_n)} + 2c(e^{q_n - q_{n+1}} + e^{q_{n-1} - q_n})$$

$$+ (-1)^n 2 e^{q_{n-1} - q_{n+1}}$$

$$= 2 \sum_{n=1}^{2N} (-1)^n e^{q_{n-1} - q_{n+1}}.$$

We thus obtain the previously mentioned invariance

$$\sum_n \frac{1}{2} p_{2n}^2 + e^{q_{2n} - q_{2n+2}} \equiv \sum_n \frac{1}{2} p_{2n-1}^2 + e^{q_{2n-1} - q_{2n+1}}. \tag{14.7}$$

This canonical transformation allows us to construct a simple particular solution.

14.3 The solitary wave

Let us start from the trivial solution to the 'odd' Hamiltonian system corresponding to static equilibrium

$$q_{2n+1} = 0, \qquad p_{2n+1} = 0.$$

We obtain from (14.6) the transformed equations of motion and corresponding trajectories

$$\frac{dq_{2n}}{dt} = p_{2n} = e^{q_{2n}} + e^{-q_{2n}} + c,$$

$$0 = e^{q_{2n}} + e^{-q_{2n+2}} + c. \tag{14.8}$$

The solution is simply

$$e^{q_{2n}} = \frac{\cosh(\kappa(n-1) + t \sinh \kappa)}{\cosh(\kappa n + t \sinh \kappa)}, \tag{14.9}$$

with the notation $c = -2 \cosh \kappa$.

This solution to the equations of motion, however, cannot represent the global evolution of a cyclic chain. The periodicity conditions $q_{2n+2N} \equiv q_{2n}$ indeed lead to the quantization of the wavenumber

$$i\kappa = \frac{\pi k}{N}; \quad k = 0, 1, 2, \ldots, N - 1. \tag{14.10}$$

This implies that, in certain regions of the motion, in n and t, one can have $e^{q_{2n}} < 0$: the trajectory becomes complex, which is mechanically inadmissible. It would also have been sufficient to notice that the total energy of this pseudo-motion was equal to zero. Solution (14.9) is nonetheless correct for an infinite chain. It describes the propagation of a *solitary wave* with potential

$$e^{q_{2n} - q_{2n+2}} = 1 + \frac{\sinh^2 \kappa}{\cosh^2(\kappa n + t \sinh \kappa)}, \tag{14.11}$$

moving (towards the left) with speed $v = 2\kappa^{-1} \sinh \kappa$. The transverse displacement of the mobile masses represents a wavefront connecting two different levels up and downstream; the height difference is 2κ, and the total energy of the *soliton* is $\sinh 2\kappa$.

A more general solution to system (14.6) can be obtained starting from the relations for quantities $r_n = q_n - q_{n+1}$, which verify the necessary but not sufficient conditions

$$\frac{dr_n}{dt} = e^{r_{n-1}} - e^{r_{n+1}}, \tag{14.12}$$

linking the even and odd systems. One looks for two different analytic functions according to the parity of n, depending on a single variable $(n + \omega t)$. The solution found depends on two parameters, k and κ.

- n odd

$$e^{r_n} = \frac{1 + k \operatorname{sn} x_n \operatorname{sn} \kappa}{1 - k \operatorname{sn} x_n \operatorname{sn} \kappa}. \tag{14.13}$$

- n even

$$e^{r_n} = \frac{\mu}{\operatorname{sn} 2\kappa} (1 - k \operatorname{sn} x_n \operatorname{sn} 2\kappa), \tag{14.14}$$

with

$$x_n = n\kappa \pm \mu t, \qquad \mu = \frac{2 \operatorname{sn} \kappa}{\operatorname{cn} \kappa \operatorname{dn} \kappa}. \tag{14.15}$$

The Jacobi elliptic functions depend on the modulus k. The potential wave has the form

$$e^{q_{2n} - q_{2n+2}} = 1 + \frac{\mu^2}{4}(\operatorname{dn} x_{2n} + k \operatorname{cn} x_{2n})^2 = 1 + \frac{(1 + k)^2 \mu^2}{4} \operatorname{dn}^2 \left(x_{2n} \frac{1 + k}{2}; \frac{2\sqrt{k}}{1 + k} \right) \tag{14.16}$$

and reduces to expression (14.11) for $k = 1$ $(2\kappa \to \kappa)$. This solution is appropriate for the infinite chain only, and not for the cyclic chain since, according to (14.16), we have $q_{2n} - q_{2n+2N}$ = function of t. The M-soliton solution was constructed by Hirota (1973), and by Flaschka (1974) using the inverse scattering method.

14.4 Complete integrability

Let us build the auxiliary Lax system by induction, starting from the necessary relations (14.12) originating from the canonical transformation

$$\dot{r}_n = e^{r_{n-1}} - e^{r_{n+1}}, \qquad r_n = q_n - q_{n+1}. \tag{14.17}$$

The relations do not imply the equations of motion

$$\ddot{q}_{2n} = e^{q_{2n} - q_{2n+2}} - e^{q_{2n-2} - q_{2n}}, \tag{14.18}$$

but simply relate two solutions of the same Hamiltonian system

$$H(p_{2n+1}, q_{2n+1}) = E \Leftrightarrow H(p_{2n}, q_{2n}) = E.$$

Let us introduce intermediate functions ρ_n such that

$$e^{q_n - \lambda t} = \rho_n \rho_{n-1} \Rightarrow e^{r_n} = \frac{\rho_{n-1}}{\rho_{n+1}}, \tag{14.19}$$

in which λ is an undetermined parameter. Relations (14.17) are implied by

$$\dot{\rho}_n = \rho_n \frac{\rho_{n-1}}{\rho_{n+1}}. \tag{14.20}$$

One can see that these would even be equivalent to (14.17) if λ was an arbitrary function of time. However, under the choice $\lambda = \text{cst}$, one can show that (14.20) implies (14.18).

Distinguishing cases according to the parity of n, we set

$$\xi_n = \rho_{2n}e^{\lambda t}, \quad \eta_n = \rho_{2n+1}^{-1} \Rightarrow e^{q_{2n}} = \frac{\xi_n}{\eta_{n-1}}. \tag{14.21}$$

We deduce from (14.20)

$$\frac{\dot{\xi}_n}{\xi_n} = \lambda + \frac{\eta_n}{\eta_{n-1}}, \quad \frac{\dot{\eta}_{n-1}}{\eta_{n-1}} = -\frac{\xi_{n-1}}{\xi_n}, \tag{14.22}$$

or further still

$$\dot{\xi}_n = \lambda \xi_n + e^{q_{2n}} \eta_n, \tag{14.23}$$

$$\dot{\eta}_n = -e^{-q_{2n+2}} \xi_n. \tag{14.24}$$

If we define the two-component vector $X_n = \begin{pmatrix} \xi_n \\ \eta_n \end{pmatrix}$, we have the auxiliary equation

$$\dot{X}_n = M_n X_n, \tag{14.25}$$

with the matrix issuing from (14.24):

$$M_n = \begin{pmatrix} \lambda & e^{q_{2n}} \\ -e^{-q_{2n+2}} & 0 \end{pmatrix}. \tag{14.26}$$

Eliminating η_n between (14.22) and (14.24),

$$\frac{d}{dt}(e^{-q_{2n}}\xi_n) = -e^{-q_{2n}}\xi_{n-1} \tag{14.27}$$

or

$$\dot{\xi}_n - \dot{p}_{2n}\xi_n + \xi_{n-1} = 0 \tag{14.28}$$

and, taking (14.23) and (14.21) into account,

$$(p_{2n} - \lambda)\xi_n - \xi_{n-1} - e^{q_{2n}-q_{2n+2}}\xi_{n+1} = 0, \tag{14.29}$$

which we can also write as a relation between X_n and X_{n-1}:

$$\begin{cases} \xi_{n-1} = (p_{2n} - \lambda)\xi_n - e^{q_{2n}}\eta_n, \\ \eta_{n-1} = e^{-q_{2n}}\xi_n, \end{cases} \tag{14.30}$$

whose coefficients depend only on variables q_{2n}, p_{2n}:

$$X_{n-1} = R_n X_n, \qquad R_n = \begin{pmatrix} p_{2n} - \lambda & -e^{q_{2n}} \\ e^{-q_{2n}} & 0 \end{pmatrix}. \qquad (14.31)$$

A sufficient compatibility condition between the linear relations (14.25) and (14.31) can be written

$$\dot{R}_n = M_{n-1} R_n - R_n M_n. \qquad (14.32)$$

As can be checked directly, this condition is equivalent to the equations of motion (14.18).

The transfer matrix of the even system

$$Z = R_1 R_2 \ldots R_N, \qquad (14.33)$$

such that we have $X_0 = Z X_N$, is of unit determinant, since $|R_n| = 1$. Its two eigenvalues are ζ and ζ^{-1}, with $\text{Tr } Z = \zeta + \zeta^{-1}$.

Let us show that the polynomial $\text{Tr } Z$ of degree N is a constant of motion for any λ. By virtue of (14.32), we have

$$\frac{d}{dt} \text{Tr } Z = \sum_{n=1}^{N} \text{Tr } R_1 \ldots R_{n-1} \left[M_{n-1} R_n - R_n M_n \right] R_{n+1} \ldots R_N$$
$$= \text{Tr } (M_0 R_1 \ldots R_N - R_1 \ldots R_N M_N) = 0, \qquad (14.34)$$

the last equality coming from the fact that the chain is cyclic. The coefficients of $\text{Tr } Z$ thus define $N - 1$ first integrals including the total momentum (coefficient of λ^{N-1}) and the total energy (coefficient of λ^{N-2}); since we can easily verify their independence, we deduce the complete integrability of the system.

If we go back to the linear system for vector X_n, it remains to impose the cyclicity relation $X_0 = X_N$ which leads to $\zeta = 1$ and $Z \equiv 1$, so $\text{Tr } Z = 2$, which is a relation between the first integrals and the parameter λ or, if one prefers, determines a spectrum of N values of λ, the former being given.

We can also go back to the problem mentioned in Section 14.2, namely that of the solution of the system of equations for the odd canonical coordinates in terms of the even ones. This problem leads directly to the fundamental linear equation (14.29) (see also (14.38)). Indeed, for given even variables, one can define u_n such that

$$e^{-q_{2n+1}} = \frac{u_{n+1}}{u_n} = e^{-q_{2n+2}} \frac{\xi_{n+1}}{\xi_n},$$

and (14.6) gives us

$$(p_{2n} - c) u_n - e^{q_{2n}} u_{n+1} - e^{-q_{2n}} u_{n-1} = 0. \qquad (14.35)$$

Parameter c is thus identical to λ. One can also see that the cyclicity condition imposes that the determinant of the matrix associated with the above linear system is identically zero in t. The characteristic polynomial in c coincides with the polynomial in λ, Tr $Z - 2$, and with the determinant $|L - \lambda|$, which shall be defined in (14.39).

14.5 The M-soliton solution for the infinite chain

Let us give here a very brief description of the inverse scattering method for the construction of the M-soliton solution according to the method of Flaschka (1974). The linear system (14.25) and (14.31) for X_n can be interpreted in an N-dimensional space as a linear relation for the vector with components ξ_n ($n = 1, 2, \ldots, N$) after elimination of η_n, as in the recurrence equation (14.29). To render the form of the latter more symmetric, we perform the change of function

$$\varphi_n = (-1)^n e^{-\frac{1}{2}(q_{2n} - \lambda t)} \xi_n \tag{14.36}$$

in order to put (14.29) in the form

$$(p_{2n} - \lambda)\varphi_n + e^{-\frac{1}{2}(q_{2n+2} - q_{2n})}\varphi_{n+1} + e^{-\frac{1}{2}(q_{2n} - q_{2n-2})}\varphi_{n-1} = 0. \tag{14.37}$$

Similarly, starting from (14.28) we have

$$\dot{\varphi}_n = \frac{1}{2}e^{\frac{1}{2}(q_{2n} - q_{2n+2})}\varphi_{n+1} - \frac{1}{2}e^{\frac{1}{2}(q_{2n-2} - q_{2n})}\varphi_{n-1}. \tag{14.38}$$

These two relations can be written

$$L\varphi = \lambda\varphi, \tag{14.39}$$

$$\dot{\varphi} = B\varphi, \tag{14.40}$$

in which L and B are two $N \times N$ matrices whose only nonzero elements are

$$L_{n,n} = p_{2n}, \tag{14.41}$$

$$L_{n+1,n} = L_{n,n+1} = -2B_{n,n+1} = 2B_{n+1,n} = e^{(q_{2n} - q_{2n+2})/2}.$$

The compatibility condition of system (14.39) and (14.40),

$$\dot{L} = [B, L],$$

is equivalent to the equations of motion and leads to the independence of the spectrum with respect to time.

From this comes the following method to construct a solution corresponding to given initial conditions in the case of the infinite chain. Equation (14.37) is a second-order difference equation, which is analogous to a Schrödinger equation for the stationary problem of a particle in a potential, the latter being determined by the functions n: q_{2n} and p_{2n}. But being given this potential (at $t = 0$, for example) allows us to determine the 'scattering data' associated with the asymptotic behaviour of φ_n, in other words the S matrix and its poles and residues. Under the hypothesis of the following behaviour at infinity:

$$q_{2n} \to \text{cst}, \qquad p_{2n} \to 0 \qquad (\lambda = z + z^{-1})$$
$$\varphi_n(z, t) \sim z^n + S(z, t)z^{-n} \qquad (n \to \infty)$$

the temporal evolution of the scattering data is very simple. The second equation (14.38) gives us

$$S(z, t) = S(z, 0)e^{-2\beta t}, \qquad 2\beta = z - z^{-1}. \tag{14.42}$$

The scattering data at time t are in principle sufficient to reconstruct the potential $q_{2n}(t)$ and $p_{2n}(t)$, which constitutes the inverse method. Its implementation in the particular case of the infinite Toda chain rests on the solution of the integral equation due to Marchenko adapted to the discrete case (Case, 1973). One obtains a solution in the form

$$e^{q_{2n} - \lambda t} = \frac{u_{2n-2}}{u_{2n}}, \qquad \rho_n = \frac{u_{n-1}}{u_n}, \tag{14.43}$$

with u_{2n} expressed as the determinant of an order-M matrix

$$u_{2n} = \text{Det}\,(1 + D^{(n)}CD^{(n)}), \tag{14.44}$$

$$C_{ij} = (1 - z_i z_j)^{-1}, \qquad D_i^{(n)} = z_i^n e^{-\beta_i t}c_i, \qquad i = 1, \ldots, M, \tag{14.45}$$

in which the c_i, z_i are arbitrary positive constants. This solution coincides with (14.9) when $M = 1$. The u_{2n} verify the nonlinear equation originating from (14.20):

$$u_n u_n'' - u_n'^2 = u_{n+2}u_{n-2} - u_n^2 \qquad (n \text{ even}). \tag{14.46}$$

It would be interesting to construct the solution associated with odd n by using the Bäcklund transformation

$$u_n' u_{n+1} - u_n u_{n+1}' = u_{n-1}u_{n+2}. \tag{14.47}$$

14.6 The quantum chain

The Hamiltonian of the cyclic quantum chain still has expression (14.2) in which q_{2n} and p_{2n} are the conjugate momenta verifying the commutation rules $[p_{2n}, q_{2m}] = -i\delta_{n,m}$, etc. We shall make use of the representation $p_{2n} = -i\dfrac{\partial}{\partial q_{2n}}$, right-acting on functions of q_{2n}. We also introduce the left-acting representation $p_{2n+1} = i\dfrac{\partial}{\partial q_{2n+1}}$.

One defines the matrix operators encountered in (14.31) for the classical system

$$R_n = \begin{pmatrix} p_n - \lambda & -e^{q_n} \\ e^{-q_n} & 0 \end{pmatrix}, \qquad p_n = (-1)^n \frac{1}{i} \frac{\partial}{\partial q_n} \tag{14.48}$$

and the transfer matrix

$$Z_{even}(\lambda) = \text{Tr } R_2 R_4 \ldots R_{2N} \equiv Z(\lambda). \tag{14.49}$$

We further define functions of $2N$ variables, even and odd,

$$\langle \{q_{2n}\}|S|\{q_{2n+1}\}\rangle = \exp i\, W(q_n) = S(q_n), \tag{14.50}$$

in which $W(q_n)$ is the generator of the classical canonical transformation defined in (14.5). We have the following

Proposition:

$$(Z_{even} - Z_{odd})S(q) = 0. \tag{14.51}$$

Proof:
According to the definitions (14.48),

$$i p_n W = (-1)^n \frac{\partial}{\partial q_n} \sum_{m=1}^{2N} (-1)^m e^{q_m - q_{m+1}} = e^{q_n - q_{n+1}} + e^{q_{n-1} - q_n} \tag{14.52}$$

and thus

$$R_n S = S \begin{pmatrix} e^{q_n - q_{n+1}} + e^{q_{n-1} - q_n} - \lambda & -e^{q_n} \\ e^{-q_n} & 0 \end{pmatrix} \equiv -\frac{1}{\lambda} S R_n R_{n+1}, \tag{14.53}$$

in which we have set

$$\hat{R}_n = \begin{pmatrix} e^{q_{n-1} - q_n} - \lambda & -e^{q_{n-1}} \\ e^{-q_n} & 1 \end{pmatrix}. \tag{14.54}$$

We thus have

$$\text{Tr}\,(R_1 R_3 \ldots R_{2N-1})\, S = S\,\text{Tr }\hat{R}_1 \hat{R}_2 \ldots \hat{R}_{2N} \left(-\frac{1}{\lambda} \right)^N = (\text{Tr } R_2 R_4 \ldots R_{2N})\, S. \quad \square \tag{14.55}$$

Corollary: If we consider the function $S(q)$ as the matrix element $\{q_{2n}\} \to \{q'_{2n}\} \equiv \{q_{2n+1}\}$ of an operator S in the $\{q_{2n}\}$ representation, the commutator of S and $Z(\lambda)$ is zero:

$$[S, Z(\lambda)] = 0. \tag{14.56}$$

S thus commutes with each of the 'first integrals' of motion, coefficients of $Z(\lambda)$. We shall show in the following section that these are indeed constants of motion of the quantum system.

14.7 The integral equation for the eigenfunctions

Since the Hamiltonian H is proportional to the coefficient of λ^{N-2} in $Z(\lambda)$, the function $S(q)$ obeys the differential equation originating from (14.51):

$$(H_{even} - H_{odd}) \cdot S(q) = 0. \tag{14.57}$$

We can then think that the eigenfunctions of H are also those of the integral kernel S. The method of proof is analogous to that one uses for the Matthieu function or spheroidal functions (Whittaker and Watson, 1927, p. 407). The eigenfunctions of the Toda Hamiltonian for $N = 2$ are precisely the Matthieu functions.

For convergence reasons which will become clear later on, let us consider the modified kernel

$$K(q_{2n}, q_{2n+1}) \equiv S\left(q_{2n}, q_{2n+1} - i\frac{\pi}{2}\right) = \exp\left\{-\sum_{n=1}^{2N} e^{q_n - q_{n+1}}\right\} \equiv K(x, y), \tag{14.58}$$

in the notation

$$x = \{x_n = q_{2n}\}, \quad y = \{y = q_{2n+1}\}, \quad \bar{x} = \{x_n^{-1}\}.$$

Let us define the circular permutation operator C such that $Cx_n = x_{n+1}$, $C^N = 1$. We have the invariance properties

$$K(x, y) \equiv K(Cx, Cy) \quad \text{or} \quad KC = CK. \tag{14.59}$$

We also have

$$K(x, y) = e^{-(x \cdot \bar{y} + y \cdot C\bar{x})} = K(y, Cx),$$

or in other words

$$KC = {}^t K = CK. \tag{14.60}$$

The symmetrizable kernel K is of Hilbert–Schmidt type on $]-\infty, +\infty[$, all of its traces are finite. We shall call $f_{k,s}(q_{2n})$ the simultaneous eigenfunctions of K and of C:

$$Cf_{k,s} = e^{2\pi ik/N} f_{k,s}, \quad k = 0, 1, \ldots, N - 1,$$

$$sf_{k,s}(q_{2n+1}) = \int_{-\infty}^{+\infty} \ldots dq_{2n} \ldots \exp\left\{-\sum_{n=1}^{2N} e^{q_n - q_{n+1}}\right\} f_{k,s}(q_{2n}), \tag{14.61}$$

associated with a momentum k and a real eigenvalue s. Owing to the differential equation (14.57) and to the convergence of the kernel to zero at infinity, the functions $f_{k,s}$ are also eigenfunctions of the Hamiltonian H:

$$-\frac{1}{2}\sum_{n=1}^{N}\frac{\partial^2 f_{k,s}}{\partial q_{2n}^2} + \left(\sum_{n=1}^{N}e^{q_{2n}-q_{2n+2}} - E_{k,s}\right)f_{k,s} = 0, \qquad (14.62)$$

and inversely. We have here a generalization of the Matthieu functions (continued to the imaginary axis). For $N = 2$, by setting $q_2 - q_4 = 2q$ and restricting ourselves to zero total momentum, we have

$$\begin{cases} sf(q) = \displaystyle\int_{-\infty}^{+\infty} dq' e^{-4\cosh q \cosh q'} f(q'), \\[2mm] -\dfrac{\partial^2 f}{\partial q^2} + 2\cosh 2q f = Ef. \end{cases} \qquad (14.63)$$

It would seem at first sight that this extension was only achieved for a certain value of the parameter called $q_{Mat.}$ in the theory of Matthieu functions (not to be confused with our argument q). We would have here $q_{Mat.} = 1/8$. In fact, we have an available parameter in the scale of time or, for the quantum case, in the constant $\hbar$.

14.8 Ternary relations and action–angle variables

From relation (14.56), $[S, Z(\lambda)] = 0$, we naturally deduce $[Z(\lambda), Z(\mu)] = 0$, $\forall \lambda, \mu$. This can be proven from the ternary relations between R matrices (14.48) which define the transition matrix (or monodromy matrix). We indeed have in the usual notation

$$R \equiv \tau \cdot R(\lambda) = \begin{pmatrix} p - \lambda & -e^q \\ e^{-q} & 0 \end{pmatrix}, \qquad (14.64)$$

$$\tau \cdot R(\mu)\tau' \cdot R(\lambda) = X(\mu - \lambda)\tau' \cdot R(\lambda)\tau \cdot R(\mu)X(\lambda - \mu). \qquad (14.65)$$

In the matrix basis

$$\tau \cdot \tau' = \begin{pmatrix} \tau_{11}\tau'_{11} & \tau_{12}\tau'_{12} & \tau_{11}\tau'_{12} & \tau_{12}\tau'_{11} \\ \tau_{21}\tau'_{21} & \tau_{12}\tau'_{22} & \tau_{11}\tau'_{22} & \tau_{12}\tau'_{21} \\ \tau_{11}\tau'_{21} & \tau_{12}\tau'_{22} & \tau_{11}\tau'_{22} & \tau_{12}\tau'_{21} \\ \tau_{21}\tau'_{11} & \tau_{22}\tau'_{12} & \tau_{21}\tau'_{12} & \tau_{22}\tau'_{11} \end{pmatrix},$$

we easily find

$$X(\mu - \lambda) = \begin{pmatrix} 1 & 0 & 0 & 0 \\ 0 & 1 & 0 & 0 \\ 0 & 0 & \dfrac{1}{1+u} & \dfrac{u}{1+u} \\ 0 & 0 & \dfrac{u}{1+u} & \dfrac{1}{1+u} \end{pmatrix}, \qquad i(\mu - \lambda) = u. \qquad (14.66)$$

According to the method exposed in Chapter 10 (see also Appendix L), we can deduce from this the algebra of the elements of the transfer matrix:

$$\left[A_{11}(\lambda), A_{11}(\lambda')\right] = 0,$$
$$\left[A_{12}(\lambda), A_{12}(\lambda')\right] = 0, \ldots \tag{14.67}$$

$$A_{11}(\lambda')A_{12}(\lambda) = \frac{1}{1+u}\left\{A_{11}(\lambda)A_{12}(\lambda') + uA_{12}(\lambda)A_{11}(\lambda')\right\},$$
$$u = i(\lambda' - \lambda) \tag{14.68}$$

or further still, and more simply,

$$\begin{cases} A_{11}(\lambda)A_{12}(\lambda') = \left(1 + \dfrac{i}{\lambda - \lambda'}\right) A_{12}(\lambda)A_{11}(\lambda') - \dfrac{i}{\lambda - \lambda'}A_{12}(\lambda')A_{11}(\lambda), \\[2mm] A_{22}(\lambda)A_{12}(\lambda') = \left(1 - \dfrac{i}{\lambda - \lambda'}\right) A_{12}(\lambda)A_{22}(\lambda') + \dfrac{i}{\lambda - \lambda'}A_{12}(\lambda')A_{22}(\lambda). \end{cases}$$
$$\tag{14.69}$$

These relations are almost identical in form to those obtained in (11.49) in the spin-1/2 fermion problem. It does not seem possible, however, to exploit these relations in an identical fashion to (11.2), since there does not exist a reference eigenstate for the Toda chain which is analogous to the vacuum of the ferromagnetic state of the spin chain. Despite this fact, something remains possible for the finite chain. Indeed, we have exactly

$$A_{11}(\lambda) = (-\lambda)^N \left\{1 - \frac{P}{\lambda} + \frac{1}{\lambda^2}\left(\frac{P^2}{2} - H_0\right) + O\left(\frac{1}{\lambda^3}\right)\right\}, \tag{14.70}$$

in which P is the total momentum and H_0 is the Hamiltonian of the open chain: $H_0 = H - e^{(q_{2N} - q_2)}$.

One can deduce from the first relation (14.57) that the coefficients of $A_{11}(\lambda)$ are constants of motion of the open chain. The first integrals being given, there exist M numbers $\lambda_1, \lambda_2, \ldots, \lambda_M$ ($M \leq N$) such that we have

$$A_{11}(\lambda_j) = 0. \tag{14.71}$$

Using exactly the same algebra as in (11.51), one defines the operator

$$F = A_{12}(\lambda_1)A_{12}(\lambda_2) \ldots A_{12}(\lambda_M) \tag{14.72}$$

and setting $Q(\lambda) = \prod_{j=1}^{M}(\lambda - \lambda_j)$ we obtain

$$A_{11}(\lambda)F = \frac{Q(\lambda + i)}{Q(\lambda)}FA_{11}(\lambda). \tag{14.73}$$

Expanding both sides in the neighbourhood of $\lambda = \infty$, one can deduce from (14.73) the following

Corollary:

$$[A_{11}(\lambda), F] = \left(\frac{Q(\lambda - i)}{Q(\lambda)} - 1 \right) F A_{11}(\lambda)$$

$$= \left(-\frac{M}{i\lambda} - \left(\frac{N(M-1)}{2} - i \sum_j \lambda_j \right) \frac{1}{\lambda^2} + \dots \right) F A_{11}(\lambda)$$

and further still

$$[P, F] = -iMF, \tag{14.74}$$

$$[H_0, F] = - \sum_{j=1}^{M} \left(\frac{1}{2} + i\lambda_j \right) F. \tag{14.75}$$

The family of operators F thus cannot be considered as creating individual excitations (whose momentum would be $-i$ and energy $-1/2 - i\lambda_j$!), but nonetheless correspond to a system of complex canonical variables, formally of action–angle type, for the integrable Toda Hamiltonian.

References

Adler, M. 1977. Some finite dimensional integrable systems and their scattering behavior. *Commun. Math. Phys.*, **55**, 195–230.

Araki, H. 1969. *Commun. Math. Phys.*, **14**, 120–157.

Babbitt, D. G. and Thomas, L. E. 1977a. Explicit Plancherel theorem for ground state representation of the Heisenberg chain. *Proc. Nat. Acad. Sci. USA*, **74**(3), 816–817.

Babbitt, D. G. and Thomas, L. E. 1977b. Ground state representation of the infinite one-dimensional Heisenberg ferromagnet. II. An explicit Plancherel formula. *Commun. Math. Phys.*, **54**, 255–278.

Babbitt, D. G. and Thomas, L. E. 1978. Ground state representation of the infinite one-dimensional Heisenberg ferromagnet. III. Scattering theory. *J. Math. Phys.*, **19**(8), 1699–1704.

Babelon, O., de Vega, H. J. and Viallet, C. M. 1983. Analysis of the Bethe ansatz equations of the XXZ model. *Nucl. Phys. B*, **220**, 13–34.

Barber, M. N. and Baxter, R. J. 1973. On the spontaneous order of the eight-vertex model. *J. Phys. C: Solid St. Phys.*, **6**(20), 2913.

Barouch, E. 1972. In: Domb, C. and Green, M. S. (eds), *Phase Transitions and Critical Phenomena*, Vol. IV. London: Academic Press.

Barouch, E., McCoy, B. M. and Dresden, M. 1970. Statistical mechanics of the XY model. I. *Phys. Rev. A*, **2**, 1075–1092.

Barut, A. O. 1977. *Nonlinear Equations in Physics and Mathematics*. Dordrecht: Reidel.

Baxter, R. J. 1969. F model on a triangular lattice. *J. Math. Phys.*, **10**(7), 1211–1216.

Baxter, R. J. 1970a. Colorings of a hexagonal lattice. *J. Math. Phys.*, **11**(3), 784–789.

Baxter, R. J. 1970b. Three-colorings of the square lattice: A hard squares model. *J. Math. Phys.*, **11**(10), 3116–3124.

Baxter, R. J. 1971a. Eight-vertex model in lattice statistics. *Phys. Rev. Lett.*, **26**, 832–833.

Baxter, R. J. 1971b. Generalized ferroelectric model on the square lattice. *Stud. App. Math.*, **50**, 51.

Baxter, R. J. 1972a. One-dimensional anisotropic Heisenberg chain. *Ann. Phys.*, **70**(2), 323–337.

Baxter, R. J. 1972b. Partition function of the eight-vertex lattice model. *Ann. Phys.*, **70**(1), 193–228.

Baxter, R. J. 1973a. Asymptotically degenerate maximum eigenvalues of the eight-vertex model transfer matrix and interfacial tension. *J. Stat. Phys.*, **8**, 25–55.

Baxter, R. J. 1973b. Eight-vertex model in lattice statistics and one-dimensional anisotropic Heisenberg chain. I. Some fundamental eigenvectors. *Ann. Phys*, **76**(1), 1–24.

Baxter, R. J. 1973c. Eight-vertex model in lattice statistics and one-dimensional anisotropic Heisenberg chain. II. Equivalence to a generalized ice-type lattice model. *Ann. Phys.*, **76**(1), 25–47.

Baxter, R. J. 1973d. Eight-vertex model in lattice statistics and one-dimensional anisotropic Heisenberg chain. III. Eigenvectors of the transfer matrix and Hamiltonian. *Ann. Phys.*, **76**(1), 48–71.

Baxter, R. J. 1973e. Spontaneous staggered polarization of the F model. *J. Phys. C: Sol. St. Phys.*, **6**(5), L94.

Baxter, R. J. 1978. Solvable eight-vertex model on an arbitrary planar lattice. *Phil. Trans. Roy. Soc. London A*, **289**(1359), 315–346.

Baxter, R. J. 1980. Hard hexagons: Exact solution. *J. Phys. A: Math. Gen.*, **13**(3), L61.

Belavin, A. A. 1979. Exact solution of the two-dimensional model with asymptotic freedom. *Phys. Lett. B*, **87**, 117–121.

Bergknoff, H. and Thacker, H. B. 1979a. Method for solving the massive Thirring model. *Phys. Rev. Lett.*, **42**, 135–138.

Bergknoff, H. and Thacker, H. B. 1979b. Structure and solution of the massive Thirring model. *Phys. Rev. D*, **19**, 3666–3681.

Bethe, H. 1931. Zur Theorie der Metalle. I. Eigenwerte und Eigenfunktionen der linearen Atomkette. *Zeit. für Physik*, **71**, 205.

Bloch, F. 1930. Zur Theorie des Ferromagnetismus. *Z. Phys.*, **61**, 206.

Boiti, M. and Pempinelli, F. 1980. Nonlinear Schrödinger equation, Bäcklund transformations and Painlevé transcendents. *Il Nuovo Cimento B*, **59**, 40–58.

Bonner, J. C. 1968. Numerical studies on the linear Ising–Heisenberg model. Ph.D. thesis, University of London.

Bonner, J. C. and Fisher, M. E. 1964. Linear magnetic chains with anisotropic coupling. *Phys. Rev.*, **135**(3A), A640–A658.

Brézin, E. and Zinn-Justin, J. 1966. Un problème à *N* corps soluble. *C. R. Acad. Sci. Paris, sér. A-B*, **263**, 670–673.

Brézin, E., Itzykson, C., Zinn-Justin, J. and Zuber, J. B. 1979. Remarks about the existence of non-local charges in two-dimensional models. *Phys. Lett. B*, **82**(3&4), 442–444.

Bychkov, Yu., Gor'kov, L. and Dzyaloshinskii, I. 1966. Possibility of superconductivity type phenomena in a one-dimensional system. *Sov. Phys. JETP*, **23**, 489–501.

Calogero, F. 1971. Solution of the one-dimensional *N*-body problems with quadratic and/or inversely quadratic pair potentials. *J. Math. Phys.*, **12**(3), 419–436.

Calogero, F. 1978. Integrable many-body problems. In: Barut, A.O. (ed.), *Nonlinear Equations in Physics and Mathematics*. Dordrecht: Reidel. Proceedings of the NATO Advanced Study Institute held in Istanbul, August 1977.

Calogero, F. 1981. Matrices, differential operators, and polynomials. *J. Math. Phys.*, **22**(5), 919–934.

Calogero, P., Ragnisco, O. and Marchioro, C. 1975. Exact solution of the classical and quantal one-dimensional many-body problems with the two-body potential $V(x) = g^2 a^2 \sinh^{-2} ax$. *Lett. al Nuovo Cimento*, **13**, 383–387.

Case, K. M. 1973. On discrete inverse scattering problems. II. *J. Math. Phys.*, **14**(7), 916–920.

Chudnovsky, D. 1979. Simplified Schlesinger's systems. *Lett. al Nuovo Cimento*, **26**, 423–427.

Chudnovsky, D. V. and Chudnovsky, G. V. 1980. Zakharov–Shabat–Mikhailov scheme of construction of two-dimensional completely integrable field theories. *Zeit. für Physik C*, **5**, 55–62.

Chudnovsky, D. and Chudnovsky, G. 1982. Hamiltonian structure of isospectral deformation equations. Elliptic curve case. In: Chudnovsky, D. and Chudnovsky, G. (eds), *The Riemann Problem, Complete Integrability and Arithmetic Applications*. Lecture Notes in Mathematics, Vol. 925. Berlin: Springer-Verlag; pp. 134–146.

Coleman, S. 1975. Quantum sine-Gordon equation as the massive Thirring model. *Phys. Rev. D*, **11**, 2088–2097.

Coxeter, H. S. M. 1948. *Regular Polytopes*. London: Methuen.

Coxeter, H. S. M. and Moser, W. O. J. 1972. *Generators and Relations for Discrete Groups*, 3rd edn. Berlin: Springer-Verlag.

Dashen, R. F., Hasslacher, B. and Neveu, A. 1975. Particle spectrum in model field theories from semiclassical functional integral techniques. *Phys. Rev. D*, **11**, 3424–3450.

Derrida, B. 1976. Solution d'un modèle à trois corps: étude de la dissusion. Ph.D. thesis, Université Paris.

des Cloizeaux, J. 1966. A soluble Fermi-gas model. Validity of transformations of the Bogoliubov type. *J. Math. Phys.*, **7**(12), 2136–2144.

des Cloizeaux, J. and Gaudin, M. 1966. Anisotropic linear magnetic chain. *J. Math. Phys.*, **7**(8), 1384–1400.

des Cloizeaux, J. and Pearson, J. J. 1962. Spin-wave spectrum of the antiferromagnetic linear chain. *Phys. Rev.*, **128**(5), 2131–2135.

Dirac, P. A. M. 1967. *The Principles of Quantum Mechanics*, 4th edn (revised). Oxford: Oxford University Press.

Dyson, F. J. 1956. General theory of spin-wave interactions. *Phys. Rev.*, **102**(5), 1217–1230.

Dyson, F. J. 1962. Statistical theory of the energy levels of complex systems. I. *J. Math. Phys.*, **3**(1), 140–156.

Dyson, F. J. 1976. Fredholm determinants and inverse scattering problems. *Commun. Math. Phys.*, **47**, 171–183.

Dzyaloshinskii, I. E. and Larkin, A. I. 1974. Correlation functions for a one-dimensional Fermi system with long-range interaction (Tomonaga model). *Sov. Phys. JETP*, **38**. Russian original: ZhETF, Vol. 65, No. 1, p. 411, January 1974.

Faddeev, L. D. 1980. Quantum completely integrable models of field theory. In: *Mathematical Physics Review Section C*, Vol. 107. New York: Harwood Academic; p. 155.

Fan, C. and Wu, F. Y. 1970. General lattice model of phase transitions. *Phys. Rev. B*, **2**, 723–733.

Fateev, V. A. 1980. The factorizable S matrix for the particles with different parity and integrable anisotropic spin chain with spin one. Technical Report CERN, Ref. TH 2963.

Fisher, M. E. 1968. In: *Proceedings of the International Conference on Statistical Mechanics*, Kyoto.

Fisher, M. E. 1969. *J. Phys. Soc. Japan Suppl.*, **26**, 87.

Flaschka, H. 1974. On the Toda lattice. II. *Prog. Theor. Phys.*, **51**(3), 703–716.

Fogedby, H. C. 1980. Solitons and magnons in the classical Heisenberg chain. *J. Phys. A: Math. Gen.*, **13**(4), 1467.

Frobenius, F. G. 1880. Über das Additions-theorem der Thetafunktionen mehrerer Variabeln. *J. reine angew. Math.*, **89**, 185.

Frobenius, F. G. 1968. *Gesammelte Abhandlungen, Band II*. Berlin: Springer-Verlag.

Gaaff, A. and Hijmans, J. 1975. Symmetry relations in the sixteen-vertex model. *Physica A*, **80**(2), 149–171.

Gaudin, M. 1967a. Étude d'une modèle à une dimension pour un système de fermions en interaction. Ph.D. thesis, Université Paris. CEA Report No. 5-3569 (1968).

Gaudin, M. 1967b. Un systeme à une dimension de fermions en interaction. *Phys. Lett. A*, **24**(1), 55–56.

Gaudin, M. 1971a. Bose gas in one dimension. I. The closure property of the scattering wavefunctions. *J. Math. Phys.*, **12**(8), 1674–1676.

Gaudin, M. 1971b. Bose gas in one dimension. II. Orthogonality of the scattering states. *J. Math. Phys.*, **12**(8), 1677–1680.

Gaudin, M. 1971c. Boundary energy of a Bose gas in one dimension. *Phys. Rev. A*, **4**, 386–394.

Gaudin, M. 1971d. Thermodynamics of the Heisenberg–Ising ring for $\Delta \geq 1$. *Phys. Rev. Lett.*, **26**(21), 1301–1304.

Gaudin, M. (1) 1972; (2) 1973. Modèles exacts en mécanique statistique: la méthode de Bethe et généralisations. Technical Report, CEA Saclay.

Gaudin, M. 1976. Diagonalisation d'une classe d'Hamiltoniens de spin. *J. Physique*, **37**, 1087–1098.

Gaudin, M. 1978. Sur le problème de deux ou trois électrons en présence d'un moment localisé. *J. Physique*, **39**, 1143–1168.

Gaudin, M. and Derrida, B. 1975. Solution exacte d'un probléme à troïs corps. *J. Physique*, **36**, 1183–1197.

Gaudin, M., McCoy, B. M. and Wu, T. T. 1981. Normalization sum for the Bethe's hypothesis wave functions of the Heisenberg–Ising chain. *Phys. Rev. D*, **23**, 417–419.

Girardeau, M. 1960. Relationship between systems of impenetrable bosons and fermions in one dimension. *J. Math. Phys.*, **1**(6), 516–523.

Green, H. S. and Hurst, C. A. 1964. *Order–Disorder Phenomena*. New York: Wiley-Interscience.

Griffiths, R. B. 1964. Magnetization curve at zero temperature for the antiferromagnetic Heisenberg linear chain. *Phys. Rev.*, **133**(3A), A768–A775.

Gutkin, E. and Sutherland, B. 1979. Completely integrable systems and groups generated by reflections. *Proc. Nat. Acad. Sci. USA*, **76**, 6057–6059.

Hammermesh, M. 1964. *Group Theory and its Application to Physical Problems*. New York: Addison-Wesley.

Heine, E. 1881. *Handbuch der Kugelfunktionen II*, 2nd edn. Berlin: Springer-Verlag; p. 472.

Heisenberg, W. 1928. Zur Theorie des Ferromagnetismus. *Z. Phys.*, **49**, 619.

Hida, K. 1981. Rigorous derivation of the distribution of the eigenstates of the quantum Heisenberg–Ising chain with XY-like anisotropy. *Phys. Lett. A*, **84**(6), 338–340.

Hirota, R. 1973. Exact N-soliton solution of a nonlinear lumped network equation. *J. Phys. Soc. Japan*, **35**(1), 286–288.

Hirota, R. and Satsuma, J. 1976. A variety of nonlinear network equations generated from the Bäcklund transformation for the Toda lattice. *Prog. Theor. Phys. Suppl.*, **59**, 64–100.

Honerkamp, J. and Weber, P. 1979. Calculation of the S-matrix from the Hamiltonian of the massive Thirring model. Technical Report, Université Fribourg.

Hulthén, L. 1938. Über das Austauschproblem eines Kristalles. *Arkiv Mat. Astron. Fysik*, **26A**, 1.

Huston, V. 1963. The circular plate condenser at small separations. *Proc. Camb. Phil. Soc.*, **59**, 211.

Iagolnitzer, D. 1978. Factorization of the multiparticle S matrix in two-dimensional space-time models. *Phys. Rev. D*, **18**, 1275–1285.

Ising, E. 1925. Beitrag zur Theorie des Ferromagnetismus. *Zeit. für Phys.*, **31**, 253–258.

Izergin, A. G. and Korepin, V. E. 1981. The inverse scattering method approach to the quantum Shabat–Mikhailov model. *Commun. Math. Phys.*, **79**, 303–316.

Jimbo, M., Miwa, T., Mori, Y. and Sato, M. 1980. Density matrix of an impenetrable Bose gas and the fifth Painlevé transcendent. *Phys. D*, **1**(1), 80–158.

Johnson, J. D. and McCoy, B. M. 1972. Low-temperature thermodynamics of the $|\Delta| \geq 1$ Heisenberg–Ising ring. *Phys. Rev. A*, **6**, 1613–1626.

Johnson, J. D., Krinsky, S. and McCoy, B. M. 1973. Vertical-arrow correlation length in the eight-vertex model and the low-lying excitations of the $X-Y-Z$ Hamiltonian. *Phys. Rev. A*, **8**(5), 2526–2547.

Jost, R. 1955. Mathematical analysis of a simple model for the stripping reaction. *Zeit. für Angew. Math. Phys.*, **6**, 316–326.

Kac, M. and Pollard, H. 1956. Partial sums of independent random variables. *Canadian J. Math.*, **2**, 375.

Kadanoff, L. P. and Wegner, F. J. 1971. Some critical properties of the eight-vertex model. *Phys. Rev. B*, **4**, 3989–3993.

Karowski, M., Thun, H. J., Truong, T. T., and Weisz, P. H. 1977. On the uniqueness of a purely elastic S-matrix in (1+1) dimensions. *Phys. Lett. B*, **67**(3), 321–322.

Kasteleyn, W. P. 1967. Graph theory and crystal physics. In: Harary, F. (ed.), *Graph Theory and Theoretical Physics*. New York: Academic Press.

Kirchhoff, G. 1877. Zur Theorice des Condensators. *Monatsb. Acad. Wiss. Berlin*, 144–162.

Korepin, V. E. 1980. The mass spectrum and the S-matrix of the massive Thirring model in the repulsive case. *Commun. Math. Phys.*, **76**, 165–176.

Korepin, V. E. 1982. Calculation of norms of Bethe wavefunctions. *Commun. Math. Phys.*, **86**, 391–418.

Kulish, P. P. 1979. Generalized Bethe ansatz and quantum inverse problem method. Preprint, Leningrad Branch of the Mathematics Institute, Leningrad.

Kulish, P. P. and Sklyanin, E. K. 1979. Quantum inverse scattering method and the Heisenberg ferromagnet. *Phys. Lett. A*, **70**(5&6), 461–463.

Kulish, P., Reshetikhin, N. and Sklyanin, E. 1981. Yang–Baxter equation and representation theory: I. *Lett. Math. Phys.*, **5**, 393–403.

Lai, C. K. 1974. Lattice gas with nearest-neighbor interaction in one dimension with arbitrary statistics. *J. Math. Phys.*, **15**(10), 1675–1676.

Lakshmanan, M. 1977. Continuum spin system as an exactly solvable dynamical system. *Phys. Lett. A*, **61**(1), 53–54.

Lax, P. D. 1968. Integrals of nonlinear equations of evolution and solitary waves. *Commun. Pure Appl. Math.*, **21**(5), 467–490.

Lenard, A. 1964. Momentum distribution in the ground state of the one-dimensional system of impenetrable bosons. *J. Math. Phys.*, **5**(7), 930–943.

Leppington, F. and Levine, H. 1970. On the capacity of the circular disc condenser at small separation. *Proc. Camb. Phil. Soc.*, **68**, 235.

Lieb, E. H. 1963. Exact analysis of an interacting Bose gas. II. The excitation spectrum. *Phys. Rev.*, **130**(4), 1616–1624.

Lieb, E. H. 1967a. Exact solution of the F model of an antiferroelectric. *Phys. Rev. Lett.*, **18**, 1046–1048.

Lieb, E. H. 1967b. Exact solution of the problem of the entropy of two-dimensional ice. *Phys. Rev. Lett.*, **18**, 692–694.

Lieb, E. H. 1967c. Exact solution of the two-dimensional Slater KDP model of a ferroelectric. *Phys. Rev. Lett.*, **19**, 108–110.

Lieb, E. H. 1967d. Residual entropy of square ice. *Phys. Rev.*, **162**, 162–172.

Lieb, E. H. and Liniger, W. 1963. Exact analysis of an interacting Bose gas. I. The general solution and the ground state. *Phys. Rev.*, **130**(4), 1605–1616.

Lieb, E. H. and Mattis, D. 1962. Theory of ferromagnetism and the ordering of electronic energy levels. *Phys. Rev.*, **125**(1), 164–172.

Lieb, E. and Mattis, D. (eds). 1966. *Mathematical Physics in One Dimension*. New York: Academic Press.

Lieb, E. H. and Wu, F. Y. 1968. Absence of Mott transition in an exact solution of the short-range, one-band model in one dimension. *Phys. Rev. Lett.*, **20**(25), 1445–1448.

Lieb, E. and Wu, F. Y. 1972. Two dimensional ferroelectric models. In: Domb, C. and Green, M. S. (eds), *Critical Phenomena and Phase Transitions*. New York: Academic Press.

Lieb, E., Schultz, T. and Mattis, D. 1961. Two soluble models of an antiferromagnetic chain. *Ann. Phys.*, **16**, 407–466.

Lipszyc, K. 1973. One-dimensional model of the rearrangement and dissociation processes. Probability amplitudes and cross-sections. *Acta Phys. Polon.*, **44**, 115–137.

Lipszyc, K. 1974. One-dimensional model of the rearrangement process and the Faddeev equations. *J. Math. Phys.*, **15**(1), 133–138.

Lipszyc, K. 1980. On the application of the Sommerfeld–Maluzhinetz transformation to some one-dimensional three-particle problems. *J. Math. Phys.*, **21**(5), 1092–1102.

Luther, A. and Peschel, I. 1975. Calculation of critical exponents in two dimensions from quantum field theory in one dimension. *Phys. Rev. B*, **12**(9), 3908–3917.

Luttinger, J. M. 1963. An exactly soluble model of a many-fermion system. *J. Math. Phys.*, **4**(9), 1154–1162.

Mattheiss, L. F. 1961. Antiferromagnetic linear chain. *Phys. Rev.*, **123**(4), 1209–1218.

Mattis, D. C. and Lieb, E. H. 1965. Exact solution of a many-fermion system and its associated boson field. *J. Math. Phys.*, **6**, 304–312.

McCoy, B. and Wu, T. 1968. Hydrogen-bonded crystals and the anisotropic Heisenberg chain. *Nuovo Cimento B*, **56**, 311–315.

McCoy, B. and Wu, T. T. 1973. *The Two-dimensional Ising Model*. Cambrdige, MA: Harvard University Press.

McGuire, J. B. 1964. Study of exactly soluble one-dimensional N-body problems. *J. Math. Phys.*, **5**(5), 622–636.

McGuire, J. B. 1965. Interacting fermions in one dimension. I. Repulsive potential. *J. Math. Phys.*, **6**(3), 432–439.

McGuire, J. B. 1966. Interacting fermions in one dimension. II. Attractive potential. *J. Math. Phys.*, **7**(1), 123–132.

McGuire, J. B. and Hurst, C. A. 1972. The scattering of three impenetrable particles in one dimension. *J. Math. Phys.*, **13**(10), 1595–1607.

Mehta, M. L. 1980. *Random Matrices*. New York: Academic Press.

Mikhlin, S. 1964. *Integral Equations*. Oxford: Pergamon Press.

Muir, T. 1966. *A Treatise on the Theory of Determinants (1933)*. New York: Dover.

Nagle, J. F. 1966. Lattice statistics of hydrogen bonded crystals. I. The residual entropy of ice. *J. Math. Phys.*, **7**(8), 1484–1491.

Nussenzveig, H. M. 1961. Soluble model of a break-up process. *Proc. Roy. Soc. London A*, **264**(1318), 408–430.

Onsager, L. 1944. Crystal statistics. I. A two-dimensional model with an order–disorder transition. *Phys. Rev.*, **65**, 117–149.

Onsager, L. and Dupuis, M. 1960. Corso: Thermodinamica dei processi irreversibili. In: *Rendiconti della Scuola internazionale di Fisica (Enrico Fermi) X*. Societa Italiana di Fisica, Bologna.

Orbach, R. 1958. Linear antiferromagnetic chain with anisotropic coupling. *Phys. Rev.*, **112**(2), 309–316.

Ovchinnikov, A. A. 1967. Complexes of several spins in a linear Heisenberg chain. *Sov. Phys. JETP Lett.*, **5**, 48.

Ovchinnikov, A. A. 1969. Excitation spectrum of an antiferromagnetic Heisenberg chain. *Sov. Phys. JETP*, **29**, 727.

Ovchinnikov, A. A. 1970. Excitation spectrum in the one-dimensional Hubbard model. *Sov. Phys. JETP*, **30**, 1160.

Pauling, L. 1960. *The Nature of the Chemical Bond*, 3rd edn. Ithaca, NY: Cornell University Press.

Percus, J. K. 1971. *Combinatorial Methods*. New York: Springer-Verlag.

Pohlmeyer, K. 1976. Integrable Hamiltonian systems and interactions through quadratic constraints. *Commun. Math. Phys.*, **46**, 207–221.

Richardson, R. W. 1965. Exact eigenstates of the pairing-force Hamiltonian. II. *J. Math. Phys.*, **6**(7), 1034–1051.

Ruijsenaars, S. N. M. 1980a. On one-dimensional integrable quantum systems with infinitely many degrees of freedom. *Ann. Phys.*, **128**(2), 335–362.

Ruijsenaars, S. N. M. 1980b. The continuum limit of the infinite isotropic Heisenberg chain in its ground state representation. *J. Funct. Anal.*, **39**(1), 75–84.

Schroer, B., Truong, T. T. and Weisz, P. 1976. Towards an explicit construction of the sine-Gordon field theory. *Phys. Lett. B*, **63**(4), 422–424.

Schwartz, L. 1950. *Théorie des distributions*. Paris: Hermann.

Siegel, C. L. 1969. *Topics in Complex Function Theory*. New York: Wiley-Interscience.

Siegel, C. L. 1973. *Topics in Complex Function Theory*, 2nd edn. New York: Wiley-Interscience.

Sklyanin, E. K., Takhtadzhyan, L. A. and Faddeev, L. D. 1979. Quantum inverse problem method. I. *Theor. Math. Phys.*, **40**, 688–706.

Slater, J. C. 1930. Cohesion in monovalent metals. *Phys. Rev.*, **35**(5), 509–529.

Sneddon, I. N. 1966. *Mixed Boundary Value Problems in Potential Theory*. Amsterdam: North-Holland.

Sutherland, B. 1968. Further results for the many-body problem in one dimension. *Phys. Rev. Lett.*, **20**(3), 98.

Sutherland, B. 1970. Two-dimensional hydrogen bonded crystals without the ice rule. *J. Math. Phys.*, **11**(11), 3183–3186.

Sutherland, B. 1971. Quantum many-body problem in one dimension: thermodynamics. *J. Math. Phys.*, **12**(2), 251–256.

Sutherland, B. 1975. Model for a multicomponent quantum system. *Phys. Rev. B*, **12**(9), 3795–3805.

Sutherland, B. 1980. Nondiffractive scattering: Scattering from kaleidoscopes. *J. Math. Phys.*, **21**(7), 1770–1775.

Sutherland, B., Yang, C. N. and Yang, C. P. 1967. Exact solution of a model of two-dimensional ferroelectrics in an arbitrary external electric field. *Phys. Rev. Lett.*, **19**, 588.

Szegö, G. 1939. *Orthogonal Polynomials*. Rhode Island: American Mathematical Society.

Takahashi, M. 1971. One-dimensional Heisenberg model at finite temperature. *Prog. Theor. Phys.*, **46**(2), 401–415.

Takahashi, M. and Suzuki, M. 1972. One-dimensional anisotropic Heisenberg model at finite temperatures. *Prog. Theor. Phys.*, **48**(6), 2187–2209.

Takhtadzhyan, L. A. and Faddeev, L. D. 1974. Essentially nonlinear one-dimensional model of classical field theory. *Theor. Math. Phys.*, **21**, 1046–1057.

Takhtajan, L. A. 1977. Integration of the continuous Heisenberg spin chain through the inverse scattering method. *Phys. Lett. A*, **64**(2), 235–237.

Thacker, H. B. 1980. Quantum inverse method for two-dimensional ice and ferroelectric lattice models. *J. Math. Phys.*, **21**(5), 1115–1117.

Thacker, H. B. 1981. Exact integrability in quantum field theory and statistical systems. *Rev. Mod. Phys.*, **53**(2), 253–285.

Thomas, L. E. 1977. Ground state representation of the infinite one-dimensional Heisenberg ferromagnet, I. *J. Math. Anal. Appl.*, **59**(2), 392–414.

Titchmarsh, E. C. 1939. *The Theory of Functions*, 2nd edn. London: Oxford University Press.

Toda, M. 1970. Waves in nonlinear lattice. *Prog. Theor. Phys. Suppl.*, **45**, 174–200.

Vaidya, H. G. and Tracy, C. A. 1979. One-particle reduced density matrix of impenetrable bosons in one dimension at zero temperature. *Phys. Rev. Lett.*, **42**, 3–6.

Walker, L. R. 1959. Antiferromagnetic linear chain. *Phys. Rev.*, **116**(5), 1089–1090.

Wannier, G. H. 1950. Antiferromagnetism. The triangular Ising net. *Phys. Rev.*, **79**, 357–364.

Weyl, H. 1935. *Elementary Theory of Invariants*. Princeton, NJ: Princeton University Press.

Whittaker, E. T. and Watson, G. N. 1927. *A Course of Modern Analysis*, 4th edn. Cambridge: Cambridge University Press.

Wiegmann, P. B. 1981. Exact solution of the s–d exchange model (Kondo problem). *J. Phys. C: Solid State Phys.*, **14**(10), 1463.

Woynarovich, F. 1982. On the $S^z = 0$ excited states of an anisotropic Heisenberg chain. *J. Phys. A.*, **15**, 2985–2996.

Yang, C. N. 1952. The spontaneous magnetization of a two-dimensional Ising model. *Phys. Rev.*, **85**, 808–816.

Yang, C. N. 1967. Some exact results for the many-body problem in one dimension with repulsive delta-function interaction. *Phys. Rev. Lett.*, **19**(23), 1312–1315.

Yang, C. N. 1968. S matrix for the one-dimensional N-body problem with repulsive or attractive δ-function interaction. *Phys. Rev.*, **168**, 1920–1923.

Yang, C. N. and Yang, C. P. 1966a. Ground-state energy of a Heisenberg–Ising lattice. *Phys. Rev.*, **147**, 303–306.

Yang, C. N. and Yang, C. P. 1966b. One-dimensional chain of anisotropic spin–spin interactions. I. Proof of Bethe's hypothesis for ground state in a finite system. *Phys. Rev.*, **150**(1), 321–327.

Yang, C. N. and Yang, C. P. 1966c. One-dimensional chain of anisotropic spin–spin interactions. II. Properties of the ground-state energy per lattice site for an infinite system. *Phys. Rev.*, **150**(1), 327–339.

Yang, C. N. and Yang, C. P. 1966d. One-dimensional chain of anisotropic spin–spin interactions. III. Applications. *Phys. Rev.*, **151**(1), 258–264.

Yang, C. N. and Yang, C. P. 1969. Thermodynamics of a one-dimensional system of bosons with repulsive delta-function interaction. *J. Math. Phys.*, **10**(7), 1115–1122.

Yang, C. P. 1967. Exact solution of a model of two-dimensional ferroelectrics in an arbitrary external electric field. *Phys. Rev. Lett.*, **19**, 586–588.

Zakharov, V. E. and Shabat, A. B. 1972. Exact theory of two-dimensional self-focusing and one-dimensional self-modulation of waves in non-linear media. *Sov. Phys. JETP*, **34**(1), 62.

Zamolodchikov, A. B. 1977. Exact two-particle S-matrix of quantum sine-Gordon solitons. *Commun. Math. Phys.*, **55**, 183–186.

Zamolodchikov, A. B. 1979. Z_4-symmetric factorized S-matrix in two space-time dimensions. *Commun. Math. Phys.*, **69**, 165–178.

Zamolodchikov, A. B. 1980. Tetrahedra equations and integrable systems in three-dimensional space. *Sov. Phys. JETP*, **52**(2), 325.

Zamolodchikov, A. B. and Zamolodchikov, Al. B. 1978. Relativistic factorized S-matrix in two dimensions having $O(N)$ isotopic symmetry. *Nucl. Phys. B*, **133**(3), 525–535.

Index

Printed in the United States
By Bookmasters

Printed in the United States
By Bookmasters